Lecture Notes in Artificial Intelligence 4911

Edited by J. G. Carbonell and J. Siekmann

Subseries of Lecture Notes in Computer Science

T0223152

Luc De Raedt   Paolo Frasconi
Kristian Kersting   Stephen Muggleton (Eds.)

Probabilistic Inductive Logic Programming

Theory and Applications

 Springer

Volume Editors

Luc De Raedt
Katholieke Universiteit Leuven
Department of Computer Science, Belgium
E-mail: Luc.DeRaedt@cs.kuleuven.be

Paolo Frasconi
Università degli Studi di Firenze
Machine Learning and Neural Networks Group,
Dipartimento di Sistemi e Informatica, Italy
E-mail: p-f@dsi.unifi.it

Kristian Kersting
Massachusetts Institute of Technology, CSAIL
E-mail: kersting@csail.mit.edu

Stephen Muggleton
Imperial College London, Department of Computing
E-mail: shm@doc.ic.ac.uk

Library of Congress Control Number: Applied for

CR Subject Classification (1998): I.2.3, I.2.6, I.2, D.1.6, F.4.1, J.3
LNCS Sublibrary: SL 7 – Artificial Intelligence

ISSN 0302-9743

ISBN 978-3-540-78651-1 Springer Berlin Heidelberg New York

Springer is a part of Springer Science+Business Media

springer.com

© Springer-Verlag Berlin Heidelberg 2008

Typesetting: Camera-ready by author, data conversion by Scientific Publishing Services, Chennai, India
Printed on acid-free paper   SPIN: 12239573   06/3180   5 4 3 2 1 0

Preface

One of the key open questions within artificial intelligence is how to combine probability and logic with learning. This question is getting an increased attention in several disciplines such as knowledge representation, reasoning about uncertainty, data mining, and machine learning simulateously, resulting in the newly emerging subfield known as statistical relational learning and probabilistic inductive logic programming. A major driving force is the explosive growth in the amount of heterogeneous data that is being collected in the business and scientific world. Example domains include bioinformatics, chemoinformatics, transportation systems, communication networks, social network analysis, link analysis, robotics, among others. The structures encountered can be as simple as sequences and trees (such as those arising in protein secondary structure prediction and natural language parsing) or as complex as citation graphs, the World Wide Web, and relational databases.

This book provides an introduction to this field with an emphasis on those methods based on logic programming principles. The book is also the main result of the successful European IST FET project no. FP6-508861 on Application of Probabilistic Inductive Logic Programming (APRIL II, 2004-2007). This project was coordinated by the Albert Ludwigs University of Freiburg (Germany, Luc De Raedt) and the partners were Imperial College London (UK, Stephen Muggleton and Michael Sternberg), the Helsinki Institute of Information Technology (Finland, Heikki Mannila), the Università degli Studi di Florence (Italy, Paolo Frasconi), and the Institut National de Recherche en Informatique et Automatique Rocquencourt (France, Francois Fages). It was concerned with theory, implementations and applications of probabilistic inductive logic programming. This structure is also reflected in the book.

The book starts with an introductory chapter to "Probabilistic Inductive Logic Programming" by De Raedt and Kersting. In a second part, it provides a detailed overview of the most important probabilistic logic learning formalisms and systems. We are very pleased and proud that the scientists behind the key probabilistic inductive logic programming systems (also those developed outside the APRIL project) have kindly contributed a chapter providing an overview of their contributions. This includes: relational sequence learning techniques (Kersting et al.), using kernels with logical representations (Frasconi and Passerini), Markov Logic (Domingos et al.), the PRISM system (Sato and Kameya), CLP($\mathcal{BN}$) (Santos Costa et al.), Bayesian Logic Programs (Kersting and De Raedt), and the Independent Choice Logic (Poole). The third part then provides a detailed account of some show-case applications of probabilistic inductive logic programming, more specifically: in protein fold discovery (Chen et al.), haplotyping (Landwehr and Mielikäinen) and systems biology (Fages and Soliman). The final part touches upon some theoretical investigations and includes chapters on behavioral comparison of probabilistic logic programming representations (Muggleton and Chen) and a model-theoretic expressivity analysis (Jaeger).

The editors would like to thank the EU (Future and Emerging Technology branch of the FP6 IST programme) for supporting the April II project as well as the partners in the consortium and all contributors to this book. We hope that you will enjoy reading this book as much as we enjoyed the process of producing it.

December 2007   Luc De Raedt
                                                               Paolo Frasconi
                                                             Kristian Kersting
                                                         Stephen H. Muggleton

Table of Contents

Theory

Probabilistic Inductive Logic Programming

Luc De Raedt[1] and Kristian Kersting[2]

[1] Departement Computerwetenschappen, K.U. Leuven
Celestijnenlaan 200A - bus 2402, B-3001 Heverlee, Belgium
Luc.DeRaedt@cs.kuleuven.be
[2] CSAIL, Massachusetts Institute of Technologie
32 Vassar Street, Cambridge, MA 02139-4307, USA
kersting@csail.mit.edu

Abstract. Probabilistic inductive logic programming aka. statistical relational learning addresses one of the central questions of artificial intelligence: the integration of probabilistic reasoning with machine learning and first order and relational logic representations. A rich variety of different formalisms and learning techniques have been developed. A unifying characterization of the underlying learning settings, however, is missing so far.

In this chapter, we start from inductive logic programming and sketch how the inductive logic programming formalisms, settings and techniques can be extended to the statistical case. More precisely, we outline three classical settings for inductive logic programming, namely *learning from entailment*, *learning from interpretations*, and *learning from proofs or traces*, and show how they can be adapted to cover state-of-the-art statistical relational learning approaches.

1 Introduction

One of the central open questions of artificial intelligence is concerned with combining expressive knowledge representation formalisms such as relational and first-order logic with principled probabilistic and statistical approaches to inference and learning. Traditionally, relational and logical reasoning, probabilistic and statistical reasoning, and machine learning are research fields in their own rights. Nowadays, they are becoming increasingly intertwined. A major driving force is the explosive growth in the amount of heterogeneous data that is being collected in the business and scientific world. Example domains include bioinformatics, transportation systems, communication networks, social network analysis, citation analysis, and robotics. They provide *uncertain* information about varying numbers of entities and relationships among the entities, i.e., about *relational* domains. Traditional machine learning approaches are able to cope either with uncertainty or with relational representations but typically not with both.

It is therefore not surprising that there has been a significant interest in integrating statistical learning with first order logic and relational representations. [14] has introduced a probabilistic variant of Comprehensive Unification Formalism (CUF). In a similar manner, Muggleton [38] and Cussens [4] have upgraded stochastic grammars towards *stochastic logic programs*; Chapter 9 presents an application of them to protein fold discovery. Sato [53] has introduced *probabilistic distributional semantics* for

L. De Raedt et al. (Eds.): Probabilistic ILP 2007, LNAI 4911, pp. 1–27, 2008.
© Springer-Verlag Berlin Heidelberg 2008

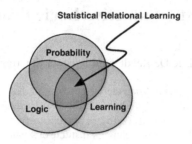

Fig. 1. Probabilistic inductive logic programming aka. statistical relational learning combines probability, logic, and learning

logic programs; latest developments of this approach are presented in Chapter 5. Taskar *et al.* [60] have upgraded Markov networks towards *relational Markov networks*, and Domingos and Richardson [12] towards *Markov logic networks* (see also Chapter 4). Another research stream includes Poole's *independent choice Logic* [49] as reviewed in Chapter 8, Ngo and Haddawy's *probabilistic-logic programs* [45], Jäger's *relational Bayesian networks* [21], Pfeffer's [47] and Getoor's [17] *probabilistic relational models*, and Kersting and De Raedt's *Bayesian logic programs* [26, see also Chapter 7], and has investigated logical and relational extensions of Bayesian networks. Neville and Jensen [44] developed relational dependency networks. This newly emerging research field is known under the name of *statistical relational learning* or *probabilistic inductive logic programming*, cf. Figure 1, and

> Deals with *machine learning* and *data mining* in *relational domains* where observations may be *missing, partially observed,* and/or *noisy.*

Employing relational and logical abstraction within statistical learning has two advantages. First, variables, i.e., placeholders for entities allow one to make abstraction of specific entities. Second, unification allows one to share information among entities. Thus, instead of learning regularities for each single entity independently, statistical relational learning aims at finding general regularities among groups of entities. The learned knowledge is declarative and compact, which makes it much easier for people to understand and to validated. Although, the learned knowledge must be recombined at run time using some reasoning mechanism such as backward chaining or resolution, which bears additional computational costs, statistical relational models are more flexible, context-aware, and offer — in principle — the full power of logical reasoning. Moreover, in many applications, there is a rich background theory available, which can efficiently and elegantly be represented as sets of general regularities. This is important because background knowledge often improves the quality of learning as it focuses learning on relevant patterns, i.e., restricts the search space. While learning, relational and logical abstraction allow one to reuse experience: learning about one entity improves the prediction for other entities; it might even generalize to objects, which have never been observed before. Thus, relational and logical abstraction can make statistical learning more robust and efficient. This has been proven beneficial in many fascinating real-world applications in citation analysis, web mining, natural language processing, robotics, bio- and chemo-informatics, electronic games, and activity recognition.

Whereas most of the existing works on statistical relational learning have started from a statistical and probabilistic learning perspective and extended probabilistic formalisms with relational aspects, we will take a different perspective, in which we will start from inductive logic programming (ILP) [43], which is often also called *multi-relational data mining* (MRDM) [13]. ILP is a research field at the intersection of machine learning and logic programming. It aims at a formal framework as well as practical algorithms for inductively learning relational descriptions (in the form of logic programs) from examples and background knowledge. However, it does not explicitly deal with uncertainty such as missing or noisy information. Therefore, we will study how inductive logic programming formalisms, settings and techniques can be extended to deal with probabilities. At the same time, it is not our intention to provide a complete survey of statistical relational learning (as [9] does), but rather to focus on the principles that underlay this new and exciting subfield of artificial intelligence.

We call the resulting framework *probabilistic* ILP. It aims at a formal framework for statistical relational learning. Dealing explicitly with uncertainty makes probabilistic ILP more powerful than ILP and, in turn, than traditional attribute-value approaches. Moreover, there are several benefits of an ILP view on statistical relational learning. First of all, classical ILP learning settings — as we will argue — naturally carry over to the probabilistic case. The probabilistic ILP settings make abstraction of specific probabilistic relational and first order logical representations and inference and learning algorithms yielding — for the first time — general statistical relational learning settings. Second, many ILP concepts and techniques such as *more–general–than*, *refinement operators*, *least general generalization*, and *greatest lower bound* can be reused. Therefore, many ILP learning algorithms such as Quinlan's FOIL and De Raedt and Dehaspe's CLAUDIEN can easily be adapted. Third, the ILP perspective highlights the importance of background knowledge within statistical relational learning. The research on ILP and on artificial intelligence in general has shown that background knowledge is the key to success in many applications. Finally, an ILP approach should make statistical relational learning more intuitive to those coming from an ILP background and should cross-fertilize ideas developed in ILP and statistical learning.

We will proceed as follows. After reviewing the basic concepts of logic programming in Section 2 and of inductive logic programming in Section 3, we will extend inductive logic programming to deal explicitly with uncertainty in Section 4. Based on this foundation, we will introduce probabilistic variants of classical ILP settings for learning from entailment, from interpretations and from proofs. Each of the resulting settings will be exemplified with different probabilistic logic representations, examples and probability distributions.

2   Logic Programming Concepts

To introduce logic programs, consider Figure 2, containing two programs, *grandparent* and *nat*. Formally speaking, we have that grandparent/2, parent/2, and nat/1 are **predicates** (with their *arity*, i.e., number of arguments listed explicitly) jef, paul and ann are **constants** and X, Y and Z are **variables** All constants and variables are also **terms** In addition, there exist structured terms such as s(X), which contains the

```
parent(jef,paul).              nat(0).
parent(paul,ann).              nat(s(X)) :- nat(X).
grandparent(X,Y) :- parent(X,Z), parent(Z,Y).
```

Fig. 2. Two logic programs, *grandparent* and *nat*

functor s/1 of arity 1 and the term X. Constants are often considered as functors of arity 0. A first order **alphabet** Σ is a set of predicate symbols, constant symbols and functor symbols. *Atoms* are predicate symbols followed by the necessary number of terms, e.g., parent(jef, paul), nat(s(X)), parent(X, Z), etc. **Literals** are atoms nat(s(X)) (positive literal) and their negations not nat(s(X)) (negative literals). We are now able to define the key concept of a **definite clause**. Definite clauses are formulas of the form

$$A :-B_1, \ldots, B_m$$

where A and the B_i are logical atoms and all variables are understood to be universally quantified. For instance, the clause c

$$c \equiv \text{grandparent}(X, Y) :-\text{parent}(X, Z), \text{parent}(Z, Y)$$

can be read as X is the grandparent of Y if X is a parent of Z and Z is a parent of Y. We call grandparent(X, Y) the head(c) of this clause, and parent(X, Z), parent(Z, Y) the body(c). Clauses with an empty body such as parent(jef, paul) are **facts**. A (definite) clause program (or **logic program** for short) consists of a set of clauses. In Figure 2, there are thus two logic programs, one defining grandparent/2 and one defining nat/1. The set of variables in a term, atom, conjunction or clause E, is denoted as Var(E), e.g., Var(c) = {X, Y, Z}. A term, atom or clause E is **ground** when there is no variable occurring in E, i.e. Var(E) = $\emptyset$. A clause c is **range-restricted** when all variables in the head of the clause also appear in the body of the clause, i.e., Var(head(c)) $\subseteq$ |$Vars$(body(c)).

A **substitution** $\theta = \{V_1/t_1, \ldots, V_n/t_n\}$, e.g. {Y/ann}, is an assignment of terms t_i to variables V_i. Applying a substitution θ to a term, atom or clause e yields the instantiated term, atom, or clause $e\theta$ where all occurrences of the variables V_i are simultaneously replaced by the term t_i, e.g. $c\theta$ is

$$c' \equiv \text{grandparent}(X, \text{ann}) :-\text{parent}(X, Z), \text{parent}(Z, \text{ann}).$$

A clause c_1 θ-subsumes[1] {head(c_2)θ} $\cup$ body(c_2)$\theta \subset$ {head(c_1)} $\cup$ body(c_1). The **Herbrand base** of a logic program P, denoted as hb(P), is the set of all ground atoms constructed with the predicate, constant and function symbols in the alphabet of P.

Example 1. The Herbrand bases of the *nat* and *grandparent* logic programs are

$$\text{hb}(nat) = \{\text{nat}(0), \text{nat}(s(0)), \text{nat}(s(s(0))), ...\}$$

$$\text{and hb}(grandparent) = \{\text{parent}(\text{ann}, \text{ann}), \text{parent}(\text{jef}, \text{jef}),$$

$$\text{parent}(\text{paul}, \text{paul}), \text{parent}(\text{ann}, \text{jef}), \text{parent}(\text{jef}, \text{ann}), ...,$$

$$\text{grandparent}(\text{ann}, \text{ann}), \text{grandparent}(\text{jef}, \text{jef}), ...\}.$$

---

[1] The definition of θ-subsumption also applies to conjunctions of literals, as these can also be defined as set of literals.

A **Herbrand interpretation** for a logic program P is a subset of hb(P). A Herbrand interpretation I is a **model** if and only if for all substitutions θ such that body$(c)\theta \subseteq I$ holds, it also holds that head$(c)\theta \in I$. The interpretation I is a model of a logic program P if I is a model of all clauses in P. A clause c (logic program P) **entails** another clause c' (logic program P'), denoted as $c \models c'$ ($P \models P'$), if and only if, each model of c (P) is also a model of c' (P'). Clearly, if clause c (program P) θ-subsumes clause c' (program P') then c (P) entails c' (P'), but the reverse is not true.

The **least Herbrand model** LH(P), which constitutes the semantics of the logic program P, consists of all facts $f \in$ hb(P) such that P logically entails f, i.e. $P \models f$. All ground atoms in the least Herbrand model are provable. **Proofs** are typically constructed using the **SLD-resolution** procedure: given a **goal** $:-G_1, G_2 \ldots, G_n$ and a clause $G:-L_1, \ldots, L_m$ such that $G_1\theta = G\theta$, applying SLD resolution yields the new goal $:-L_1\theta, \ldots, L_m\theta, G_2\theta \ldots, G_n\theta$. A *successful* refutation, i.e., a proof of a goal is then a sequence of resolution steps yielding the empty goal, i.e. $:-$. *Failed* proofs do not end in the empty goal.

Example 2. The atom grandparent(jeff, ann) is true because of

$$:-\text{grandparent}(\text{jeff}, \text{ann})$$
$$:-\text{parent}(\text{jeff}, \text{Z}), \text{parent}(\text{Z}, \text{ann})$$
$$:-\text{parent}(\text{paul}, \text{ann})$$
$$:-$$

Resolution is employed by many theorem provers (such as Prolog). Indeed, when given the goal grandparent(jeff, ann), Prolog would compute the above successful resolution refutation and answer that the goal is true.

For a detailed introduction to logic programming, we refer to [33], for a more gentle introduction, we refer to [15], and for a detailed discussion of Prolog, see [58].

3 Inductive Logic Programming (ILP) and Its Settings

Inductive logic programming is concerned with finding a hypothesis H (a logic program, i.e. a definite clause program) from a set of positive and negative examples *Pos* and *Neg*.

Example 3 (Adapted from Example 1.1 in [32]). Consider learning a definition for the daughter/2 predicate, i.e., a set of clauses with head predicates over daughter/2, given the following facts as learning examples

Pos	daughter(dorothy, ann).
	daughter(dorothy, brian).
Neg	daughter(rex, ann).
	daughter(rex, brian).

Additionally, we have some general knowledge called background knowledge B, which describes the family relationships and sex of each person:

mother(ann, dorothy). female(dorothy).   female(ann).
mother(ann, rex).   father(brian, dorothy). father(brian, rex).

From this information, we could induce H

$$\text{daughter}(C, P) \; : - \; \text{female}(C), \text{mother}(P, C).$$
$$\text{daughter}(C, P) \; : - \; \text{female}(C), \text{father}(P, C).$$

which perfectly explains the examples in terms of the background knowledge, i.e., *Pos* are entailed by H together with B, but *Neg* are not entailed.

More formally, ILP is concerned with the following learning problem.

Definition 1 (ILP Learning Problem). **Given** *a set of positive and negative examples* Pos *and* Neg *over some language* $\mathcal{L}_E$, *a background theory* B, *in the form of a set of definite clauses, a hypothesis language* $\mathcal{L}_H$, *which specifies the clauses that are allowed in hypotheses, and a covers relation* $covers(e, H, B) \in \{0, 1\}$, *which basically returns the classification of an example* e *with respect to* H *and* B, **find** *a hypothesis* H *in* $\mathcal{H}$ *that covers (with respect to the background theory* B*) all positive examples in* Pos *(completeness) and none of the negative examples in* Neg *(consistency).*

The language $\mathcal{L}_E$ chosen for representing the examples together with the *covers* relation determines the inductive logic programming setting. Various settings have been considered in the literature [7]. In the following, we will formalize learning from *entailment* [48] and from *interpretations* [20,8]. We further introduce a novel, intermediate setting, which we call learning from *proofs*. It is inspired on the seminal work by [55].

3.1 Learning from Entailment

Learning from entailment is by far the most popular ILP setting and it is addressed by a wide variety of well-known ILP systems such as FOIL [50], PROGOL [37], and ALEPH [56].

Definition 2 (Covers Relation for Learning from Entailment). *When learning from entailment, the examples are definite clauses and a hypothesis* H *covers an example* e *with respect to the background theory* B *if and only if* $B \cup H \models e$, *i.e., each model of* $B \cup H$ *is also a model of* e.

In many well-known systems, such as FOIL, one requires that the examples are ground facts, a special form of clauses. To illustrate the above setting, consider the following example inspired on the well-known mutagenicity application [57].

Example 4. Consider the following facts in the background theory B, which describe part of molecule 225.

molecule(225).	bond(225, f1_1, f1_2, 7).
logmutag(225, 0.64).	bond(225, f1_2, f1_3, 7).
lumo(225, −1.785).	bond(225, f1_3, f1_4, 7).
logp(225, 1.01).	bond(225, f1_4, f1_5, 7).
nitro(225, [f1_4, f1_8, f1_10, f1_9]).	bond(225, f1_5, f1_1, 7).
atom(225, f1_1, c, 21, 0.187).	bond(225, f1_8, f1_9, 2).
atom(225, f1_2, c, 21, −0.143).	bond(225, f1_8, f1_10, 2).
atom(225, f1_3, c, 21, −0.143).	bond(225, f1_1, f1_11, 1).

$$\texttt{atom}(225, \texttt{f1\_4}, \texttt{c}, 21, -0.013). \qquad \texttt{bond}(225, \texttt{f1\_11}, \texttt{f1\_12}, 2).$$
$$\texttt{atom}(225, \texttt{f1\_5}, \texttt{o}, 52, -0.043). \qquad \texttt{bond}(225, \texttt{f1\_11}, \texttt{f1\_13}, 1).$$

. . .

$$\texttt{ring\_size\_5}(225, [\texttt{f1\_5}, \texttt{f1\_1}, \texttt{f1\_2}, \texttt{f1\_3}, \texttt{f1\_4}]).$$
$$\texttt{hetero\_aromatic\_5\_ring}(225, [\texttt{f1\_5}, \texttt{f1\_1}, \texttt{f1\_2}, \texttt{f1\_3}, \texttt{f1\_4}]).$$

. . .

Consider now the positive example $\texttt{mutagenic}(225)$. It is covered by H

$$\texttt{mutagenic}(\texttt{M}) \; : - \; \texttt{nitro}(\texttt{M}, \texttt{R1}), \texttt{logp}(\texttt{M}, \texttt{C}), \texttt{C} > 1.$$

together with the background knowledge B, because $H \cup B$ entails the example. To see this, we unify $\texttt{mutagenic}(225)$ with the clause's head. This yields

$$\texttt{mutagenic}(225) \; : - \; \texttt{nitro}(225, \texttt{R1}), \texttt{logp}(225, \texttt{C}), \texttt{C} > 1.$$

Now, $\texttt{nitro}(225, \texttt{R1})$ unifies with the fifth ground atom (left-hand side column) in B, and $\texttt{logp}(225, \texttt{C})$ with the fourth one. Because $1.01 > 1$, we found a proof of $\texttt{mutagenic}(225)$.

3.2 Learning from Interpretations

The learning from interpretations setting [8] upgrades boolean concept-learning in computational learning theory [61].

Definition 3 (Covers Relational for Learning from Interpretations). *When learning from interpretations, the examples are Herbrand interpretations and a hypothesis H covers an example e with respect to the background theory B if and only if e is a model of $B \cup H$.*

Recall that Herbrand interpretations are sets of true ground facts and they completely describe a possible situation.

Example 5. Consider the interpretation I, which is the union of B

$$B = \{ \texttt{father}(\texttt{henry}, \texttt{bill}), \texttt{father}(\texttt{alan}, \texttt{betsy}), \texttt{father}(\texttt{alan}, \texttt{benny}),$$
$$\texttt{father}(\texttt{brian}, \texttt{bonnie}), \texttt{father}(\texttt{bill}, \texttt{carl}), \texttt{father}(\texttt{benny}, \texttt{cecily}),$$
$$\texttt{father}(\texttt{carl}, \texttt{dennis}), \texttt{mother}(\texttt{ann}, \texttt{bill}), \texttt{mother}(\texttt{ann}, \texttt{betsy}),$$
$$\texttt{mother}(\texttt{ann}, \texttt{bonnie}), \texttt{mother}(\texttt{alice}, \texttt{benny}), \texttt{mother}(\texttt{betsy}, \texttt{carl}),$$
$$\texttt{mother}(\texttt{bonnie}, \texttt{cecily}), \texttt{mother}(\texttt{cecily}, \texttt{dennis}), \texttt{founder}(\texttt{henry}),$$
$$\texttt{founder}(\texttt{alan}), \texttt{founder}(\texttt{ann}), \texttt{founder}(\texttt{brian}), \texttt{founder}(\texttt{alice}) \}$$

and

$$C = \{ \texttt{carrier}(\texttt{alan}), \texttt{carrier}(\texttt{ann}), \texttt{carrier}(\texttt{betsy}) \}.$$

The interpretation I is covered by the clause c

$$\texttt{carrier}(\texttt{X}) \; : - \; \texttt{mother}(\texttt{M}, \texttt{X}), \texttt{carrier}(\texttt{M}), \texttt{father}(\texttt{F}, \texttt{X}), \texttt{carrier}(\texttt{F}).$$

because I is a model of c, i.e., for all substitutions θ such that $\mathrm{body}(c)\theta \subseteq I$, it holds that $\mathrm{head}(c)\theta \in I$.

The key difference between learning from interpretations and learning from entailment is that interpretations carry much more — even complete — information. Indeed, when learning from entailment, an example can consist of a *single* fact, whereas when learning from interpretations, all facts that hold in the example are known. Therefore, learning from interpretations is typically easier and computationally more tractable than learning from entailment, cf. [7].

3.3   Learning from Proofs

Because learning from entailment (with ground facts as examples) and interpretations occupy extreme positions with respect to the information the examples carry, it is interesting to investigate intermediate positions. Shapiro's [55] Model Inference System (MIS) fits nicely within the learning from entailment setting where examples are facts. However, to deal with missing information, Shapiro employs a clever strategy: MIS queries the users for missing information by asking them for the truth-value of facts. The answers to these queries allow MIS to reconstruct the trace or the proof of the positive examples. Inspired by Shapiro, we define the learning from proofs setting.

Definition 4 (Covers Relation for Learning from Proofs). *When learning from proofs, the examples are ground proof-trees and an example e is covered by a hypothesis H with respect to the background theory B if and only if e is a proof-tree for H ∪ B.*

At this point, there exist various possible forms of proof-trees. Here, we will — for reasons that will become clear later — assume that the proof-tree is given in the form of a ground and-tree where the nodes contain ground atoms. More formally:

Definition 5 (Proof Tree). *A tree t is a proof-tree for a logic program T if and only if t is a rooted tree where for every node $n \in t$ with* children(n) *satisfies the property that there exists a substitution θ and a clause $c \in T$ such that $n =$* head$(c)\theta$ *and* children$(n) =$ body$(c)\theta$.

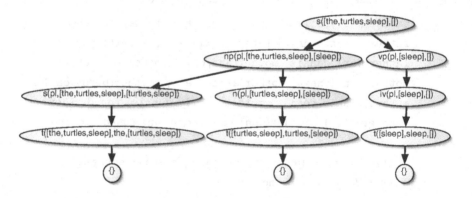

Fig. 3. A proof tree, which is covered by the definite clause grammar in Example 6. Symbols are abbreviated.

Example 6. Consider the following *definite clause grammar.*

$$\text{sentence(A, B)}: -\text{noun\_phrase(C, A, D)}, \text{verb\_phrase(C, D, B)}.$$
$$\text{noun\_phrase(A, B, C)}: -\text{article(A, B, D)}, \text{noun(A, D, C)}.$$
$$\text{verb\_phrase(A, B, C)}: -\text{intransitive\_verb(A, B, C)}.$$
$$\text{article(singular, A, B)}: -\text{terminal(A, a, B)}.$$
$$\text{article(singular, A, B)}: -\text{terminal(A, the, B)}.$$
$$\text{article(plural, A, B)}: -\text{terminal(A, the, B)}.$$
$$\text{noun(singular, A, B)}: -\text{terminal(A, turtle, B)}.$$
$$\text{noun(plural, A, B)}: -\text{terminal(A, turtles, B)}.$$
$$\text{intransitive\_verb(singular, A, B)}: -\text{terminal(A, sleeps, B)}.$$
$$\text{intransitive\_verb(plural, A, B)}: -\text{terminal(A, sleep, B)}.$$
$$\text{terminal([A|B], A, B)}.$$

It covers the proof tree shown in Figure 3.

Proof-trees contain — as interpretations — a lot of information. Indeed, they contain instances of the clauses that were used in the proofs. Therefore, it may be hard for the user to provide this type of examples. Even though this is generally true, there exist specific situations for which this is feasible. Indeed, consider tree banks such as the UPenn Wall Street Journal corpus [35], which contain parse trees. These trees directly correspond to the proof-trees we talk about.

4 Probabilistic ILP Settings

Indeed, ILP has been developed for coping with relational data. It does not, however, handle uncertainty in a principled way. In the reminder of this paper, we will show how to extend the ILP settings to the probabilistic case. To do so, we shall quickly review some of the key concepts and notation that will be used[2].

Let X be a random variable with a finite domain domain$(X) = \{x_1, x_2, \ldots, x_n\}$ of possible states. We use the notation $\mathbf{P}(X)$ to denote the probability distribution over domain(X), and $P(X = x_i)$ or $P(x_i)$ to denote the probability that the random variable X takes the value $x_i \in$ domain X. For instance, consider the random variable (or proposition) earthquake with domain $\{true, false\}$. Then, $\mathbf{P}(\text{earthquake})$ denotes the probability distribution, and $P(\text{earthquake} = false)$ (or, in shorthand notation, $P(\neg\text{earthquake})$) denotes the probability that earthquake is *false*. The distribution $\mathbf{P}(X_1, \cdots, X_n)$ over a set of random variables $\{X_1, \cdots, X_n\}$, $n > 1$, is called **joint probability distributions**. For instance, we may be interested in the joint probability that earthquake = *true* and burglary = *true* at the same time. Generalizing the above notation, we will be using the notation $\mathbf{P}(\text{earthquake}, \text{burglary})$ and $P(\text{earthquake} = true, \text{burglary} = true)$ respectively.

---

[2] The reader not familiar with the basics of probability theory is encouraged to consult [52] for an excellent overview from an artificial intelligence perspective.

Some useful definitions and properties of probability theory can now be listed. The **conditional probability** is defined as $\mathbf{P}(\mathtt{X}|\mathtt{Y}) = \mathbf{P}(\mathtt{X},\mathtt{Y})/\mathbf{P}(\mathtt{Y})$ if $\mathbf{P}(\mathtt{Y}) > 0$. Note that the use of $\mathbf{P}$ in equalities is a short hand notation that denotes that the equality is valid *for all* possible states of the involved random variables. The **chain rule** says that $\mathbf{P}(\mathtt{X}_1,\cdots,\mathtt{X}_n) = \mathbf{P}(\mathtt{X}_1) \prod_{i=2}^{n} \mathbf{P}(\mathtt{X}_i|\mathtt{X}_{i-1},\cdots,\mathtt{X}_1)$. The **law of Bayes** states that $\mathbf{P}(\mathtt{X}|\mathtt{Y}) = \mathbf{P}(\mathtt{Y}|\mathtt{X}) \cdot \mathbf{P}(\mathtt{X})/\mathbf{P}(\mathtt{Y})$. The distribution of $\mathtt{X}$ is related to the joint distribution $\mathbf{P}(\mathtt{X},\mathtt{Y})$ by $\mathbf{P}(\mathtt{X}) = \sum_{y \in \mathrm{domain}(\mathtt{Y})} \mathbf{P}(\mathtt{X},y)$, which is called **marginalization**. Finally, two random variables $\mathtt{X}$ and $\mathtt{Y}$ are **conditionally independent** given a third random variable $\mathtt{Z}$ if and only if $\mathbf{P}(\mathtt{X},\mathtt{Y}|\mathtt{Z}) = \mathbf{P}(\mathtt{X}|\mathtt{Z}) \cdot \mathbf{P}(\mathtt{Y}|\mathtt{Z})$. In case that the property holds without needing to condition on variables $\mathtt{Z}$, we say that $\mathtt{X}$ and $\mathtt{Y}$ are **independent**.

When working with probabilistic ILP representations, there are essentially two changes:

1. Clauses are annotated with probabilistic information such as conditional probabilities, and
2. The covers relation becomes probabilistic.

A probabilistic covers relation softens the hard covers relation employed in traditional ILP and is defined as the probability of an example given the hypothesis and the background theory.

Definition 6 (Probabilistic Covers Relation). *A probabilistic covers relation takes as arguments an example e, a hypothesis H and possibly the background theory B, and returns the probability value* $\mathbf{P}(e \mid H, B)$ *between 0 and 1 of the example e given H and B, i.e.,* $\mathrm{covers}(e, H, B) = \mathbf{P}(e \mid H, B)$.

Using the probabilistic covers relation of Definition 6, our first attempt at a definition of the probabilistic ILP learning problem is as follows.

Preliminary Definition 1 (Probabilistic ILP Learning Problem)
Given *a probabilistic-logical language* $\mathcal{L}_H$ *and a set E of examples over some language* $\mathcal{L}_E$, **find** *the hypothesis* H^* *in* $\mathcal{L}_H$ *that maximizes* $\mathbf{P}(E \mid H^*, B)$.

Under the usual **i.i.d.** assumption, i.e., examples are sampled independently from identical distributions, this results in the maximization of

$$\mathbf{P}(E \mid H^*, B) = \prod_{e \in E} \mathbf{P}(e \mid H^*, B) = \prod_{e \in E} \mathrm{covers}(e, H^*, B).$$

Similar to the ILP learning problem, the language $\mathcal{L}_E$ selected for representing the examples together with the probabilistic covers relation determines different learning setting. In ILP, this lead to learning from interpretations, from proofs, and from entailment. It should therefore be no surprise that this very same distinction also applies to probabilistic knowledge representation formalisms. Indeed, Bayesian networks [46] essentially define a probability distribution over *interpretations* or *possible worlds*, and stochastic grammars [34] define a distribution over *proofs*, derivations or traces.

Guided by Definition 1, we will now introduce three probabilistic ILP settings, which extend the purely logical ones sketched before. Afterwards, we will refine Definition 1 in Definition 7.

4.1 Probabilistic Learning from Interpretations

In order to integrate probabilities in the learning from interpretations setting, we need to find a way to assign probabilities to interpretations covered by an annotated logic program. In the past decade, this issue has received a lot of attention and various different approaches have been developed. The most popular propositional frameworks are Bayesian network and Markov networks. Later on, these propositional frameworks have been extended to the relational case such probabilistic-logic programs [45], probabilistic relational models [47], relational Bayesian networks [21], and Bayesian logic programs [24,25].

In this book, the two most popular propositional formalisms, namely Bayesian networks and Markov networks, are considered, as well as their relational versions. The present chapter focuses on Bayesian networks, and their extension towards Bayesian logic programs [27, more details in Chapter 7], whereas Chapter 6 by Santos Costas *et al.* discusses an integration of Bayesian networks and logic programs called CLP($\mathcal{BN}$) and Chapter 9 by Domingos *et al.* focuses on Markov networks and their extension to Markov Logic [12].

Bayesian Networks. The most popular formalism for defining probabilities on possible worlds is that of Bayesian networks. As an example of a Bayesian network, consider Judea Pearl's famous alarm network graphically illustrated in Figure 4. Formally speaking, a **Bayesian network** is an augmented, directed acyclic graph, where each node corresponds to a random variable X_i and each edge indicates a *direct influence* among the random variables. It represents the joint probability distribution $\mathbf{P}(X_1, \ldots, X_n)$. The influence is quantified with a conditional probability distribution $cpd(X_i)$ associated to each node X_i. It is defined in terms of the parents of the node X, which we denote by $\mathbf{Pa}(X_i)$, and specifies $\mathrm{cpd}(X_i) = \mathbf{P}(X_i \mid \mathbf{Pa}(X_i))$.

Example 7. Consider the Bayesian network in Figure 4. It contains the random variables `alarm`, `earthquake`, `marycalls`, `johncalls` and `alarm`. The CPDs associated to each of the nodes are specified in Table 1. They include the CPDs $\mathbf{P}(\text{alarm} \mid \text{earthquake}, \text{burglary})$, and $\mathbf{P}(\text{earthquake})$, etc.

The Bayesian network thus has two components: a qualitative one, i.e. the directed acyclic graph, and a quantitative one, i.e. the conditional probability distributions. Together they specify the joint probability distribution.

As we will – for simplicity – assume that the random variables are all boolean, i.e., they can have the domain $\{true, false\}$, this actually amounts to specifying a probability distribution on the set of all possible interpretations. Indeed, in our alarm example, the Bayesian network defines a probability distribution over truth-assignments to $\{$alarm, earthquake, marycalls, johncalls, burglary$\}$.

The qualitative component specifies a set of conditional independence assumptions. More formally, it stipulates the following conditional independency assumption:

Assumption 1. *Each node X_i in the graph is conditionally independent of any subset $\mathbf{A}$ of nodes that are not descendants of X_i given a joint state of $\mathbf{Pa}(X_i)$, i.e.* $\mathbf{P}(X_i \mid \mathbf{A}, \mathbf{Pa}(X_i)) = \mathbf{P}(X_i \mid \mathbf{Pa}(X_i))$.

Fig. 4. The Bayesian alarm network. Nodes denote random variables and edges denote direct influences among the random variables.

Table 1. The conditional probability distributions associated with the nodes in the alarm network, cf. Figure 4

P(burglary)	P(earthquake)
$(0.001, 0.999)$	$(0.002, 0.998)$

burglary	earthquake	P(alarm)
true	*true*	$(0.95, 0.05)$
true	*false*	$(0.94, 0.06)$
false	*true*	$(0.29, 0.71)$
false	*false*	$(0.001, 0.999)$

alarm	P(johncalls)
true	$(0.90, 0.10)$
false	$(0.05, 0.95)$

alarm	P(marycalls)
true	$(0.70, 0.30)$
false	$(0.01, 0.99)$

For example, alarm is conditionally independent of marycalls) given a joint state of its parents {earthquake, burglary}. Because of the conditional independence assumption, we can write down the joint probability density as

$$\mathbf{P}(X_1, \ldots, X_n) = \prod_{i=1}^{n} \mathbf{P}(X_i \mid \mathbf{Pa}(X_i)) \qquad (1)$$

by applying the independency assumption and the chain rule to the joint probability distribution.

Bayesian Logic Programs. The idea underlying Bayesian logic programs is to view ground atoms as random variables that are defined by the underlying definite clause programs. Furthermore, two types of predicates are distinguished: deterministic and probabilistic ones. The former are called *logical*, the latter *Bayesian*. Likewise we will also speak of Bayesian and logical atoms. A Bayesian logic program now consists of a set of of Bayesian (definite) clauses, which are expressions of the form A | $A_1, \ldots, A_n$ where A is a Bayesian atom, $A_1, \ldots, A_n$, $n \geq 0$, are Bayesian and logical atoms and all variables are (implicitly) universally quantified. To quantify probabilistic dependencies, each Bayesian clause c is annotated with its conditional probability distribution

$cpd(c) = \mathbf{P}(A \mid A_1, \ldots, A_n)$, which quantifies as a macro the probabilistic dependency among ground instances of the clause.

Let us illustrate Bayesian logic programs on an example inspired on Jensen's stud farm example [22], which describes the processes underlying a life threatening hereditary disease.

Example 8. Consider the following Bayesian clauses:

$$\text{carrier}(X) \mid \text{founder}(X). \tag{2}$$
$$\text{carrier}(X) \mid \text{mother}(M, X), \text{carrier}(M), \text{father}(F, X), \text{carrier}(F). \tag{3}$$
$$\text{suffers}(X) \mid \text{carrier}(X). \tag{4}$$

They specify the probabilistic dependencies governing the inheritance process. For instance, clause (3) says that the probability for a horse being a carrier of the disease depends on its parents being carriers. In this example, the mother, father, and founder are logical, whereas the other ones, such as carrier and suffers, are Bayesian. The logical predicates are then defined by a classical definite clause program which constitute the background theory for this example. It is listed as interpretation B in Example 5. Furthermore, the conditional probability distributions for the Bayesian clauses are

$P(\text{carrier}(X) = true)$
0.6

carrier(X)	$P(\text{suffers}(X) = true)$
$true$	0.7
$false$	0.01

carrier(M)	carrier(F)	$P(\text{carrier}(X) = true)$
$true$	$true$	0.6
$true$	$false$	0.5
$false$	$true$	0.5
$false$	$false$	0.0

Observe that logical atoms, such as mother(M, X), do not affect the distribution of Bayesian atoms, such as carrier(X), and are therefore not considered in the conditional probability distribution. They only provide variable bindings, e.g., between carrier(X) and carrier(M).

By now, we are able to define the covers relation for Bayesian logic programs. A Bayesian logic program together with the background theory induces a Bayesian network. The random variables A of the Bayesian network are the Bayesian ground atoms in the least Herbrand model I of the annotated logic program. A Bayesian ground atom, say carrier(alan), influences another Bayesian ground atom, say carrier(betsy), if and only if there exists a Bayesian clause c such that

1. carrier(alan) $\in body(c)\theta \subseteq I$, and
2. carrier(betsy) $\equiv head(c)\theta \in I$.

Each node A has $cpd(c\theta)$ as associated conditional probability distribution. If there are multiple ground instances in I with the same head, a *combining rule* combine$\{\cdot\}$ is used to quantified the combined effect. A combining rule is a function that maps finite sets of conditional probability distributions onto one (*combined*) conditional probability distribution. Examples of combining rules are *noisy-or*, and *noisy-and*, see e.g. [22].

Note that we assume that the induced network is acyclic and has a finite branching factor. The probability distribution induced is now

$$\mathbf{P}(I|H) = \prod_{\text{Bayesian atom } A \in I} \text{combine}\{\text{cpd}(c\theta) \,|body(c)\theta \subseteq I, head(c)\theta \equiv A\}. \quad (5)$$

Let us illustrate this fro the stud farm example.

Example 9. Using the above definition, the probability of the interpretation

> {carrier(henry) = $false$, suffers(henry) = $false$, carrier(ann) = $true$,
> suffers(ann) = $false$,   carrier(brian) = $false$, suffers(brian) = $false$,
> carrier(alan) = $false$,   suffers(alan) = $false$,   carrier(alice) = $false$,
> suffers(alice) = $false$,   ...}

can be computed using a standard Bayesian network inference engine because the facts together with the program induce the Bayesian network shown in Figure 5. Thus (5) defines a probabilistic coverage relation. In addition, various types of inference would be possible. One might, for instance, ask for the probability P(suffers(henry)|

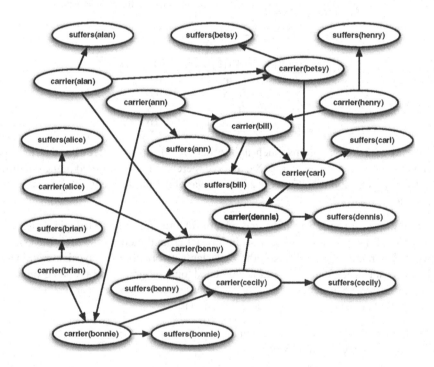

Fig. 5. The structure of the Bayesian network induced by the Stud farm Bayesian logic program. For the ease of comprehensibility, we have omitted the *logical* Bayesian atoms over `founder/1`, `father/2`, and `mother/2`.

carrier(henry) $= true$), which can again be computed using a standard Bayesian network inference engine.

4.2 Probabilistic Proofs

To define probabilities on proofs, ICL [49, Chapter 8], PRISMs [53,54, Chapter 5], and stochastic logic programs [14,38,5, an application can be found in Chapter 9] attach probabilities to facts (respectively clauses) and treat them as stochastic choices within resolution. Relational Markov models [2] and logical hidden Markov models [28], which we will briefly review in Chapter 2, can be viewed as a simple fragment of them, where heads and bodies of clauses are singletons only, so-called iterative clauses. We will illustrate probabilistic learning from proofs using stochastic logic programs. For a discussion of the close relationship among stochastic logic programs, ICL, and PRISM, we refer to [6].

Stochastic logic programs are inspired on stochastic context free grammars [1,34]. The analogy between context free grammars and logic programs is that

- Grammar rules correspond to definite clauses,
- Sentences (or strings) to atoms, and
- Productions to derivations.

Furthermore, in stochastic context-free grammars, the rules are annotated with probability labels in such a way that the sum of the probabilities associated to the rules defining a non-terminal is 1.0.

Eisele and Muggleton have exploited this analogy to define stochastic logic programs. These are essentially definite clause programs, where each clause c has an associated probability label p_c such that the sum of the probabilities associated to the rules defining any predicate is 1.0 (though [4] considered less restricted versions as well).

This framework allows ones to assign probabilities to proofs for a given predicate q given a stochastic logic program $H \cup B$ in the following manner. Let D_q denote the set of all possible ground proofs for atoms over the predicate q. For simplicity reasons, it will be assumed that there is a finite number of such proofs and that all proofs are finite (but again see [4] for the more general case). Now associate to each proof $t_q \in D_q$ the probability

$$v_t = \prod_c p_c^{n_{c,t}}$$

where the product ranges over all clauses c and $n_{c,t}$ denotes the number of times clause c has been used in the proof t_q. For stochastic context free grammars, the values v_t correspond to the probabilities of the production. However, the difference between context free grammars and logic programs is that in grammars two rules of the form $n \to q, n_1, ..., n_m$ and $q \to q_1, ..., q_k$ always 'resolve' to give $n \to q_1, ..., q_k, n_1, ..., n_m$ whereas resolution may fail due to unification. Therefore, the probability of a proof tree t in D_q, i.e., a successful derivation is

$$P(t \mid H, B) = \frac{v_t}{\sum_{s \in D_q} v_s}. \qquad (6)$$

The probability of a ground atom a is then defined as the sum of all the probabilities of all the proofs for that ground atom.

$$P(\text{a} \mid H, B) = \sum_{\substack{s \in D_q \\ s \text{ is a proof for } \alpha}} v_s. \tag{7}$$

Example 10. Consider a stochastic variant of the definite clause grammar in Example 6 with uniform probability values for each predicate. The value v_u of the proof (tree) u in Example 6 is $v_u = \frac{1}{3} \cdot \frac{1}{2} \cdot \frac{1}{2} = \frac{1}{12}$. The only other ground proofs s_1, s_2 of atoms over the predicate sentence are those of

$$\text{sentence}([\text{a}, \text{turtle}, \text{sleeps}], [])$$

$$\text{and sentence}([\text{the}, \text{turtle}, \text{sleeps}], []).$$

Both get the value $v_{s_1} = v_{s_2} = \frac{1}{12}$. Because there is only one proof for each of the sentences,

$$P(\text{sentence}([\text{the}, \text{turtles}, \text{sleep}], [])) = v_u = \frac{1}{3}.$$

For stochastic logic programs, there are at least two natural learning settings.

Motivated by Equation (6), we can learn them from proofs. This makes structure learning for stochastic logic programs relatively easy, because proofs carry a lot of information about the structure of the underlying stochastic logic program. Furthermore, the learning setting can be considered as an extension of the work on learning stochastic grammars from proof-banks. It should therefore also be applicable to learning unification based grammars. We will present a probabilistic ILP approach within the learning from proofs setting in Section 5.4.

On the other hand, we can use Equation (7) as covers relation and, hence, employ the learning from entailment setting. Here, the examples are ground atoms entailed by the target stochastic logic program. Learning stochastic logic programs from atoms only is much harder than learning them from proofs because atoms carry much less information than proofs. Nevertheless, this setting has been studied by [5] and by [54], who solves the parameter estimation problem for stochastic logic programs respectively PRISM programs, and by [39,41], who presents an approach to structure learning of stochastic logic programs: adding one clause at a time to an existing stochastic logic program. In the following section, we will introduce the probabilistic learning from entailment. Instead of considering stochastic logic programs, however, we will study a Naïve Bayes framework, which has a much lower computational complexity.

4.3   Probabilistic Learning from Entailment

In order to integrate probabilities in the entailment setting, we need to find a way to assign probabilities to clauses that are entailed by an annotated logic program. Since most ILP systems working under entailment employ ground facts for a single predicate as examples, and the authors are unaware of any existing probabilistic ILP formalisms that implement a probabilistic covers relation for definite clauses as examples in general, we will restrict our attention to assign probabilities to facts for a single predicate. It remains an open question as how to formulate more general frameworks for working with entailment.

More formally, let us annotate a logic program H consisting of a set of clauses of the form $p \leftarrow b_i$, where p is an atom of the form $p(V_1, ..., V_n)$ with the V_i different variables, and the b_i are different bodies of clauses. Furthermore, we associate to each clause in H the probability values $\mathbf{P}(b_i \mid p)$; they constitute the conditional probability distribution that for a random substitution θ for which $p\theta$ is ground and true (resp. false), the query $b_i\theta$ succeeds (resp. fails) in the knowledge base B.[3] Furthermore, we assume the prior probability of p is given as $\mathbf{P}(p)$, it denotes the probability that for a random substitution θ, $p\theta$ is true (resp. false). This can then be used to define the covers relation $\mathbf{P}(p\theta \mid H, B)$ as follows (we delete the B as it is fixed):

$$\mathbf{P}(p\theta \mid H) = \mathbf{P}(p\theta \mid b_1\theta, ..., b_k\theta) = \frac{\mathbf{P}(b_1\theta, ..., b_k\theta \mid p\theta) \times \mathbf{P}(p\theta)}{\mathbf{P}(b_1\theta, ..., b_k\theta)} \tag{8}$$

For instance, applying the naïve Bayes assumption yields

$$\mathbf{P}(p\theta \mid H) = \frac{\prod_i \mathbf{P}(b_i\theta \mid p\theta) \times \mathbf{P}(p\theta)}{\mathbf{P}(b_1\theta, ..., b_k\theta)} \tag{9}$$

Finally, since $P(p\theta \mid H) + P(\neg p\theta \mid H) = 1$, we can compute $P(p\theta \mid H)$ without $P(b_1\theta, ..., b_k\theta)$ through normalization.

Example 11. Consider again the mutagenicity domain and the following annotated logic program:

$(0.01, 0.21) : \mathtt{mutagenetic(M)} \leftarrow \mathtt{atom(M, \_, \_, 8, \_)}$

$(0.38, 0.99) : \mathtt{mutagenetic(M)} \leftarrow \mathtt{bond(M,, A, 1), atom(M, A, c, 22, \_), bond(M, A,, 2)}$

We denote the first clause by b_1 and the second one by b_2. The vectors on the left-hand side of the clauses specify $P(b_i\theta = true \mid p\theta = true)$ and $P(b_i\theta = true \mid p\theta = false)$ respectively. The covers relation (assuming the Naïve Bayes assumption) assigns probability 0.97 to example 225 because both features fail for $\theta = \{M \leftarrow 225\}$. Hence,

$$\begin{aligned} P(\, \mathtt{mutagenetic(225)} = true&, b_1\theta = false, \ b_2\theta = false) \\ = \ &P(\, b_1\theta = false \mid \mathtt{mutagenetic(225)} = true \,) \\ &\cdot P(\, b_2\theta = false \mid \mathtt{mutagenetic(225)} = true \,) \\ &\cdot P(\, \mathtt{mutagenetic(225)} = true \,) \\ = \ &0.99 \cdot 0.62 \cdot 0.31 \approx 0.19 \end{aligned}$$

and $P(\, \mathtt{mutagenetic(225)} = false, b_1\theta = false, \ b_2\theta = false) = 0.79 \cdot 0.01 \cdot 0.68 \approx 0.005$. This yields

$$P(\, \mathtt{muta(225)} = true \mid b_1\theta = false, \ b_2\theta = false\}) = \frac{0.19}{0.19 + 0.005} \approx 0.97.$$

5   Probabilistic ILP: A Definition and Example Algorithms

Guided by Definition 1, we have introduced several probabilistic ILP settings for statistical relational learning. The main idea was to lift traditional ILP settings by associating

---

[3] The query q succeeds in B if there is a substitution σ such that $B \models q\sigma$.

probabilistic information with clauses and interpretations and by replacing ILP's deterministic covers relation by a probabilistic one. In the discussion, we made one trivial but important observation:

Observation 1. *Derivations might fail.*

The probability of a failure is zero and, consequently, failures are never observable. Only succeeding derivations are observable, i.e., the probabilities of such derivations are greater zero. As an extreme case, recall the negative examples *Neg* employed in the ILP learning problem definition 1. They are supposed to be not covered, i.e., $P(Neg|H, B) = 0$.

Example 12. Reconsider Example 3. Rex is a male person; he cannot be the daughter of ann. Thus, daughter(rex, ann) was listed as a negative example.

Negative examples conflict with the usual view on learning examples in statistical learning. In statistical learning, we seek to find that hypothesis H^*, which is most likely given the learning examples:

$$H^* = \arg\max_H P(H|E) = \arg\max_H \frac{P(E|H) \cdot P(F)}{P(E)} \quad \text{with} \quad P(E) > 0.$$

Thus, examples E are observable, i.e., $P(E) > 0$. Therefore, we refine the preliminary probabilistic ILP learning problem definition 1. In contrast to the purely logical case of ILP, we do not speak of *positive* and *negative* examples anymore but of *observed* and *unobserved* ones.

Definition 7 (Probabilistic ILP Problem). **Given** *a set* $E = E_p \cup E_i$ *of observed and unobserved examples* E_p *and* E_i *(with* $E_p \cap E_i = \emptyset$*) over some example language* $\mathcal{L}_E$, *a probabilistic covers relation* covers$(e, H, B) = P(e \mid H, B)$, *a probabilistic logical language* $\mathcal{L}_H$ *for hypotheses, and a background theory* B, **find** *a hypothesis* H^* *in* $\mathcal{L}_H$ *such that* $H^* = \arg\max_H$ score(E, H, B) *and the following constraints hold:* $\forall e_p \in E_p$: covers$(e_p, H^*, B) > 0$ *and* $\forall e_i \in E_i$: covers$(e_i, H^*, B) = 0$. *The score is some objective function, usually involving the probabilistic covers relation of the observed examples such as the observed likelihood* $\prod_{e_p \in E_p}$ covers(e_p, H^*, B) *or some penalized variant thereof.*

The probabilistic ILP learning problem of Definition 7 unifies ILP and statistical learning in the following sense: using a deterministic covers relation (,which is either 1 or 0) yields the classical ILP learning problem, see Definition 1, whereas sticking to propositional logic and learning from *observed* examples, i.e., $P(E) > 0$, only yields traditional statistical learning.

To come up with algorithms solving probabilistic ILP learning problems, say for density estimation, one typically distinguishes two subtasks because $H = (L, \lambda)$ is essentially a logic program L annotated with probabilistic parameters λ:

1. *Parameter estimation* where it is assumed that the underlying logic program L is fixed, and the learning task consists of estimating the parameters λ that maximize the likelihood.
2. *Structure learning* where both L and λ have to be learned from the data.

Below, we will sketch basic parameter estimation and structure learning techniques, and illustrate them for each setting. In the remainder of the thesis, we will then discuss selected probabilistic ILP approaches for learning from interpretations and probabilistic learning from traces in detail. A more complete survey of learning probabilistic logic representations can be found in [9] and in the related work sections of this thesis.

5.1 Parameter Estimation

The problem of parameter estimation is thus concerned with estimating the values of the parameters λ of a fixed probabilistic program $H = (L, \lambda)$ that best explains the examples E. So, λ is a set of parameters and can be represented as a vector. As already indicated above, to measure the extent to which a model fits the data, one usually employs the likelihood of the data, i.e. $P(E \mid L, \lambda)$, though other scores or variants could be used as well.

When all examples are fully observable, maximum likelihood reduces to frequency counting. In the presence of missing data, however, the maximum likelihood estimate typically cannot be written in closed form. It is a numerical optimization problem, and all known algorithms involve nonlinear optimization The most commonly adapted technique for probabilistic logic learning is the Expectation-Maximization (EM) algorithm [11,36]. EM is based on the observation that learning would be easy (i.e., correspond to frequency counting), if the values of all the random variables would be known. Therefore, it estimates these values, maximizes the likelihood based on the estimates, and then iterates. More specifically, EM assumes that the parameters have been initialized (e.g., at random) and then iteratively performs the following two steps until convergence:

(E-Step). On the basis of the observed data and the present parameters of the model, it computes a distribution over all possible completions of each partially observed data case.

(M-Step). Treating each completion as a fully observed data case weighted by its probability, it computes the improved parameter values using (weighted) frequency counting.

The frequencies over the completions are called the *expected counts*. Examples for parameter estimation of probabilistic relational models in general can be found in Chapters 2 and 10 for sequential relational models, in Chapter 4 for Markov logic, in Chapter 5 for PRISM, in Chapter 6 for CLP($\mathcal{BN}$), in Chapter 7 for Bayesian logic programs, and in Chapters 9 and 11 for stochastic logic programs and variants.

5.2 Structure Learning

The problem is now to learn both the structure L and the parameters λ of the probabilistic program $H = (L, \lambda)$ from data. Often, further information is given as well. As in ILP, the additional knowledge can take various different forms, including a *language bias* that imposes restrictions on the syntax of L, and an *initial hypothesis* (L, λ) from which the learning process can start.

Nearly all (score-based) approaches to structure learning perform a heuristic search through the space of possible hypotheses. Typically, hill-climbing or beam-search is applied until the hypothesis satisfies the logical constraints and the score(H, E) is no longer improving. The steps in the search-space are typically made using refinement operators, which make small, syntactic modification to the (underlying) logic program.

At this points, it is interesting to observe that the logical constraints often require that the observed examples are covered in the logical sense. For instance, when learning stochastic logic programs from entailment, the observed example clauses must be entailed by the logic program, and when learning Markov logic networks, the observed interpretations must be models of the underlying logic program. Thus, for a probabilistic program $H = (L_H, \lambda_H)$ and a background theory $B = (L_B, \lambda_B)$ it holds that $\forall e_p \in E_p : P(e|H, B) > 0$ if and only if covers$(e, L_H, L_B) = 1$, where L_H (respectively L_B) is the underlying logic program (logical background theory) and covers(e, L_H, L_B) is the purely logical *covers* relation, which is either 0 or 1.

Let us now sketch for each probabilistic ILP setting one learning approach.

5.3   Learning from Probabilistic Interpretations

The large majority of statistical relational learning techniques proposed so far fall into the learning from interpretations setting including parameter estimation of probabilistic logic programs [30], learning of probabilistic relational models [18], parameter estimation of relational Markov models [60], learning of object-oriented Bayesian networks [3], learning relational dependency networks [44], and learning logic programs with annotated disjunctions [62,51]. This book provides details on learning sequential relational models in Chapter 2 and 10, on learning Markov logic programs in Chapter 4, and on learning CLP($\mathcal{BN}$) in Chapter 6.

As an example, which will be discussed in detail in Chapter 7, consider learning Bayesian logic programs. SCOOBY [26] is a greedy hill-climbing approach for learning Bayesian logic programs. SCOOBY takes the initial Bayesian logic program $H = (L, \lambda)$ as starting point and computes the parameters maximizing $score(L, \lambda, E)$. Then, refinement operators generalizing respectively specializing H are used to to compute all legal neighbors of H in the hypothesis space, see Figure 6. Each neighbor is scored. Let

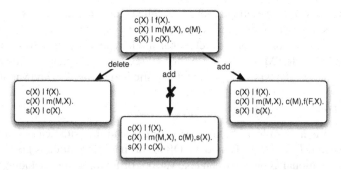

Fig. 6. The use of refinement operators during structural search within the framework of Bayesian logic programs. We can add an atom or delete an atom from the body of a clause. Candidates crossed out are illegal because they are cyclic. Other refinement operators are reasonable such as adding or deleting logically valid clauses.

$H' = (L', \lambda')$ be the legal neighbor scoring best. If $score(L, \lambda, E) < score(L', \lambda', E)$ then SCOOBY takes H' as new hypothesis. The process is continued until no improvements in score are obtained.

SCOOBY is akin to theory revision approaches in inductive logic programming, which also form the basis for learning biochemical reaction models in Chapter 11. In case that only propositional clauses are considered, SCOOBY coincides with greedy hill-climbing approaches for learning Bayesian networks [19].

5.4   Learning from Probabilistic Proofs

Given a training set E containing ground proofs as examples, one possible approach to learning from observed proofs only combines ideas from the early ILP system GOLEM [42] that employs Plotkin's [48] least general generalization (LGG) with bottom-up generalization of grammars and hidden Markov models [59]. The resulting algorithm employs the likelihood of the proofs $score(L, \lambda, E)$ as the scoring function. It starts by taking as L_0 the set of ground clauses that have been used in the proofs in the training set and scores it to obtain λ_0. After initialization, the algorithm will then repeatedly select a pair of clauses in L_i, and replace the pair by their LGG to yield a candidate L'. The candidate that scores best is then taken as $H_{i+1} = (L_{i+1}, \lambda_{i+1})$, and the process iterates until the score no longer improves. One interesting issue is that strong logical constraints can be imposed on the LGG. These logical constraints directly follow from the fact that the example proofs should still be valid proofs for the logical component L of all hypotheses considered. Therefore, it makes sense to apply the LGG only to clauses that define the same predicate, that contain the same predicates, and whose (reduced) LGG also has the same length as the original clauses.

Preliminary results with a prototype implementation are promising. In one experiment, we generated from the target stochastic logic program

$$1 : s(A, B) \leftarrow np(Number, A, C), vp(Number, C, B).$$
$$1/2 : np(Number, A, B) \leftarrow det(A, C), n(Number, C, B).$$
$$1/2 : np(Number, A, B) \leftarrow pronom(Number, A, B).$$
$$1/2 : vp(Number, A, B) \leftarrow v(Number, A, B).$$
$$1/2 : vp(Number, A, B) \leftarrow v(Number, A, C), np(D, C, B).$$
$$1 : det(A, B) \leftarrow term(A, the, B).$$
$$1/4 : n(s, A, B) \leftarrow term(A, man, B).$$
$$1/4 : n(s, A, B) \leftarrow term(A, apple, B).$$
$$1/4 : n(pl, A, B) \leftarrow term(A, men, B).$$
$$1/4 : n(pl, A, B) \leftarrow term(A, apples, B).$$
$$1/4 : v(s, A, B) \leftarrow term(A, eats, B).$$
$$1/4 : v(s, A, B) \leftarrow term(A, sings, B).$$
$$1/4 : v(pl, A, B) \leftarrow term(A, eat, B).$$
$$1/4 : v(pl, A, B) \leftarrow term(A, sing, B).$$
$$1 : pronom(pl, A, B) \leftarrow term(A, you, B).$$
$$1 : term([A|B], A, B) \leftarrow$$

(independent) training sets of 50, 100, 200, and 500 proofs. For each training set, 4 different random initial sets of parameters were tried. We ran the learning algorithm on

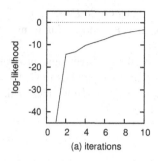
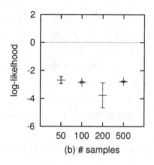

(a) iterations   (b) # samples

Fig. 7. Experimental results on learning stochastic logic programs from proofs. (a) A typical learning curve. (b) Final log-likelihood averaged over 4 runs. The error bars show the standard deviations.

each data set starting from each of the initial sets of parameters. The algorithm stopped when a limit of 200 iterations was exceeded or a change in log-likelihood between two successive iterations was smaller than 0.0001.

Figure 7 (a) shows a typical learning curve, and Figure 7 (b) summarizes the overall results. In all runs, the original structure was induced from the proof-trees. Moreover, already 50 proof-trees suffice to rediscover the structure of the original stochastic logic program. Further experiments with 20 and 10 samples respectively show that even 20 samples suffice to learn the given structure. Sampling 10 proofs, the original structure is rediscovered in one of five experiments. This supports that the *learning from proof trees* setting carries a lot information. Furthermore, our methods scales well. Runs on two independently sampled sets of 1000 training proofs yield similar results: -4.77 and -3.17, and the original structure was learned in both cases. More details can be found in [10].

Other statistical relational learning frameworks that have been developed within the learning from proofs setting are relational Markov models [2] and logical hidden Markov models [28,29, see also Chapters 2 and 10].

5.5 Probabilistic Learning from Entailment

This setting has been investigated for learning stochastic logic programs [39,40,5,41] and for parameter estimation of PRISM programs [54,23] from observed examples only, cf. Chapters 5 and 11. Here, we will illustrate a promising, alternative approach with less computational complexity, which adapts FOIL [50] with the conditional likelihood as described in Equation (9) as the scoring function $score(L, \lambda, E)$. This idea has been followed with NFOIL, see [31] for more details.

Given a training set E containing positive and negative examples (i.e. true and false ground facts), this algorithm stays in the learning from observed examples only to induce a probabilistic logical model to distinguish between the positive and negative examples. It computes Horn clause features $b_1, b_2, \ldots$ in an outer loop. It terminates when no further improvements in the score are obtained, i.e, when $score(\{b_1, \ldots, b_i\}, \lambda_i, E) < score(\{b_1, \ldots, b_{i+1}\}, \lambda_{i+1}, E)$, where λ denotes the maximum likelihood parameters. A major difference with FOIL is, however, that the

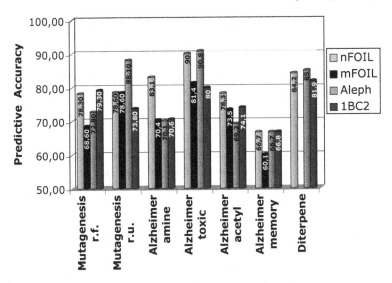

Fig. 8. Cross-validated accuracy results of NFOIL on ILP benchmark data sets. For **Mutagenesis r.u.**, leave-one-out cross-validated accuracies are reported because of the small size of the data set. For all other domains, 10-fold cross-validated results are given. **mFOIL** [32] and **Aleph** [56] are standard ILP algorithms. **1BC2** [16] is a first order logical variant of Naïve Bayes. For **1BC2**, we do not test significance because the results on **Mutagenesis** are taken from [16]. **Diterpene** is a multiclass problem but **mFOIL** has been developed for two-class problems only. Therefore, we do not report results for **mFOIL** on Diterpene.

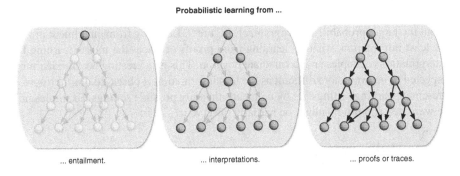

Fig. 9. The level of information on the target probabilistic program provided by probabilistic ILP settings: shaded parts denote unobserved information. Learning from entailment provides the least information. Only roots of proof tree are observed. In contrast, learning from proofs or traces provides the most information. All ground clauses and atoms used in proofs are observed. Learning from interpretations provides an intermediate level of information. All ground atoms but not the clauses are observed.

covered positive examples are *not* removed. The inner loop is concerned with inducing the next feature b_{i+1} top-down, i.e., from general to specific. To this aim it starts with a clause with an empty body, e.g., muta(M) ←. This clause is then specialized by repeatedly adding atoms to the body, e.g., $\texttt{muta(M)} \leftarrow \texttt{bond(M, A, 1)}$, $\texttt{muta(M)} \leftarrow$ $\texttt{bond(M, A, 1)}, \texttt{atom(M, A, c, 22, \_)}$, etc. For each refinement b'_{i+1} we then compute the maximum-likelihood parameters λ'_{i+1} and $score(\{b_1, \ldots, b'_{i+1}\}, \lambda'_{i+1}, E)$. The refinement that scores best, say b''_{i+1}, is then considered for further refinement and the refinement process terminates when $score(\{b_1, \ldots, b_{i+1}\}, \lambda_{i+1}, E) <$ $score(\{b_1, \ldots, b''_{i+1}\}, \lambda''_{i+1}, E)$. As Figure 8 shows, NFOIL performs well compared to other ILP systems on traditional ILP benchmark data sets. MFOIL and ALEPH, two standard ILP systems, were never significantly better than NFOIL (paired sampled t-test, $p = 0.05$). NFOIL achieved significantly higher predictive accuracies than MFOIL on Alzheimer amine, toxic, and acetyl. Compared to ALEPH, NFOIL achieved significantly higher accuracies on Alzheimer amine and acetyl (paired sampled t-test, $p = 0.05$). For more details, we refer to [31].

6   Conclusions

This chapter has defined the formal framework of *probabilistic ILP* for statistical relational learning and presented three learning setting settings: *probabilistic learning from entailment, from interpretations*, and *from proofs*. They differ in their representation of examples and the corresponding covers relation. The probabilistic ILP settings and learning approaches are by no means the only possible settings for probabilistic ILP. Nevertheless, two of the settings have – to the best of our knowledge – not been introduced before. Furthermore, we have sketched how the settings combine and generalize ILP and statistical learning. Finally, we have shown how state-of-the-art SRL frameworks fit into these learning settings

At present, it is still an open question as to what the relation among these different settings is. It is, however, apparent that they provide different levels of information about the target probabilistic program, cf. Figure 9. Learning from entailment provides the least information, whereas learning from proofs or traces the most. Learning from interpretations occupies an intermediate position. This is interesting because learning is expected to be even more difficult as the less information is observed. Furthermore, the presented learning settings are by no means the only possible settings. Examples might be weighted and proofs might be partially observed.

Acknowledgements

This work was supported by the European Union, contract number FP6-508861, Applications of Probabilistic Inductive Logic Programming II.

References

1. Abney, S.P.: Stochastic Attribute-Value Grammars. Computational Linguistics 23(4), 597–618 (1997)
2. Anderson, C.R., Domingos, P., Weld, D.S.: Relational Markov Models and their Application to Adaptive Web Navigation. In: Hand, D., Keim, D., Ng, R. (eds.) Proceedings of the Eighth International Conference on Knowledge Discovery and Data Mining (KDD 2002), Edmonton, Canada, July 2002, pp. 143–152. ACM Press, New York (2002)

3. Bangsø, O., Langseth, H., Nielsen, T.D.: Structural learning in object oriented domains. In: Russell, I., Kolen, J. (eds.) Proceedings of the Fourteenth International Florida Artificial Intelligence Research Society Conference (FLAIRS 2001), Key West, Florida, USA, pp. 340–344. AAAI Press, Menlo Park (2001)

4. Cussens, J.: Loglinear models for first-order probabilistic reasoning. In: Laskey, K.B., Prade, H. (eds.) Proceedings of the Fifteenth Annual Conference on Uncertainty in Artificial Intelligence (UAI 1999), Stockholm, Sweden, pp. 126–133. Morgan Kaufmann, San Francisco (1999)

5. Cussens, J.: Parameter estimation in stochastic logic programs. Machine Learning Journal 44(3), 245–271 (2001)

6. Cussens, J.: Integrating by separating: Combining probability and logic with ICL, PRISM and SLPs. Technical report, APrIL Projetc (January 2005)

7. De Raedt, L.: Logical settings for concept-learning. Artificial Intelligence Journal 95(1), 197–201 (1997)

8. De Raedt, L., Džeroski, S.: First-Order jk-Clausal Theories are PAC-Learnable. Artificial Intelligence Journal 70(1-2), 375–392 (1994)

9. De Raedt, L., Kersting, K.: Probabilistic Logic Learning. ACM-SIGKDD Explorations: Special issue on Multi-Relational Data Mining 5(1), 31–48 (2003)

10. De Raedt, L., Kersting, K., Torge, S.: Towards learning stochastic logic programs from proof-banks. In: Veloso, M., Kambhampati, S. (eds.) Proceedings of the Twentieth National Conference on Artificial Intelligence (AAAI 2005), Pittsburgh, Pennsylvania, USA, July 9–13, 2005, pp. 752–757. AAAI (2005)

11. Dempster, A.P., Laird, N.M., Rubin, D.B.: Maximum likelihood from incomplete data via the EM algorithm. Journal of the Royal Statistical Society B 39, 1–39 (1977)

12. Domingos, P., Richardson, M.: Markov Logic: A Unifying Framework for Statistical Relational Learning. In: Dietterich, T.G., Getoor, L., Murphy, K. (eds.) Proceedings of the ICML-2004 Workshop on Statistical Relational Learning and its Connections to Other Fields (SRL 2004), Banff, Alberta, Canada, July 8, 2004, pp. 49–54 (2004)

13. Džeroski, S., Lavrač, N. (eds.): Relational data mining. Springer, Berlin (2001)

14. Eisele, A.: Towards Probabilistic Extensions of Contraint-based Grammars. In: Dörne, J. (ed.) Computational Aspects of Constraint-Based Linguistics Decription-II, DYNA-2 deliverable R1.2.B (1994)

15. Flach, P.: Simply Logical: Intelligent Reasoning by Example. John Wiley, Chichester (1994)

16. Flach, P.A., Lachiche, N.: Naive Bayesian classification of structured data. Machine Learning Journal 57(3), 233–269 (2004)

17. Getoor, L.: Learning Statistical Models from Relational Data. PhD thesis, Stanford University, USA (June 2001)

18. Getoor, L., Friedman, N., Koller, D., Taskar, B.: Learning Probabilistic Models of Link Structure. Journal of Machine Leaning Research (JMLR) 3, 679–707 (2002)

19. Heckerman, D.: A Tutorial on Learning with Bayesian Networks. Technical Report MSR-TR-95-06, Microsoft Research (1995)

20. Helft, N.: Induction as nonmonotonic inference. In: Brachman, R.J., Levesque, H.J. (eds.) Proceedings of the First International Conference on Principles of Knowledge Representation and Reasoning(KR 1989), Toronto, Canada, May 15-18, 1989, pp. 149–156. Morgan Kaufmann, San Francisco (1989)

21. Jäger, M.: Relational Bayesian Networks. In: Laskey, K.B., Prade, H. (eds.) Proceedings of the Thirteenth Conference on Uncertainty in Artificial Intelligence (UAI 1997), Stockholm, Sweden, July 30–August 1, 1997, pp. 266–273. Morgan Kaufmann, San Francisco (1997)

22. Jensen, F.V.: Bayesian networks and decision graphs. Springer, Heidelberg (2001)

23. Kameya, Y., Sato, T., Zhou, N.-G.: Yet more efficient EM learning for parameterized logic programs by inter goal sharing. In: de Mantaras, R.L., Saitta, L. (eds.) Proceedings of the 16th European Conference on Artificial Intelligence (ECAI 2004), Valencia, Spain, August 22-27, 2004, pp. 490–494. IOS Press, Amsterdam (2004)

24. Kersting, K.: Bayes'sche-logische Programme. Master's thesis, Institute for Computer Science, University of Freiburg (2000)

25. Kersting, K., De Raedt, L.: Bayesian logic programs. Technical Report 151, Institute for Computer Science, University of Freiburg, Freiburg, Germany (April 2001)

26. Kersting, K., De Raedt, L.: Towards Combining Inductive Logic Programming with Bayesian Networks. In: Rouveirol, C., Sebag, M. (eds.) ILP 2001. LNCS (LNAI), vol. 2157, pp. 118–131. Springer, Heidelberg (2001)

27. Kersting, K., De Raedt, L.: Bayesian Logic Programming: Theory and Tool. In: Getoor, L., Taskar, B. (eds.) An Introduction to Statistical Relational Learning, pp. 291–321. MIT Press, Cambridge (2007)

28. Kersting, K., De Raedt, L., Raiko, T.: Logial Hidden Markov Models. Journal of Artificial Intelligence Research (JAIR) 25, 425–456 (2006)

29. Kersting, K., Raiko, T.: 'Say EM' for Selecting Probabilistic Models for Logical Sequences. In: Bacchus, F., Jaakkola, T. (eds.) Proceedings of the Twenty-First Conference on Uncertainty in Artificial Intelligence (UAI 2005), Edinburgh, Scotland, July 26-29, 2005, pp. 300–307 (2005)

30. Koller, D., Pfeffer, A.: Learning probabilities for noisy first order rules. In: Pollack, M. (ed.) Proceedings of the Fifteenth International Joint Conference on Artificial Intelligence (IJCAI 1997), Nagoya, Japan, pp. 1316–1321. Morgan Kaufmann, San Francisco (1997)

31. Landwehr, N., Kersting, K., De Raedt, L.: nFOIL: Integrating Naïve Bayes and Foil. In: Veloso, M., Kambhampati, S. (eds.) Proceedings of the Twentieth National Conference on Artificial Intelligence (AAAI 2005), Pittsburgh, Pennsylvania, USA, July 9–13, 2005, pp. 795–800. AAAI Press, Menlo Park (2005)

32. Lavrač, N., Džeroski, S.: Inductive Logic Programming. Ellis Horwood (1994)

33. Lloyd, J.W.: Foundations of Logic Programming, 2nd edn. Springer, Berlin (1989)

34. Manning, C.H., Schütze, H.: Foundations of Statistical Natural Language Processing. MIT Press, Cambridge (1999)

35. Marcus, M., Kim, G., Marcinkiewicz, M.A., MacIntyre, R., Bies, A., Ferguson, M., Katz, K., Schasberger, B.: The Penn treebank: Annotating predicate argument structure. In: Weinstein, C.J. (ed.) In ARPA Human Language Technology Workshop, Plainsboro, NJ, USA, March 8–11, 1994, pp. 114–119 (1994)

36. McLachlan, G., Krishnan, T.: The EM Algorithm and Extensions. Wiley, New York (1997)

37. Muggleton, S.H.: Inverse Entailment and Progol. New Generation Computing Journal, 245–286 (1995)

38. Muggleton, S.H.: Stochastic logic programs. In: De Raedt, L. (ed.) Advances in Inductive Logic Programming, pp. 254–264. IOS Press, Amsterdam (1996)

39. Muggleton, S.H.: Learning Stochastic Logic Programs. Electronic Transactions in Artificial Intelligence 4(041) (2000)

40. Muggleton, S.H.: Learning stochastic logic programs. In: Getoor, L., Jensen, D. (eds.) Working Notes of the AAAI-2000 Workshop on Learning Statistical Models from Relational Data (SRL 2000), Austin, Texas, July 31, 2000, pp. 36–41. AAAI Press, Menlo Park (2000)

41. Muggleton, S.H.: Learning Structure and Parameters of Stochastic Logic Programs. In: Matwin, S., Sammut, C. (eds.) ILP 2002. LNCS (LNAI), vol. 2583, Springer, Heidelberg (2003)

42. Muggleton, S.H., Feng, C.: Efficient Induction of Logic Programs. In: Muggleton, S.H. (ed.) Inductive Logic Programming, Acadamic Press (1992)

43. Muggleton, S.H., De Raedt, L.: Inductive Logic Programming: Theory and Methods. Journal of Logic Programming 19(20), 629–679 (1994)
44. Neville, J., Jensen, D.: Dependency Networks for Relational Data. In: Rastogi, R., Morik, K., Bramer, M., Wu, X. (eds.) Proceedings of The Fourth IEEE International Conference on Data Mining (ICDM 2004), Brighton, UK, November 1–4, 2004, pp. 170–177. IEEE Computer Society Press, Los Alamitos (2004)
45. Ngo, L., Haddawy, P.: Answering Queries from Context-Sensitive Probabilistic Knowledge Bases. Theoretical Computer Science 171, 147–177 (1997)
46. Pearl, J.: Reasoning in Intelligent Systems: Networks of Plausible Inference, 2nd edn. Morgan Kaufmann, San Francisco (1991)
47. Pfeffer, A.J.: Probabilistic Reasoning for Complex Systems. PhD thesis, Computer Science Department, Stanford University (December 2000)
48. Plotkin, G.D.: A note on inductive generalization. In: Machine Intelligence, vol. 5, pp. 153–163. Edinburgh University Press (1970)
49. Poole, D.: Probabilistic Horn abduction and Bayesian networks. Artificial Intelligence Journal 64, 81–129 (1993)
50. Quinlan, J.R., Cameron-Jones, R.M.: Induction of logic programs: FOIL and related systems. New Generation Computing, 287–312 (1995)
51. Riguzzi, F.: Learning logic programs with annotated disjunctions. In: Camacho, R., King, R., Srinivasan, A. (eds.) ILP 2004. LNCS (LNAI), vol. 3194, pp. 270–287. Springer, Heidelberg (2004)
52. Russell, S.J., Norvig, P.: Artificial Intelligence: A Modern Approach, 2nd edn. Prentice-Hall, Inc., Englewood Cliffs (2002)
53. Sato, T.: A Statistical Learning Method for Logic Programs with Distribution Semantics. In: Sterling, L. (ed.) Proceedings of the Twelfth International Conference on Logic Programming (ICLP 1995), Tokyo, Japan, pp. 715–729. MIT Press, Cambridge (1995)
54. Sato, T., Kameya, Y.: Parameter Learning of Logic Programs for Symbolic-Statistical Modeling. Journal of Artificial Intelligence Research (JAIR) 15, 391–454 (2001)
55. Shapiro, E.: Algorithmic Program Debugging. MIT Press, Cambridge (1983)
56. Srinivasan, A.: The Aleph Manual (1999), Available at: http://www.comlab.ox.ac.uk/oucl/~research/areas/machlearn/Aleph/
57. Srinivasan, A., Muggleton, S.H., King, R.D., Sternberg, M.J.E.: Theories for Mutagenicity: A Study of First-Order and Feature based Induction. Artificial Intelligence Journal 85, 277–299 (1996)
58. Sterling, L., Shapiro, E.: The Art of Prolog: Advanced Programming Techniques. MIT Press, Cambridge (1986)
59. Stolcke, A., Omohundro, S.: Hidden Markov model induction by Bayesian model merging. In: Hanson, S.J., Cowan, J.D., Giles, C.L. (eds.) Advances in Neural Information Processing Systems, vol. 5, pp. 11–18. Morgan Kaufmann, San Francisco (1993); (Proceedings of NIPS-92, Denver, Colorado, USA, November 30–December 3 (1992)
60. Taskar, B., Abbeel, P., Koller, D.: Discriminative Probabilistic Models for Relational Data. In: Darwiche, A., Friedman, N. (eds.) Proceedings of the Eighteenth Conference on Uncertainty in Artificial Intelligence (UAI 2002), Edmonton, Alberta, Canada, August 1-4, 2002, pp. 485–492 (2002)
61. Valiant, L.G.: A theory of the Learnable. Communications of the ACM 27(11), 1134–1142 (1984)
62. Vennekens, J., Verbaeten, S., Bruynooghe, M.: Logic Programs with Annotated Disjunctions. In: Demoen, B., Lifschitz, V. (eds.) Proceedings of 20th International Conference on Logic Programming (ICLP 2004), Saint-Malo, France, September 6-10, 2004, pp. 431–445 (2004)

Relational Sequence Learning

Kristian Kersting[1], Luc De Raedt[2], Bernd Gutmann[2],
Andreas Karwath[3], and Niels Landwehr[3]

[1] CSAIL, Massachusetts Institute of Technology
32 Vassar Street, Cambridge, MA 02139-4307, USA
`kersting@csail.mit.edu`
[2] Departement Computerwetenschappen, K.U. Leuven
Celestijnenlaan 200A - bus 2402, B-3001 Heverlee, Belgium
{`Luc.DeRaedt,Bernd.Gutmann`}`@cs.kuleuven.be`
[3] Machine Learning Lab, Institute for Computer Science, University of Freiburg
Georges-Koehler Allee, Building 079, 79110 Freiburg, Germany
{`landwehr,karwath`}`@informatik.uni-freiburg.de`

Abstract. Sequential behavior and sequence learning are essential to
intelligence. Often the elements of sequences exhibit an internal structure that can elegantly be represented using relational atoms. Applying
traditional sequential learning techniques to such relational sequences
requires one either to ignore the internal structure or to live with a combinatorial explosion of the model complexity. This chapter briefly reviews
relational sequence learning and describes several techniques tailored towards realizing this, such as local pattern mining techniques, (hidden)
Markov models, conditional random fields, dynamic programming and
reinforcement learning.

1   Introduction

The activities humans perform are by their very nature sequential. Sequences
occur in many activities ranging from reasoning to language understanding, in
everyday skills as well as in complex problem solving. The ability to learn from
such sequences is therefore essential to artificial intelligence and sequence learning techniques can be applied in many domains such as planning, reasoning,
robotics, user modeling, natural language processing, speech recognition, adaptive control, activity recognition, information extraction, and computational biology. This explains why learning from sequences has received a lot of attention
in the past few decades. Learning tasks investigated include classification, prediction, local pattern mining, labeling, alignment, transduction, and density and
policy estimation.

One major dimension along which to differentiate sequential learning techniques is the complexity of the language they employ to describe sequences and
models. At one extreme are learning approaches that assume a propositional
language. The simplicity of a propositional language allows such methods to
represent the model in matrix form: cells typically denote the transition probabilities among symbols. In turn, efficient matrix operations can be used to devise

L. De Raedt et al. (Eds.): Probabilistic ILP 2007, LNAI 4911, pp. 28–55, 2008.

efficient algorithms. At the other end of the spectrum, (probabilistic) relational systems accept descriptions of complex, structured sequence elements and generate relationally structured models. They typically have access to background knowledge and allow for a more compact description of the entities, but often have also a higher computational complexity. This chapter reviews several relational sequences learning techniques that build on ideas developed on both sides of the spectrum. They fill an interesting, intermediate position on the expressiveness scale by using sequences of relational atoms. Such sequences are more expressive than their propositional counter part, while at the same time they are less expressive (and hence, often more efficient to deal with) than the fully relational models discussed in the rest of this book.

The following section briefly reviews sequential learning. After illustrating the limitations of propositional languages, the chapter introduces the more complex data model of relational sequences. In the remaining sections, we will then present methods for dealing with relational sequences. This includes: local pattern mining [22], relational alignment [14], r-grams [20] logical hidden Markov models [16], logical conditional random fields [10] and relational Markov decision processes [15].

2 Sequential Learning

Consider sequences of UNIX commands. They typically tell a lot about the user's behavior since users tend to respond in a similar manner to similar situations, leading to repeated sequences of actions.

Example 1. For instance, LATEX users frequently run EMACS to edit their LATEX files and afterwards compile the edited file using LATEX:

$$\text{emacs rsl.tex}, \text{ls}, \text{latex dvips.tex}, \text{dvips rsl} \ldots \tag{1}$$

The existence of command alias mechanisms in many UNIX command interpreters also supports the idea that users tend to enter many repeated sequences of commands. Thus, UNIX command sequences carry a lot information, which can be used to automatically construct user profiles, which in turn can be used to predict the next command, to identify the current user, etc.

In general, sequence learning considers essentially strings (command logs) $s = \mathtt{w}_1, \mathtt{w}_2, \ldots, \mathtt{w}_T$ ($T > 0$) of symbols $\mathtt{w}_i$ (UNIX commands) over an alphabet Σ. We now identify various sequence learning problems (in a somewhat simplified form): In *sequence prediction* the aim is to predict elements (commands) of a sequence based on preceding elements (commands), i.e., $\mathtt{w}_{t-k}, \mathtt{w}_{t-k+1}, \ldots, \mathtt{w}_t \to \mathtt{w}_{t+1}$. In *frequent sequence mining* the aim is to compute the (sub)sequences frequently occurring in a set of sequences. In *sequence classification* one aims at predicting a single class label (user or user type) $\mathtt{c}$ that applies to an entire sequence s, i.e., $s \to \mathtt{c}$. In *sequence labeling*, the goal is to assign a (class) labels (shell sessions) $\mathtt{c}_i$ to each sequence element $\mathtt{w}_i$, i.e., $\mathtt{w}_1, \mathtt{w}_2, \ldots, \mathtt{w}_T \to \mathtt{c}_1, \mathtt{c}_2, \ldots, \mathtt{c}_T$. *Sequential decision making* involves selecting sequences of actions to accomplish a goal or to maximize the future reward function (for instance to optimally organize email folders). In addition, there are also other issues related to these sequence learning tasks. For example, we may want to segment a sequence, cluster sequences, align two or more sequences, or compute a general mapping between sequences realizing transduction.

Another dimension along which sequential learning can be characterized is the learning paradigm employed. Learning tasks can be supervised, unsupervised, or reinforced. In all cases, however, sequence learning methods essentially rely on models for "legitimate" sequences (in the form of production rules, Markov chains, hidden Markov models, or some other form), which can typically be developed from data using grammar induction, expectation-maximization, gradient descent, policy iteration or some other form of machine learning algorithm. The motivation for this chapter is that all prominent types of models that have been investigated over the last decades share a principal weakness stemming from a lack of expressive power in the language used to described sequences and models, as we shall show in the next Section.

3   Moving to More Complex Sequences

Prominent sequence learning techniques such as (hidden) Markov models assume atomic representations, which essentially amounts to explicitly enumerating all unique configurations or states. It might then be possible to learn, for example, that state state234 follows (with high probability) state654321. Atomic representations are simple and learning can be implemented using efficient matrix operations. These matrices, however, can become intractably large as they scale quadratically in the size of the language, i.e. the number of states.

In many applications, sequence elements are indeed structured and can elegantly be represented as relational *ground atoms*.

Example 2. Using ground atoms, the UNIX command sequence of Example 1 can be represented as

$$emacs(rsl, tex), ls, latex(rsl, tex), dvips(rsl, dvi) \ldots$$

Here, emacs/2, ls/0, latex/1, dvips/2 are *predicates* (of arity 2, 0 and 1 respectively) that identify relations. Lower-case strings like rsl, tex, and dvi are *constants*. *Ground atoms* are now predicates together with their arguments, for example emacs(rsl, tex) and ls. In principle, symbols can even be described propositionally, i.e., conjunctions of symbols (the ground atoms) that ignore the structure in. Using a propositional representation, each of the possible ground atoms becomes a symbol. For instance, the conjunction file(f1), name(f1, rsl), suffix(f1, tex) describes that there is a file with name rsl and suffix tex. Though this propositional representation allows for some opportunities for generalization, it ignores the structure exhibited in the domain, for instance, that latex takes two arguments, and also, each of the symbols contains the names and identifiers of specific entities such as f1. This prevents generalization over

several entities such as emacs(X, tex). The *abstract symbol* emacs(X, tex) is —
by definition — a logical atom, i.e., a predicate together with it arguments,
where an argument can now be a placeholder X, Y, . . . for some constant. It is abstract in that it represents the set of all ground, i.e., variable-free atoms such as
emacs(rsl, tex), emacs(rsl, dvi), emacs(april, tex) etc. Moreover, unification
allows one to share information between subsequent symbols. For example, the
(abstract) sub-sequence latex(X, tex), dvips(X, dvi) describes that a user, after
compiling a LATEXfile into a DVI file, turns the DVI into a POSTSCRIPT file, without stating the name of the file. This is especially important when generalizing
patterns across filenames as the objects referred to will typically be different,
and the precise identifiers do not matter but the relationships they occur in, do.

Having specified a more complex language to describe sequences, the next step
is to develop sequential learning methods capable of using these representations.
This is what relational sequence learning is about. In the remaining sections,
we will review several relational sequence learning and mining methods that
have been proven successful in applications. Their underlying idea is to make
use of relational abstraction: similar symbols are grouped together by means of
logical variables and knowledge is shared across abstract symbols by means of
unification. More precisely, we will discuss relational sequence mining, alignment,
Markov models, and reinforcement learning in turn.

4   Mining Logical Sequences

Many of the traditional data mining tasks can be phrased as that of finding the
set of patterns $Th(\mathcal{L}, D, q) = \{\phi \in \mathcal{L} | q(\phi, D)$ holds $\}$, cf. [23]. Here, $\mathcal{L}$ is the
space or language of all possible patterns, D is a set of observations, and q is a
predicate or constraint that characterizes the solutions to the data mining task.

The MineSeqLog algorithm of [22] (see also [4]) is a constraint based pattern
mining system for logical sequences. The basic component is a frequent pattern
miner, which makes the following choices in $Th(\mathcal{L}, D, q)$:

  – D is a set of ground logical sequences over an alphabet Σ
  – $\mathcal{L}$ consists of the abstract sequences over Σ, (in which variables can occur)
  – q is a constraint of the form $freq(\phi, D) \geq t$ expressing that the pattern ϕ
    must cover at least t of the sequences in D

This formulation makes some simplifications, in that MineSeqLog can also cope
with sequences with gaps as well as with other constraints than a minimum
frequency threshold, cf. [22].

The key constraint is the minimum frequency threshold. The *frequency*,
$freq(\phi, D)$, of a pattern (in the form of an abstract sequence) ϕ is the number of sequences s in D for which ϕ subsumes s. A sequence $s = w_1, w_2, \ldots, w_T$ is
subsumed by a pattern $\phi = p_1, \ldots, p_k$ if and only if there exists a substitution
θ and natural numbers i, j such that $p_1\theta = w_i, p_2\theta = w_{i+1}, \ldots, p_k\theta = w_j$. For instance, the pattern latex(File, tex), dvipdf(File, dvi) subsumes the concrete
sequence cd(april), latex(par, tex), dvipdf(par, dvi), lpr(par, pdf) with substitution $\theta = \{File/par\}$. We sometimes will say that ϕ is more general than s,

or vice versa, that s is more specific than ϕ. The subsumption relation induces a partial order on the language $\mathcal{L}$, which is used in order to structure the search for frequent patterns. The subsumption ordering can be exploited because the minimum frequency constraint is anti-monotonic. More formally, a constraint q is *anti-monotonic* if and only if $\forall$ sequences x: $\left(x \text{ subsumes } y \wedge p(y) \rightarrow p(x)\right)$. It is easy to see that this holds for frequency because the frequency can only decrease when refining patterns. The anti-monotonicity property implies that there is a border of maximally specific sequences satisfying the constraint.

The set $Th(\mathcal{L}, D, freq(\phi, D))$ can now be computed by instantiating the traditional level-wise algorithm of Mannila and Toivonen [23], which is essentially a breadth-first general-to-specific search algorithm. To generate more specific patterns from more general ones, a refinement operator ρ is employed. A refinement operator ρ is an operator that maps each sequence s to a set of specializations of it, i.e. $\rho(s) \subseteq \{s' \in \mathcal{L} \mid s \text{ subsumes } s'\}$. Furthermore, to avoid generating the same pattern more than once, the operator should be *optimal*, i.e.

Complete. By applying the operator ρ on ϵ, the empty sequence (possibly with repetitions), it must be possible to generate all other queries in $\mathcal{L}$. This requirement guarantees that we will not miss any queries that may satisfy the constraints.

Single path. Given pattern p, there should exist exactly one sequence of patterns $p_0 = \epsilon, p_1, \ldots, p_T = p$ such that $p_{i+1} \in \rho(p_i)$ for all i. This requirement helps ensuring that no query is generated more than once, i.e. there are no duplicates.

The following operator satisfies these requirements. $\rho(\mathbf{s}_1, \ldots, \mathbf{s}_l)$ is obtained by applying one of the following operations.

Add an atom $\mathbf{s}_{l+1}$ to the right of the query such that $\mathbf{s}_{l+1}$ is an atom whose arguments are different variables not yet occurring in $\mathbf{s}_1, \ldots, \mathbf{s}_l$

Apply a substitution of the form $\theta = \{X/c\}$, where X is a variable, c a constant such that there are no constants occurring to the right of X in $\mathbf{s}_1, \ldots, \mathbf{s}_l$, and all variables in $\mathbf{s}_1, \ldots, \mathbf{s}_l$ are different

Unify two variables X and Y such that X occurs only once, all variables to the right of X occur only once, and X occurs to the right of Y.

This operator can then be integrated in the standard level-wise algorithm for frequent pattern mining. This algorithm is sketched below. It starts from the empty sequence and repeatedly generates candidates (on C_i) to determine afterwards (using F_i) whether they are frequent. To generate candidates the refinement operator ρ is applied. Furthermore, only frequent sequences are refined due to the anti-monotonicity property. This process continues until no further candidates are frequent.

The MineSeqLog algorithm as described by [22] cannot only cope with anti-monotonic constraints, but also with monotonic ones, and even conjunctions of the two. A maximum frequency threshold, which is of the form $freq(\phi, D) < f$, is a monotonic constraint. [22] also report on experiments with MineSeqLog using the Unix-command data set of [9] and constraints of the form $(freq(\phi, ClassA) \geq f_1) \wedge (freq(\phi, ClassB) < f_2)$.

Algorithm 1. Computing the frequent sequences

$i := 0;\ C_0 := \{\epsilon\};\ F_0 := \emptyset$
while $C_i \neq \emptyset$ **do**
$\quad F_i := \{p \in C_i \mid freq(p, D) \geq t\}$
$\quad$**output** F_i
$\quad C_{i+1} := \{p \mid p \in \rho(p'), p' \in F_i\}$
$\quad i := i + 1$

5   Relational Alignments

The need to measure sequence similarity arises in many application domains and often coincides with sequence alignment: the more similar two sequences are, the better they can be aligned. Aligning sequences not only shows how similar sequences are, it also shows where there are differences and correspondences between the sequences. As an example, consider aligning proteins, which is the major application area for sequence alignment in bioinformatics. One common approach is, given the amino acid sequence of an unknown protein (query sequence) to scan an existing database of other amino acid sequences (containing proteins with more or less known function) and extract the most similar ones with regard to the query sequence. The result is usually a list, ordered by some score, with the best hits at the top of this list. The common approach for biologists, is now to investigate these top scoring alignments or hits to conclude about the function, shape, or other features of query sequence.

5.1   Sequence Alignment Algorithms

One of the earliest alignment algorithms, based on dynamic programming, is that by Needleman and Wunsch [26] in 1970 for global alignment, i.e., an alignment that spans the entire length of sequences to be aligned. The algorithm, as shown in algorithm 2, is based on dynamic programming, and is able to find the alignment of two sequences A and B with the maximal overall similarity w.r.t. a given pairwise similarity model. More precisely, the algorithm proceeds as follows: initially, for two sequences of length l and k, a matrix with $l + 1$ columns and $k + 1$ rows is created. There is one column for each symbol in B and one row for each symbol in sequence A. The matrix then is filled with the maximum scores as follows:

$$M_{i,j} = \max \begin{cases} M_{i-1,j-1} + S_{i,j} & : \text{a match or mismatch} \\ M_{i,j-1} + w & : \text{a gap in sequence } B \\ M_{i-1,j} + w & : \text{a gap in sequence } A \end{cases} \tag{2}$$

where $S_{i,j}$ is the pairwise similarity of amino acids and w reflects a linear gap (insert or deletion step) penalty. So, as the algorithm proceeds, the $M_{i,j}$ will be assignedto the optimal score for the alignment of the first i symbols in A and the first j symbols in B. After filling the matrix, the maximum score for

---

Algorithm 2. Core procedure of the Needleman and Wunsch algorithm for global sequence alignment. The inputs are two sequences A and B and the output is the M matrix. For more details, see text.

---

for $i = 1$ *to length*$(A) - 1$ **do**
 | $M(0, i) = w \cdot i$;
for $i = 1$ *to length*$(B) - 1$ **do**
 | $M(0, j) = w \cdot j$;
for $i = 1$ *to length*(A) **do**
  for $j = 1$ *to length*(B) **do**
   $M(i, j) =$

$$\max \left(\underbrace{M(i - 1, j - 1) + S\left(A(i), B(j)\right)}_{(mis-)match}, \underbrace{M(i - 1, j) + w}_{gap\ B}, \underbrace{M(i, j - 1) + w}_{gap\ A} \right);$$

---

any alignment can be found in cell $M_{l,k}$. To compute which alignment actually gives this score, one start from cell $M_{l,k}$, and compares the value with the three possible sources ((mis-)match, gap in A, gap in B) to see which it came from. If (mis)-match, then A_i and B_j are aligned. Otherwise, A_i resp. B_j are aligned with a gap.

In the biological domain, this similarity model $S_{i,j}$ is typically represented by pair-wise similarity or dissimilarity scores of pairs of amino acids. These scores are commonly specified by a so-called similarity matrix, like the PAM [5] or BLOSUM [11] families of substitution matrices. The scores, or costs, associated with a match or mismatch between two amino acids, are sometimes interpreted as the probability that this change in amino acids might have occurred during evolution.

The Needleman-Wunsch algorithm computes global alignments, i.e. it attempts to align every element in one sequence with an element in the other one By contrast, local alignments identify regions of similarity within long sequences that are often widely divergent overall. To calculate the best *local* alignment of two sequences, one often employs the Smith-Waterman local alignment algorithm [33]. The main difference of this algorithm as compared to the Needleman-Wunsch algorithm is that all negative scores are set to zero. To compute the highest scoring local alignment, one now start with the maximal scoring matrix cell and proceed as before until a cell with score zero is encountered. For large data sets, the exhaustive Smith-Waterman approach is too computational intensive. For this reason, the BLAST algorithm [1] uses a heuristic approach, searching for small identical elements in both sequences and extending those using a dynamic programming approach similar to the Smith-Waterman algorithm. The BLAST family of algorithms is somewhat less accurate than the full Smith-Waterman algorithm, but much more efficient.

In general, the alignments resulting from a global or local alignment show the more *conserved* regions between two sequences. To enhance the detection of these conserved regions, multiple sequence alignments can be constructed. Given a number of sequences belonging to the same class, i.e. in biological terms believed to belong to the same family, fold, or other classes, so-called profiles are constructed that align all sequences together. A common approach for the construction of a multiple alignment proceeds in three steps: First, all pairwise alignments are constructed. Second, using this information as starting point a phylogenetic tree is created as *guiding tree*. Third, using this tree, sequences are joined consecutively into one single alignment according to their similarity. This approach is known as the neighbor joining approach [30].

5.2   Moving Towards the Alignment of Relational Sequences

The alignment algorithms discussed in the previous paragraphs assume a given similarity measure $S_{i,j}$. Typically, this similarity measure is a propositional one as the considered sequences consist of propositional symbols. Many sequences occurring in real-world problems, however, can elegantly be represented as relational sequences. A *relational sequence alignment* simply denotes the alignment of sequences of such structured terms.

One attractive way to solve this problem is to use a standard alignment algorithm but to replace the propositional similarity measure $S_{i,j}$ in Equation (2) by a structured one. In [14] we proposed the use of one of the many distance measures developed within the field of Inductive Logic Programming [25]. As an example, consider one of the most basic measures proposed by Nienhuys-Cheng [27][1]. It treats ground structured terms as hierarchies, where the top structure is most important and the deeper, nested sub-structures are less important. Let S denote the set of all symbols, then Nienhuys-Cheng's distance d is inductively defined as follows:

$$\forall \mathsf{c}/0 \in S : \qquad\qquad d(\mathsf{c},\mathsf{c}) = 0$$
$$\forall \mathsf{p}/u, \mathsf{q}/v \in S : \mathsf{p}/u \neq \mathsf{q}/v : d(\mathsf{p}(\mathsf{t}_1,\ldots,\mathsf{t}_u), \mathsf{q}(\mathsf{s}_1,\ldots,\mathsf{s}_v)) = 1$$
$$\forall \mathsf{p}/u \in S : \qquad\qquad d(\mathsf{p}(\mathsf{t}_1,\ldots,\mathsf{t}_u), \mathsf{p}(\mathsf{s}_1,\ldots,\mathsf{s}_u)) = \tfrac{1}{2u} \sum_{i=1}^{u} d(\mathsf{t}_i, \mathsf{s}_i)$$

For different symbols the distance is one; however, when the symbols are the same, the distance linearly decreases with the number of arguments that have different values and is at most 0.5. The intuition is that longer tuples are more error-prone and that multiple errors in the same tuple are less likely. To solve the corresponding relational alignment problem, one simply sets $S_{i,j} = 1 - d(\mathsf{x}_i, \mathsf{y}_i)$ in Equation (2).

5.3   Relational Information Content

Now that we have introduced relational sequence alignments, we will investigate how informative they are. Following Gorodkin *et al.* [7], the information

---

[1] For sequences of more complex logical objects such as interpretations and queries, a different, appropriate similarity function could be chosen. We refer to Jan Ramon's PhD Thesis [29] for a nice review of them.

content I_i of position i of a relational sequence alignment is $I_i = \sum_{k \in G} I_{ik} = \sum_{k \in G} q_{ik} \log_2 \left(\frac{q_{ik}}{p_k} \right)$, where G is the Herbrand base over the language of the aligned sequences including gaps (denoted as '$-$') and q_{ik} is the fraction of ground atoms k at position i. When k is not a *gap*, we interpret p_k as the *a priori* distribution of the ground atom. Following Gorodkin *et al.*, we set $p_- = 1.0$, since then $q_{i-} \log_2(q_{i-}/p_-)$ is zero for q_{i-} equal to zero or one. We choose $p_k = 1/ (|G| - 1)$ when $k \neq -$. The intuition is as follows: *if I_{ik} is negative, we observe fewer copies of ground atom k at position i than expected, and vice versa, if I_{ik} is positive, we observe more that expected.*

The total information content becomes $I = \sum_{i=1}^{T} I_i$ (where T is the length of the alignment) and can be used to evaluate relational sequence alignments. So far, however, we have defined the information content at the most informative level only, namely the level of ground atoms. Relational sequences exhibit a rich internal structure and, therefore, multiple abstraction levels can be explored by using variables to make abstraction of specific symbols. To compute the information content at higher abstraction levels, i.e., of an atom a replacing all covered ground atoms k at position i, we view q_{ia} (resp. p_a) as the sum of q_{ik} (resp. p_k) of the ground atoms k covered by a. Figure 1 shows the (cumulative)

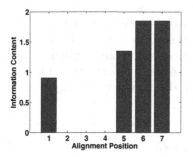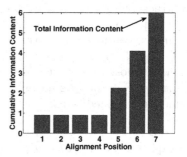

Fig. 1. Information content (IC) for the *Unix command line* example in table 1. The bar graph on the left-hand side shows the IC at each position in the alignment. The bar graph on the right-hand side shows the cumulative IC up to each position in the alignment.

Table 1. The multiple alignment of five arbitrary Unix command line sequences using gap opening cost 1.5, gap extension cost 0.5, and padding cost 0.25. The '-' denotes a gap in a sequence. Clearly one can see the aligned commands for xdvi, dvipdf, and pdfview. In sequence four, the corresponding vi and latex commands are not properly aligned due to the gap opening costs, as the proper alignment would require two gaps instead of the single one employed here.

1: -	vi(ch2,tex)	ls		latex(ch2,tex)	xdvi(ch2,dvi)	dvipdf(ch2,dvi)	pdfview(ch2,pdf)
2: cd(thesis)	vi(ch1,tex)	bibtex(ch1)	latex(ch1,tex)	xdvi(ch1,dvi)	dvipdf(ch1,dvi)	pdfview(ch1,pdf)
3: -	-	-	-	xdvi(pap2,dvi)	dvipdf(pap2,dvi)	pdfview(pap2,pdf)
4: cd(pap1)	-	-	vi(pap1,tex)	latex(pap1,tex)	dvipdf(pap1,dvi)	pdfview(pap1,pdf)
5: -	vi(rsl,tex)	-	latex(rsl,tex)	dvips(rsl,dvi)	-	-

information content for our running *Unix command line* example. As prior we use the empirical frequencies over all five sequences.

The information content is a significant concept as it allows one to evaluate alignments and to find common motifs in relational sequences. Moreover, it allows one to represent alignments graphically by so-called *relational sequence logos*.

5.4   Relational Sequence Logos

Reconsider the alignment in Table 1. It consists of several lines of information. This makes it for longer sequences difficult – if not impossible – to read off information such as the general consensus of the sequences, the order of predominance of the symbols at every position, their relative frequencies, the amount of information present at every position, and significant locations within the sequences. In contrast, the corresponding sequence logo as shown in Figure 2 concentrates all of this into a single graphical representation. In other words, 'a logo says more than a thousand lines alignment'.

Each position i in a *relational sequence logo* is represented by a stack consisting of the atoms at position i in the corresponding alignment. The height of the stack at position i indicates the information content I_i. The height h_{ik} of each atom k at position i is proportional to its frequency relative to the expected frequency, i.e., $h_{ik} = \alpha_i \cdot \left(\frac{q_{ik}}{p_k}\right) \cdot I_i$, where α_i is a normalization constant. The atoms are sorted according to their heights. If I_{ik} is negative, a simple bar is shown using the absolute value of the I_{ik}.

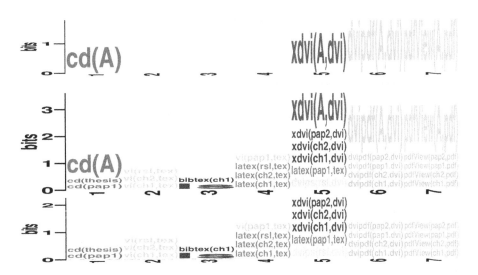

Fig. 2. Sequence logos for the *Unix command line* example in table 1 (from bottom to top: ground, relational, and abstract). For positions 1 as well as positions 5 -7 the abstract logo contributes substantially towards a conserved region.

Sequence logos at lower abstraction levels can become quite complex. Relational abstraction can be used to straighten them up. Reconsider Figure 2. It also shows the logo at the highest abstraction level, where we considered as symbols the *least general generalization* of all ground atoms over the same predicate at each position in the alignment only. Because the prior probabilities change dramatically, the *abstract logo* looks very different from the ground one. It actually highlights the more conserved region of the sequences at the end (positions 5-7). Both views provide relevant information. *Relational logos* combine these views by putting at each position the individual stack items together and sort them in ascending order of heights.

To summarize, relational sequence logos illustrate that while relational alignments can be quite complex, they exhibit rich internal structures which, if exploited, can lead to new insights not present in flat alignments. For applications to information extraction from MedLine abstracts and to protein fold description, we refer to [14].

Both *frequent sequence mining* and *sequence alignment* are nonparametric methods in that they do not assume an underlying model. We will now turn to probabilisitc model-based sequence learning methods, which are appealing as they take uncertainty explicitly into account.

6   Relational Grams

Relational Grams extend *n-gram* models to sequences of logical atoms. In a nutshell, *n*-gram models are smoothed Markov chains: they model the probability of a sequence $s = w_1...w_m$ as a mixture of Markov distributions of different orders. For $k \in \mathbb{N}$, a k-th order Markov chain estimates the probability for s as

$$P_k(w_1...w_m) = \prod_{i=1}^{m} P_k(w_i \mid w_{i-k+1}...w_{i-1}) \tag{3}$$

In the most basic case, the conditional probabilities are estimated from a set S of training sequences in terms of "gram" counts, i.e., counts of short patterns of symbols such as $w_{i-k+1}...w_{i-1}$:

$$P_k(w_i \mid w_{i-k+1} \ldots w_{i-1}) = \frac{C(w_{i-k+1} \ldots w_i)}{C(w_{i-k+1} \ldots w_{i-1})} \tag{4}$$

where $C(w_{i-k+1}...w_{i-1})$ is the number of times $w_{i-k+1}...w_{i-1}$ appeared as a subsequence in any $s \in S$. This is the maximum likelihood estimate for the model described by Equation 3.

In this model, the gram order k defines the trade-off between reliability of probability estimates and discriminatory power of the model. For larger k, many probability estimates will be zero due to data sparseness, which can deteriorate model accuracy. *n*-grams combine models of different order [24], and estimate the conditional probabilities as

$$P(w_i \mid w_{i-n+1} \ldots w_{i-1}) = \sum_{k=1}^{n} \alpha_k P_k(w_i \mid w_{i-k+1} \ldots w_{i-1}) \tag{5}$$

where the $\alpha_1, ..., \alpha_n$ are suitable weights with $\sum_{k=1}^{n} \alpha_k = 1$, and the distributions $P_k(w_i \mid w_{i-k+1} ... w_{i-1})$ are estimated according to Equation 4. This effectively smoothes the probability estimates of the higher-order models with the more robust estimates of lower-order models and thereby avoids the data sparseness problem. More advanced smoothing techniques have also been proposed (cf. [24]), but are beyond the scope of this chapter. Despite their simplicity, n-grams have proven to be a powerful tool for sequence classification and probability estimation.

By generalizing the sequence elements w_i to first-order logical atoms, relational grams (or *r-grams*) inherit the power and simplicity of the n-gram method. However, they go beyond a simple relational upgrade of n-grams in two ways:

1. **Relational Smoothing.** In addition to smoothing by shorter histories (as for n-grams), relational grams can smooth probability estimates by relational generalization of grams. For example, the gram $emacs(rsl, tex)$, $latex(rsl, tex)$ could be generalized by shortening it to $emacs(rsl, tex)$, but also by logical abstraction to $emacs(F, tex), latex(F, tex)$. Both generalizations avoid the data sparseness problem by estimating probabilities from a larger sample though they represent a different bias. In the second case, we model a pattern indicating that a user first runs $emacs$ and then $latex$ on the same file, independently of a particular filename.

2. **Abstraction from Identifiers.** Some arguments in the predicates used to define sequence elements which should never be grounded. Consider, for example, filenames in the Unix user modeling domain described above. File names, in contrast to file extensions, are just names—except for some system files, they are chosen more or less arbitrarily to describe the particular content of a file. Accordingly, it does not make sense to estimate distributions over filenames, especially if we want the induced models to generalize across different users, who typically name their files differently. Relational grams therefore provide a mechanism to abstract from such identifiers, and define distributions over ground sequences modulo identifier renaming.

More formally, r-gram models can be defined as follows. Let Σ denote a typed relational alphabet, for which the set of types is partitioned into *constants* and *identifiers*. We will also talk about *constant-variables* and *identifier-variables* depending on the variable's type. Let $\bar{\Sigma}$ denote the subset of Σ where no arguments of identifier types are grounded.

Definition 1 (r-gram model). *An **r-gram model** R of order n over an alphabet Σ is a set of relational grams*

$$l_n^1 \vee ... \vee l_n^d \leftarrow l_1...l_{n-1}$$

where

1. *$\forall i : l_1...l_{n-1}l_n^i \in \bar{\Sigma}^*$;*
2. *$\forall i : l_n^i$ contains no constant-variables;*

3. $\forall i : l_n^i$ is annotated with probability values
   $P_r(l_n^i \mid l_1...l_{n-1})$ such that $\sum_{i=1}^{d} P_r(l_n^i \mid l_1...l_{n-1}) = 1$
4. $\forall i \neq j : l_1...l_{n-1}l_n^i \not\preceq l_1...l_{n-1}l_n^j$; i.e. the heads are mutually exclusive. Here, the operator $\preceq$ implements subsumption[2]

Example 3. An example of an order 2 relational gram in the Unix user domain is:

$$\left.\begin{array}{l} 0.4 \; latex(F, tex) \\ 0.1 \; latex(F', tex) \\ 0.1 \; emacs(F', tex) \\ ... \\ 0.05 \; cd(Dir) \end{array}\right\} \leftarrow emacs(F, tex)$$

It states that after editing a file with emacs, a user is more likely to use latex on that file than she is to use latex on a different file or execute another command.

We still need to show how an r-gram model R defines a distribution over relational sequences. We first discuss a basic model by analogy to an unsmoothed n-gram, before extending it to a smoothed one in analogy to Equation 5.

A Basic Model. In the basic r-gram model, for any ground sequence $g_1...g_{n-1}$ there is exactly one gram $l_n^1 \vee ... \vee l_n^d \leftarrow l_1...l_{n-1}$ with $l_1...l_{n-1} \preceq_\theta g_1...g_{n-1}$. Its body $l_1...l_{n-1}$ is the most specific sequence subsuming $g_1...g_{n-1}$. According to Equation 3, we start by defining a probability $P_R(g \mid g_1...g_{n-1})$ for any ground atom g given a sequence $g_1...g_{n-1}$ of ground literals. Let g be a ground literal and consider the above gram subsuming $g_1...g_{n-1}$. If there is an $i \in \{1, ..., d\}$ such that $l_1...l_{n-1}l_n^i \preceq_\theta g_1...g_{n-1}g$ it is unique and we define

$$P_R(g \mid g_1...g_{n-1}) := P_r(g \mid g_1...g_{n-1}) := P_r(l_n^i \mid l_1...l_{n-1})$$

Otherwise, $P_R(g \mid g_1...g_{n-1}) = 0$. From $P_R(g \mid g_1...g_{n-1})$, a sequence probability $P_R(g_1...g_m)$ can be derived as in Equation 3.

In this way, the model assigns a probability value to any ground sequence s over the alphabet Σ. If two sequences are identical up to local identifier renaming, the model will assign them the same probability value. For example, the same probability is assigned to $emacs(chapter1, tex), latex(chapter1, tex)$ and $emacs(chapter2, tex), latex(chapter2, tex)$. We have therefore modeled patterns of object identifiers (the fact that the same file name is used in both commands) without referring to any concrete identifiers. As the model does not distinguish between sequences that are identical up to identifier renaming, the sum of probability estimates over all ground sequences is larger than one. However, the model defines a proper probability distribution over the set of equivalence classes modulo local identifier renaming. More details can be found in [20].

---

[2] We actually use a variant of the subsumption relation introduced above. It is called *object identity* as in [31] and it requires that each variable is instantiated to a different constant that does not occur in the pattern.

Smoothing r-Grams. In the basic model, there was exactly one gram $r \in R$ subsuming a given ground subsequence $g_1...g_{n-1}$, namely the most specific one. As for n-grams, the problem with this approach is that there is a large number of such grams and the amount of training data needed to reliably estimate all of their frequencies is prohibitive unless n is very small. The basic idea behind smoothing in r-grams is to generalize grams logically, and mix the resulting distributions, i.e., $P_R(g \mid g_1...g_{n-1}) = \sum_{r \in \hat{R}} \frac{\alpha_r}{\alpha} P_r(g \mid g_1...g_{n-1})$ where $P_r(g \mid g_1...g_{n-1})$ is the probability defined by r as explained above, $\hat{R}$ is the subset of grams in R subsuming $g_1...g_{n-1}$, and α is a normalization constant, i.e. $\alpha = \sum_{r \in \hat{R}} \alpha_r$. The more general r, the more smooth the probability estimate $P_r(g \mid g_1...g_{n-1})$ will be. The actual degree and characteristic of the smoothing is defined by the set of matching r-grams together with their relative weights α_r.

To summarize, r-grams upgrade n-grams to deal with sequences of logical atoms. As n-grams, they combine simple Markov models with powerful smoothing techniques. Furthermore, they allow us to make abstract of the specific identifiers in the data. As for n-grams, learning r-grams is straightforward, and basically amounts to counting frequencies of first-order patterns in the data. These could be determined efficiently, e.g., by a first-order sequential pattern miner such as SeqLog [21], see also Section 4. Furthermore, r-grams need a user-defined *language bias*, which constrains the allowed patterns in terms of types and determines which types are treated as identifiers.

R-grams have been successfully applied to structured sequential problems in Unix user modeling, protein fold prediction, and mobile phone user pattern analysis (see [20]).

7   Logical Hidden Markov Models

In r-grams and Markov models in general, the (structured) states are directly visible, and therefore the transition probabilities among states are the only parameters. In hidden Markov models [28], the states are not directly observable, but only by means of variables (called observations) influenced by the state.

Definition 2. Abstract transitions *are expressions of the form* $p : H \xleftarrow{0} B$ *where* $p \in [0, 1]$, *and* H, B *and* 0 *are atoms. The atoms* H *and* B *are abstract states and* 0 *represents an abstract output symbol. All variables are implicitly assumed to be universally quantified, i.e., the scope of variables is a single abstract transition.*

Consider Figure 3. Here, the gray node emacs(File, tex) denotes that a LATEX user edits a file F using emacs. That the user is indeed a LATEXuser, however, is not directly observable but only through a sequence of observations such as emacs(F) and latex(F) specified in terms of abstract transitions (arrows) such as $c \equiv 0.6 : \texttt{latex(File, tex)} \xleftarrow{\texttt{emacs(File)}} \texttt{emacs(File, tex)}$. Assume now that we are in state emacs(rsl, tex), i.e. $\theta_B = \{\texttt{File/rsl}\}$. Then c specifies that there is a probability of 0.6 that the next state will be subsumed by latex(rsl, tex) and that one of the symbols represented by emacs(rsl) will be emitted. This

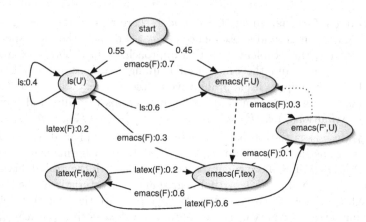

Fig. 3. A logical hidden Markov model. Abstract states are represented by gray nodes. Arrows between nodes denote abstract transitions. The abstract emissions and transition probabilities are associated with the arrows. Dotted arrows denote 'must-follow' links; dashed arrows the 'more-general-than' relation.

was a simple example for an abstract transition because θ_H and θ_O were both empty. In general, the resulting state and output symbol sets are not singletons. For instance, for $0.6 : \mathtt{emacs(File',U)} \xleftarrow{\mathtt{latex(File)}} \mathtt{latex(File,tex)}$ the resulting state set is the set of subsumed ground states of $\mathtt{emacs(File',U)}$ such as $\mathtt{emacs(rsl,tex)}$, $\mathtt{emacs(rsl,dvi)}$, $\mathtt{emacs(lohmm,tex)}$ etc. We therefore need a way to assign probabilities to these possible alternatives.

Definition 3. *The selection distribution μ specifies for each abstract state and observation symbol* A *over the alphabet Σ a distribution $\mu(\cdot \mid A)$ over all ground atoms subsumed by* A.

In our example, assume a selection probability

$$\mu(\mathtt{emacs(rsl,tex)} \mid \mathtt{emacs(File',U)}) = 0.4,$$
$$\mu(\mathtt{emacs(april,tex)} \mid \mathtt{emacs(File',U)}) = 0.6$$
$$\mu(\mathtt{emacs(rsl,word)} \mid \mathtt{emacs(File',U)}) = 0.5,$$
$$\mu(\mathtt{emacs(april,word)} \mid \mathtt{emacs(File',U)}) = 0.5$$

Then there would be a probability of $0.4 \times 0.6 = 0.24$ that the next state is $\mathtt{emacs(rsl,tex)}$. Taking μ into account, the meaning of an abstract transition $p : \mathtt{H} \xleftarrow{O} \mathtt{B}$ can be summarized as follows: *Let* $\mathtt{B}\theta_B, \mathtt{H}\theta_B\theta_H$ *and* $\mathtt{O}\theta_B\theta_H\theta_O$ *be ground atoms. Then the model makes a transition from* $\mathtt{O}\theta_B\theta_H\theta_O$ *with probability*

$$p \cdot \mu(\mathtt{H}\theta_B\theta_H \mid \mathtt{H}\theta_B) \cdot \mu(\mathtt{O}\theta_B\theta_H\theta_O \mid \mathtt{O}\theta_B\theta_H). \tag{6}$$

To represent μ, any probabilistic representation can – in principle – be used, e.g. a Bayesian network or a Markov chain. In [16], we show how to use a *naïve Bayes* approach to reduce the model complexity.

Thus far the semantics of a single abstract transition has been defined. A logical hidden Markov model usually consists of multiple abstract transitions, which makes things a little bit more complicated. Reconsider Figure 3, where *dotted edges* indicate that two abstract states behave in exactly the same way. If we follow a transition to an abstract state with an outgoing dotted edge, we will automatically follow that edge making appropriate unifications. Furthermore, *dashed edges* encode a preference order among abstract states used as conflict resolution strategy, which is needed because multiple abstract transitions can match a given ground state. Consider the dashed edge in Figure 3 connecting emacs(File, U) and emacs(File, tex). For the state emacs(rsl, tex) the matching abstract transitions do not sum to 1.0. To resolve this, we only consider the maximally specific transitions (with respect to the body parts B) that apply to a state in order to determine the successor states. The rationale behind this is that if there exists a substitution θ such that $B_2\theta = B_1$, i.e., B_2 subsumes B_1, then the first transition can therefore be regarded as more informative than the second one.

Finally, in order to specify a prior distribution over states, we assume a finite set Υ of clauses of the form $p : H \leftarrow$ start using a distinguished start symbol such that p is the probability of the logical hidden Markov model to start in a ground instance of the some ground state of start.

In [16] it is proven that logical hidden Markov models specify a unique probability measure over sequences of ground atoms over Σ. Moreover all algorithms for hidden Markov models such as the forward, the Viterbi and the Baum-Welch algorithms carry over to the relational case. Thus they can be used for sequence prediction, sequence classification and sequence labeling tasks. Here, we would like to exemplify the practical relevance of logical Hidden Markov models on two bioinformatics domains [18,16]: protein fold classification and mRNA signal structure detection.

Protein fold classification is concerned with how proteins fold in nature, which is related to their three-dimensional structures. More precisely, the task is to predict the fold class of unknown proteins. Each fold contains a set of proteins which fold in a similar way. Fold class predication is an important problem as the biological functions of proteins depend on the way they fold. As already shown in the section on relational alignments, the secondary structure of proteins can elegantly be represented as logical sequences. Here, however, we used a more fine grained discretization of lengths of helices and strands. For instance, the Ribosomal protein L4 is represented as follows:

st(null, 2), he(h(right, alpha), 6), st(plus, 2), he(h(right, alpha), 4), . . .

The task was to predict one of the five most populated SCOP [13] folds of alpha and beta proteins (a/b): TIM beta/alpha-barrel, NAD(P)-binding Rossmann-fold domains, Ribosomal protein L4, Cysteine hydrolase, and Phosphotyrosine protein phosphatases I-like. We took the 816 corresponding proteins from the ASTRAL dataset version 1.65 (similarity cut 95%) consisting of 22210 ground atoms. The 10-fold cross-validated accuracy was 76%. This is in a similar range as Turcotte *et al.*'s [36] 75% accuracy for a similar task. What is important to

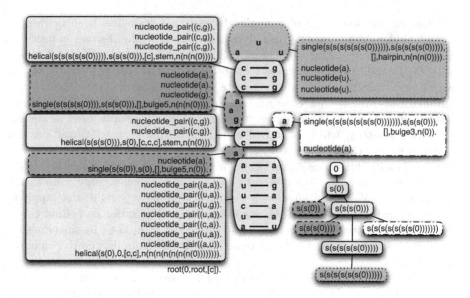

Fig. 4. The *tree* representation of a mRNA secondary structure (SECIS signal structure). (a) The logical sequence, i.e., the sequence of ground atoms representing the tree. The ground atoms are ordered clockwise starting with root(0, root, [c]) in the lower left-hand side corner. (b) The tree formed by the secondary structure elements. For more details, we refer to [16].

realize when comparing the Logical HMM approach for this application to the propositional HMM is to that the logical hidden Markov models were by an order of magnitude smaller than the number of the equivalent hidden Markov models (120 parameters vs. approx. 62000). This clearly shows the benefit of learning in a relational setting.

The secondary structure of mRNA contains special subsequences called signal structures that are responsible for special biological functions, such as RNA-protein interactions and cellular transport. In contrast to the secondary structure of proteins, however, it does not form linear chains, but trees and hence, cannot be represent using ordinary hidden Markov models, cf. Figure 4. We performed leave-one-out cross-validation experiments on a similar data set as used in [12]: 15 and 5 SECIS (Selenocysteine Insertion Sequence), 27 IRE (Iron Responsive Element), 36 TAR (Trans Activating Region) and 10 histone stemloops structures constituting five classes. The logical hidden Markov models achieved an accuracy of 99%, which is in the range of Horvath *et al.*'s relational instance-based learning approach.

8   Relational Conditional Random Fields

(Logical) HMMs model a sequence X by assuming that there is an underlying sequence of states Y drawn from a finite set of states S. To model the joint

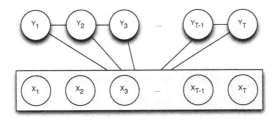

Fig. 5. Graphical representation of linear-chain CRF

distribution $P(X, Y)$ tractably, HMMs make two independency assumptions: each state depends only on its immediate predecessor and each observation/input x_j depends only on the current state y_j. The downside of these assumptions is that they make it relatively cumbersome to model arbitrary dependencies in the input space, i.e., in the space of X.

For the sequence labeling task, conditional random fields [19] (CRFs) are an alternative to (logical) HMMs that make it relatively easy to model arbitrary dependencies in the input space. They have become popular in language processing, computer vision, and information extraction. They have outperformed HMMs on language processing tasks such as information extraction and shallow parsing. CRFs are undirected graphical models that represent the conditional probability distribution $P(Y|X)$. Instead of the generatively trained (Lo)HMM, the discriminatively trained CRF is designed to handle non-independent input features, which can be beneficial in complex domains.

When used for sequences, the graph structure of a CRF is a first-order chain as shown in Figure 5. Normalization by the $Z(X)$ ensures that the defined function returns a probability:

$$P(Y|X) = \frac{1}{Z(X)} \exp \sum_{t=1}^{T} \Psi_t(y_t, X) + \Psi_{t-1,t}(y_{t-1}, y_t, X). \qquad (7)$$

In contrast to a Markov Random Field, both the normalization factor $Z(X)$ and the potential functions Ψ are conditioned on the input nodes X. For the sequential learning setting, the potentials are typically represented as a linear combination of feature functions $\{f_k\}$, which are given and fixed:

$$\Psi(y_t, X) = \sum \alpha_k g_k(y_t, X) \ \text{ and } \ \Psi(y_{t-1}, y_t, X) = \sum \beta_k f_k(y_{t-1}, y_t, X). \ (8)$$

The model parameters are then a set of real-valued weights α_k, β_k, one weight for each feature. In linear-chain CRFs (see Figure 5), a first-order Markov assumption is made on the hidden variables. In this case, there are features for each label transition. Feature functions can be arbitrary such as a binary test that has value 1 if and only if y_{t-1} has the label a.

So far, CRFs have mainly been applied on propositional input sequences. In the next subsection, we will show how to lift them to the relational sequences case. A more detailed description can be found in [10].

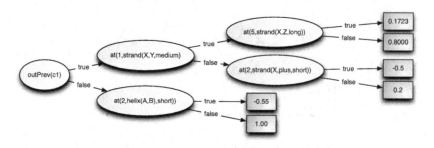

Fig. 6. A relational regression tree; used in TildeCRF to represent the potential function. The inner nodes are logical tests.

8.1   TildeCRF

The only parts of a CRF that access the input sequence are the potential functions. Therefore, CRFs can easily be lifted to the relational sequences case by representing the potential function F as a sum of relational regression trees learned by a relational regression tree learner such as Tilde [3]. Each regression tree stands for a gradient and the sum of all for the potential function. We adapted Dietterich *et al.*'s Gradient Tree Boosting approach, called TreeCRF, to learn the trees. Following their notation, we define $F^{y_t}(y_{t-1}, X) = \Psi(y_t, X) + \Psi(y_{t-1}, y_t, X)$. We do not make any assumptions about F; it can be any function and not only a linear combination of features.

The gradient of $\frac{\partial \log P(Y|X)}{\partial F^v(u, w_d(X))}$ can be evaluated quite easily for every training sequence and position:

$$\frac{\partial \log P(Y|X)}{\partial F^v(u, w_d(X))} = I(y_{d-1} \prec u, y_d \prec v) - P(y_{d-1} \prec u, y_d \prec v | w_d(X)) \qquad (9)$$

where I is the identity function, the symbol $\prec$ denotes that u θ-subsumes y, and $P(y_{d-1} \prec u, y_d \prec v | w_d(X))$ is the probability that class labels u, v fit the class labels at positions $d, d-1$. It is calculated as shown in GenExamples in Alg. 3.

By evaluating the gradient at every known position in our training data and fitting a regression model to this values, we get an approximation of the expectation of the gradient. In order to simplify the derivation of the gradient and afterwards the evaluation, we do not use the complete input X but a window $w_d(X) = x_{d-s}, \ldots, x_d, \ldots, x_{d+s}$, where s is a fixed window size. This is exactly the learning setting of Tilde: each window, i.e., each regression example is a (weighted) set of ground atoms.

All other issues of Dietterich *al.*'s original approach, TreeCRF, remain unchanged. That is, we can use the forward-backward algorithm as proposed by [6] to compute $Z(X)$. The forward recursion is defined as $\alpha(k, 1) = \exp F^k(\bot, w_1(X))$ and $\alpha(k, t) = \sum_{k' \in K} \left[\exp F^k(k', w_t(X)) \right] \cdot \alpha(k', t-1)$. The backward recursion is defined as $\beta(k, T) = 1$ and $\beta(k, t) = \sum_{k' \in K} \left[\exp F^{k'}(k, W_{t+1}(X)) \right] \cdot \beta(k', t+1)$.

Algorithm 3. Gradient Tree Boosting

Function TREECRF($Data, L$)/* $Data = \{(X_i, Y_i)|i = 1, 2, \ldots, N\}$ */
begin
 For each label k, initialize $F_0^k \equiv 0$ **for** $1 \leq m \leq M$ **do**
  for $1 \leq k \leq K$ /* `Iterate over all labels` */
  do
   $Sk :=$ GENEXAMPLES($k, Data, Pot_{m-1}$)
   /* `where` $Pot_{m-1} = \{F_{m-1}^k \mid k = 1, 2, \ldots, K\}$ */
   $\Delta_m(k) :=$ FITRELREGRESSTREE($S(k), L$)/* `at most` L `leaves` */
   $F_m^k := F_{m-1}^k + \Delta_m(k)$

 Return Pot_M
end
Function GENEXAMPLES($k, Data, Pot_m$)
begin
 $S := \emptyset$
 for $(X_i, Y_i) \in Data$ **do**
  $(\alpha, \beta, Z(X_i)) :=$ FORWARDBACKWARD(X_i, T_i, K)
  /* `run the forward-backward algorithm on` (X_i, Y_i) `to get` $\alpha(k, t)$
  `and` $\beta(k, t)$ `for all` k `and` t */
  for $1 \leq t \leq T_i$ **do**
   for $1 \leq k' \leq K$ **do**
    /* `Compute gradient at position` t `for label` k */
    $P(y_{t-1} = k', y_t = k|X_i) :=$
     $\dfrac{\alpha(k', t-1) \cdot \exp(F_m^k(k', w_t(X))) \cdot \beta(k, t)}{Z(X_i)}$
    $\Delta(k, k', t) := I(y_{t-1} \prec k', y_t \prec k) - P(y_{t-1} \prec k', y_t \prec k|X_i)$
    /* `add example to set of regression examples` */
    $S := S \cup \{((w_t(X_i), k'), \Delta(k, k', t))\}$

 Return S
end

8.2 Making Predictions

There are several ways for obtaining a classifier from a trained CRF. We can predict the output sequence Y with the highest probability: $H(X) = \arg\max_Y P(Y|X)$, which can be computed using the Viterbi algorithm [28]. Another option is to predict every atom y_t in the output sequence individually. This makes sense when we want to maximize the number of correctly tagged input atoms: $H_t(X) = \arg\max_{k \in K} P(y_t = k|X)$. Finally, one can also use a CRF for sequence classification, i.e., to predict a single label for the entire sequence. To do so, we can simply make a kind of majority vote. That is, we first predict $H(X)$. Next, we count the number of times each class atom was predicted, i.e.,

$count(c, Y) := |\{i \in \{1, \ldots, T\} \mid y_i = c\}|$. Then, the sequence X is assigned to class c with probability $P(c|X) = T^{-1} \cdot count(c, H(X))$.

We employed TildeCRF to the protein fold classification problem. We used the same subset on the subset of the SCOP database [8] that was mentioned in 7. We have performed a 10-fold cross-validation and obtained an overall accuracy of 92.62%, whereas the LoHMMs achieved only 75% accuracy. This difference is significant (one-tailored t-test, $p = 0.05$).

To summarize, the previous sections have shown that it is indeed possible to lift many prominent sequence learning techniques to the relational case. Before concluding, we will show that this also holds for relational sequential decision making, i.e., for relational sequence generation through actions.

9   Relational Sequential Decision Making

In the machine learning community, learning about stimuli and actions solely on the basis of the rewards and punishments associated with them is called reinforcement learning [34]. It is the problem faced by an agent that acts in an environment and occasionally receives some reward based on the state the agent is in and the action(s) the agent took. The agent's learning task is then to find a policy for action selection that maximizes its expected reward over time. This task requires not only choosing those actions that are associated with high rewards in the current state but also looking ahead by choosing actions that will lead the agent to more lucrative parts of the state space. Thus, in contrast to supervised sequence learning tasks such as sequence labeling, reinforcement learning is minimally supervised because agents are not told explicitly which actions to take in particular situations, but must discover this themselves on the basis of the received rewards.

9.1   Markov Decision Processes

Consider an agent acting in the *blocks world* [32]. The domain consists of a surface, called the *floor*, on which there are *blocks*. Blocks may be on the floor or on top of other blocks. They are said to pile up in stacks, each of which is on the floor. Valid relations are $on(X, Y)$, i.e., block X is on Y, and $cl(Z)$, i.e., block Z is *clear*. At each time, the agent can move a clear (and movable) block X onto another clear block Y. The $move(X, Y, Z)$ action is probabilistic, i.e., it may not always succeed. For instance, with probability p_1 the action succeeds, i.e. X will be on top of Y. With probability $1 - p_1$, however, the action fails. More precisely, with probability p_2 the block X remains at its current position, and with probability p_3 (with $p_1 + p_2 + p_3 = 1$) it falls on some clear block Z.

A natural formalism to represent the utilities and uncertainties is the formalism of Markov decision processes. A Markov decision process (MDP) is a tuple $\mathbf{M} = (S, A, \mathbf{T}, \lambda)$. Here, S is a set of system states such as

$$z \equiv cl(a), on(a, b), on(b, floor), block(a), block(b)$$

describing the blocks world consisting of two blocks a and b where a is on top of b. The agent has available a finite set of actions $A(z) \subseteq A$ for each state $z \in S$, which cause stochastic state transitions, for instance, move(a, floor) moving a on the floor. For each $z, z' \in S$ and $a \in A(z)$ there is a transition T in $\mathbf{T}$, i.e., $z' \xleftarrow{p:r:a} z$. The transition denotes that with probability $P(z, a, z') := p$ action a causes a transition to state z' when executed in state z. For instance $z' \equiv$ cl(a), cl(b), on(a, floor), on(b, floor), block(a), block(b). For each $z \in S$ and $a \in A(z)$ it holds $\sum_{z' \in S} P(z, a, z') = 1$. The agent gains a reward when entering a state, denoted as $R(z) := r$. For instance, on the blocks world we could have $R(z') = 10$.

A solution of a (relational) Markov decision process is a policy $\pi : S \mapsto A$ mapping state to actions. Essentially policy can be viewed as sets of expressions of the form $a \leftarrow z$ for each $z \in S$ where $a \in A(z)$ such as move(a, floor) $\leftarrow$ cl(a), on(a, b), on(b, floor), block(a), block(b). It denotes a particular course of actions to be adopted by an agent, with $\pi(z) := a$ being the action to be executed whenever the agent is in state z.

Assuming that the sequence of rewards after step t is $r_{t+1}, r_{t+2}, r_{t+3}, \ldots$, the agent's goal is to find a policy that maximizes the expected reward $E[R]$ for each step t. Typically, future rewards are discounted by $0 \leq \lambda < 1$ so that the expected return basically becomes $\sum_{k=0}^{\infty} \lambda^k \cdot r_{t+k+1}$. To achieve this, most techniques employ value functions. More precisely, given some MDP $M = \langle S, A, T, R \rangle$, a policy π for M, and a *discount factor* $\gamma \in [0, 1]$, the *state value function* $V^\pi : S \to \mathbb{R}$ represents the value of being in a state following policy π with respect to the expected reward. In other words, the value $V^\pi(z)$ of a state z is the expected return starting from that state, which depends on the agent's policy π. A policy π' is better than or equal to another policy π, $\pi' \geq \pi$, if and only if $\forall s \in S : V^{\pi'}(s) \geq V^\pi(s)$. Thus, a policy π^* is optimal, i.e., it maximizes the expected return for all states if $\pi^* \geq \pi$ for all π. Optimal value functions are denoted V^*. Bellman's [2] *optimality equation* states: $V^*(s) = \max_a \sum_{s'} T(s, a, s')[R(s, a, s') + \gamma V^*(s')]$. From this equation, all model-based and model-free methods for solving MDPs can be derived.

9.2   Abstract Policies

A policy over ground states is propositional in the sense that it specifies for each ground state separately which action to execute. In turn, specifying such policies for Markov decision programs with large state spaces is cumbersome and learning them will require much effort. This motivates the introduction of *abstract policies*.

An abstract policy $\boldsymbol{\pi}$ intentionally specifies the action to take for sets of states, i.e., for an abstract state.

Definition 4. *An abstract policy $\boldsymbol{\pi}$ over Σ is a finite set of decision rules of the form* a $\leftarrow$ L, *where* a *is an abstract action and* L *is an abstract state. We assume* a *to be applicable in* L, *i.e.,* vars(a) $\subseteq vars(L)$.

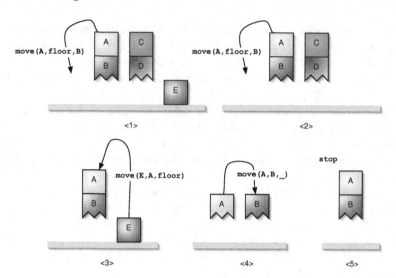

Fig. 7. The decision rules of the *unstack-stack* policy. In the figure, the decision rules are ordered from left to right, i.e., a rule fires only if no rule further to the left fires.

The meaning of a single decision rule $a \leftarrow L$ is as follows: *If the agent is in a state Z such that $a \leq_\theta L$, then the agent performs action $a\theta$ with probability $1/|\theta|$, i.e., uniformly with respect to number of possible instantiations of action a in Z.* Usually, π consists of multiple decision rules. We assume a total order $\prec^\pi$ among the decision rules in π and use the first matching decision rule.

Consider the following *unstack-stack* abstract policy:

$\langle 1 \rangle$ move(A, floor, B) ← on(A, B), on(C, D), on(E, floor),cl(A), cl(C), cl(E).
$\langle 2 \rangle$ move(A, floor, B) ← on(A, B), on(C, D), cl(A), cl(C).
$\langle 3 \rangle$ move(E, A, floor) ← on(A, B), on(E, floor), cl(A), cl(E).
$\langle 4 \rangle$ move(A, B, floor) ← cl(A), cl(B).
$\langle 5 \rangle$ stop ← on(A, B), cl(A).

where the start action adds the **absorbing** propositions, i.e., it encodes that we enter an absorbing state[3]. For instance in state z (see before), only decision rule $\langle 3 \rangle$ fires.

The policy, which is graphically depicted in Figure 7, is interesting for several reasons. First, it is close to the *unstack-stack* strategy, which is well known in the planning community [32]. Basically, the strategy amounts to first putting all blocks on the table and then building the goal state by stacking all blocks from the floor onto one single stack. No block is moved more than twice. Second, it perfectly generalizes to all other blocks worlds, no matter how many blocks there are. Finally, it cannot be learned in a propositional setting because here

---

[3] For ease of exposition, we have omitted the **absorbing** state in front and statements that variables refer to different blocks.

Algorithm 4. Relational TD(0) where α is the learning rate and $\widehat{V}(\mathbb{L})$ is the approximation of $V(\mathbb{L})$

Let π be an abstract policy over abstract states $\mathbb{L}$
Initialize $\widehat{V}_0(L)$ arbitrarily for each L in $\mathbb{L}$
repeat
    Pick a ground state $\mathtt{Z}$ of the underlying (relational) MDP M
    repeat
        Choose action $\mathtt{a}$ in $\mathtt{Z}$ based on π, i.e., (1) select first decision rule $\mathtt{a} \leftarrow \mathtt{L}$
        in π that matches according to $\prec^{\pi}$, (2) select $\mathtt{a}\theta$ uniformally among
        induced ground actions
        Take $\mathtt{a}\theta$, observe immediate reward r and successor state $\mathtt{Z}'$, i.e., (1) select
        with probability p_i the i-th outcome of $\mathtt{a}\theta$, (2) compute $\mathtt{Z}'$ as $[\mathtt{b} \setminus \mathtt{B}\theta] \cup \mathtt{H}_i\theta$
        Let $\mathtt{L}'$ in $\mathbb{L}$ be the abstract state first matching $\mathtt{Z}'$ according to $\prec^{\pi}$
        $\widehat{V}(L) := \widehat{V}(L) + \alpha \cdot (r + \lambda \cdot \widehat{V}(L') - \widehat{V}(L))$
        Set $\mathtt{Z} := \mathtt{Z}'$
    until $\mathtt{Z}$ *is terminal, i.e., absorbing*
until *converged or some maximal number of episodes exceeded*

---

the optimal, propositional policy would encode the number of states and the optimal number of moves.

9.3 Relational Temporal Difference Learning

The crucial question for (relational) Markov decision programs and for relational reinforcement learning is how one can learn abstract policies. Almost all relational MDP solvers and reinforcement learning systems follow the so called *generalized relational policy iteration* scheme. It consists of three interacting processes: *policy evaluation*, *policy improvement*, and *policy refinement*. Here, evaluating a policy refers to computing a performance measure of the current policy; policy improvement refers to computing a new policy based on the current value function; and *policy refinement* makes small modifications to an abstract policy such as adding rules.

Here, we will focus on model-free approaches, i.e., we do not assume any model of the world. For a model-based approach, we refer to [17]. Moreover, we will focus on the *relational evaluation problem*, which considers how to compute the state-value function V^{π} for an arbitrary abstract policy π: **Given** an abstract policy π, **find** the state-value function V^{π} from experiences $\langle S_t, a_t, S_{t+1}, r_t \rangle$ only, where action a_t leads from state S_t to state S_{t+1} receiving reward r_t.

The basic idea is to define the value of an abstract state L_i (i.e., a body of a decision rule) to be the average expected value for all the states subsumed by that state. This is a good model because if we examine each state subsumed, we make contradictory observations of rewards and transition probabilities. The best model is the average of these observations given no prior knowledge of the model. For ease of explanation, we will focus on a *TD(0)* approach, see e.g. [34]. Results for general *TD(λ)* can be obtained by applying Tsitsiklis and van Roy's [35] results.

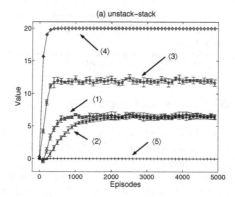

Fig. 8. Relational TD(0)'s learning curves on the evaluation problem for the *unstack-stack* policy. The predicted values are shown as a function of the number of episodes. These data are averages over 5 runs; the error bars show the standard deviations.

Relational $TD(0)$ as shown in Algorithm 4 sketches the resulting approach. Given some experience following an abstract policy π, RTD(0) updates its estimate $\widehat{V}$ of V. If the estimate is not changing considerably, the algorithm stops. If an absorbing state is reached, an episode ends and a new "starting" state is selected. If a nonabsorbing state is visited, then it updates its estimate based on what happens after that visit. Instead of updating the estimate at the level of states, $RTD(0)$ updates its estimate at the abstract states of π only.

RTD(0) can be proven to converge, see e.g. [15]. Figure 8 shows the performance of RTD(0) when evaluating the *unstack-stack* abstract policy (see above). We randomly generated 100 blocks world states for 6 blocks, for 8 blocks, and for 10 blocks using the procedure described by [32]. This set of 300 states constituted the set *Start* of starting states in all experiments. Note that for 10 blocks a traditional MDP would have to represent $58,941,091$ states of which $3,628,800$ are goal states. The result of each experiment is an average of five runs of 5000 episodes, where for each new episode we randomly selected one state from *Start* as starting state. For each run, the value function was initialized to zero. Furthermore, we used a discount factor λ of 0.9 and a learning rate α of 0.015. The learning curves show that the values of the abstract states converged and, hence, $RTD(0)$ converged. Note that the value of abstract state $\langle 5 \rangle$ remained 0. The reason for this is that, by accident, no state with all blocks on the floor was in *Start*. Furthermore, the values converged to similar values in all runs. The values basically reflect the nature of the policy. It is better to have a single stack than multiple ones.

10   Conclusions

Relational sequence learning problems arise in many applications. This chapter has defined various relational sequence learning tasks and reviewed some methods for solving them. In contrast to learning with propositional sequences, relational sequence learning assumes the elements of the sequences to be structured in terms of (ground) atoms. Using relational abstraction and unification, relational sequence models cannot only compress traditional models significantly, but also make abstraction of identifiers. This in turn is important because it reduces the number of parameters of the resulting models, which results in easier learning tasks.

Acknowledgments

The authors would like to thank Sau Dan Lee for his work on the SeqLog system, Sarah Finkel for setting up the REAListic web server for relational sequence alignments, Tapani Raiko for his involvement in the Logical HMMs, Ingo Thon for implementing Logical (H)MMs in Java, Johannes Horstmann for valuable work on applying Logical HMMs to the protein data, and Angelika Kimmig for valuable comments on an earlier draft. This work was supported by the European Union, contract number FP6-508861, Applications of Probabilistic Inductive Logic Programming II.

References

1. Altschul, S.F., Gish, W., Miller, W., Myers, E.W., Lipman, D.J.: Basic local alignment search toll. Journal of Molecular Biology 215(3), 403–410 (1990)
2. Bellman, D.P.: Dynamic Programming. Princeton University Press, Princeton (1957)
3. Blockeel, H., De Raedt, L.: Top-down Induction of First-order Logical Decision Trees. Artificial Intelligence 101(1–2), 285–297 (1998)
4. Bruynooghe, M., De Raedt, L., Lee, S.D., Troncon, R.: Mining logical sequences. Technical report, Department of Computer Science, Katholieke Universiteit Leuven (forthcoming, 2007)
5. Dayhoff, M.O., Schwartz, R.M., Orcutt, B.C.: A model of evolutionary change in proteins. In: Dayhoff, M.O. (ed.) Atlas of Protein Sequence and Structure, vol. 5, ch. 22 pp. 345–352. Nat. Biomedical Research Foundation (1978)
6. Dietterich, T., Ashenfelter, A., Bulatov, Y.: Training conditional random fields via gradient tree boosting. In: Proc. 21st International Conf. on Machine Learning, pp. 217–224. ACM Press, New York (2004)
7. Gorodkin, J., Heyer, L.J., Brunak, S., Stormo, G.D.: Displaying the information contents of structural RNA alignments: The structure logos. CABIOS 13(6), 583–586 (1997)
8. Gough, J., Karplus, K., Hughey, R., Chothia, C.: Assignment of homology to genome sequences using a library of hidden markov models that represent all proteins of known structure. JMB 313(4), 903–919 (2001)
9. Greenberg, S.: Using unix: Collected traces of 168 users. Research Report 88/333/45, Department of Computer Science, University of Calgary, Calgary, Canada (1988)
10. Gutmann, B., Kersting, K.: Tildecrf: Conditional random fields for logical sequences. In: Fürnkranz, J., Scheffer, T., Spiliopoulou, M. (eds.) ECML 2006. LNCS (LNAI), vol. 4212, pp. 174–185. Springer, Heidelberg (2006)

11. Henikoff, S., Henikoff, J.G.: Amino acid substitution matrices from protein blocks. Proc. Natl Acad. Sci. 89, 10915–10919 (1992)
12. Horváth, T., Wrobel, S., Bohnebeck, U.: Relational Instance-Based learning with Lists and Terms. Machine Learning Journal 43(1/2), 53–80 (2001)
13. Hubbard, T., Murzin, A., Brenner, S., Chotia, C.: SCOP: A structural classification of proteins database. NAR 27(1), 236–239 (1997)
14. Karwath, A., Kersting, K.: Relational sequences alignments and logos. In: Muggleton, S., Otero, R., Tamaddoni-Nezhad, A. (eds.) ILP 2006. LNCS (LNAI), vol. 4455, pp. 290–304. Springer, Heidelberg (2007)
15. Kersting, K., De Raedt, L.: Logical Markov Decision Programs and the Convergence of Logical TD(λ). In: Camacho, R., King, R., Srinivasan, A. (eds.) ILP 2004. LNCS (LNAI), vol. 3194, pp. 180–197. Springer, Heidelberg (2004)
16. Kersting, K., De Raedt, L., Raiko, T.: Logical Hidden Markov Models. Journal of Artificial Intelligence Research (JAIR) 25, 425–456 (2006)
17. Kersting, K., Van Otterlo, M., De Raedt, L.: Bellman goes Relational. In: Greiner, R., Schuurmans, D. (eds.) Proceedings of the Twenty-First International Conference on Machine Learning (ICML 2004), Banff, Alberta, Canada, July 4–8, 2004, pp. 465–472 (2004)
18. Kersting, K., Raiko, T., Kramer, S., De Raedt, L.: Towards discovering structural signatures of protein folds based on logical hidden markov models. In: Proceedings of the Pacific Symposium on Biocomputing (PSB 2003), pp. 192–203 (2003)
19. Lafferty, J., McCallum, A., Pereira, F.: Conditional random fields: Probabilistic models for segmenting and labeling sequence data. In: Proc. 18th International Conf. on Machine Learning, pp. 282–289. Morgan Kaufmann, San Francisco (2001)
20. Landwehr, N., De Raedt, L.: r-grams: Relational Grams. In: Proceedings of the Twentieth Joint International Conference on Artificial Intelligence (IJCAI 2007), AAAI Press, Menlo Park (2007)
21. Lee, S.D., De Raedt, L.: Mining Logical Sequences Using SeqLog. In: Meo, R., Lanzi, P.L., Klemettinen, M. (eds.) Database Support for Data Mining Applications. LNCS (LNAI), vol. 2682, Springer, Heidelberg (2004)
22. Lee, S.D., De Raedt, L.: Constraint based mining of first order sequences in seqlog. In: Meo, R., Lanzi, P.L., Klemettinen, M. (eds.) Database Support for Data Mining Applications. LNCS (LNAI), vol. 2682, pp. 154–173. Springer, Heidelberg (2004)
23. Mannila, H., Toivonen, H.: Levelwise search and borders of theories in knowledge discovery. Data Mining and Knowledge Discovery 1(3), 241–258 (1997)
24. Manning, C.H., Schütze, H.: Foundations of Statistical Natural Language Processing. MIT Press, Cambridge (1999)
25. Muggleton, S.H., De Raedt, L.: Inductive Logic Programming: Theory and Methods. Journal of Logic Programming 19(20), 629–679 (1994)
26. Needleman, S., Wunsch, C.: A general method applicable to the search for similarities in the amino acid sequence of two proteins. J. Mol. Bio. 48(3), 443–453 (1970)
27. Nienhuys-Cheng, S.-H.: Distance between Herbrand interpretations: A measure for approximations to a target concept. In: Džeroski, S., Lavrač, N. (eds.) ILP 1997. LNCS, vol. 1297, pp. 250–260. Springer, Heidelberg (1997)
28. Rabiner, L.R.: A Tutorial on Hidden Markov Models and Selected Applications in Speech Recognition. Proceedings of the IEEE 77(2), 257–286 (1989)
29. Ramon, J.: Clustering and instance based learning in first order logic. PhD thesis, Department of Computer Science, K.U. Leuven, Leuven, Belgium (October 2002)
30. Saitou, N., Nei, M.: The neighbor-joining method: a new method for reconstructing phylogenetic trees. Mol. Evol. Biol. 4(4), 406–425 (1987)

31. Semeraro, G., Esposito, F., Malerba, D.: Ideal Refinement of Datalog Programs. In: Proceedings of the 5th Intternational Workshop on Logic Programming Synthesis and Transformation (1995)
32. Slaney, J., Thiébaux, S.: Blocks World revisited. Artificial Intelligence Journal 125, 119–153 (2001)
33. Smith, T.F., Waterman, M.S.: Identification of common molecular subsequences. Journal of Molecular Biology 147, 195–197 (1981)
34. Sutton, R.S., Barto, A.G.: Reinforcement Learning: An Introduction. MIT Press, Cambridge (1998)
35. Tsitsiklis, J.N., Van Roy, B.: An analysis of temporal-difference learning with function approximation. IEEE Transactions of Automatic Control 42, 674–690 (1997)
36. Turcotte, M., Muggleton, S.H., Sternberg, M.J.E.: The Effect of Relational Background Knowledge on Learning of Protein Three-Dimensional Fold Signatures. Machine Learning Journal 43(1/2), 81–95 (2001)

Learning with Kernels and Logical Representations

Paolo Frasconi and Andrea Passerini

Machine Learning and Neural Networks Group
Dipartimento di Sistemi e Informatica
Università degli Studi di Firenze, Italy
http://www.dsi.unifi.it/neural/

Abstract. In this chapter, we describe a view of statistical learning in
the inductive logic programming setting based on kernel methods. The
relational representation of data and background knowledge are used
to form a kernel function, enabling us to subsequently apply a number
of kernel-based statistical learning algorithms. Different representational
frameworks and associated algorithms are explored in this chapter. In
kernels on Prolog proof trees, the representation of an example is obtained by recording the execution trace of a program expressing background knowledge. In *declarative kernels*, features are directly associated
with mereotopological relations. Finally, in *kFOIL*, features correspond
to the truth values of clauses dynamically generated by a greedy search
algorithm guided by the empirical risk.

1 Introduction

Kernel methods are one of the highly popular state-of-the-art techniques in machine learning [1,2]. They make it possible to design generic learning algorithms,
abstracting away details about data types, a trait that makes them especially
appealing in relational domains for dealing with structured objects such as sequences [3,4,5,6], trees [7,8], or graphs [9,10,11,12,13,14]. When using kernel machines, instances are mapped to a Hilbert space commonly called the *feature
space*, where the kernel function is the inner product. In principle, there is no
need to explicitly represent feature vectors as an intermediate step, as it happens
for example with many propositionalization schemes [15,16]. This trick has often
been exploited from the algorithmic point of view when the kernel function can
be computed efficiently in spite of very high-dimensional feature spaces.

In the simplest supervised learning settings (such as classification and regression with independent examples), all representational issues are dealt with by
the kernel, whereas the learning algorithm has mainly a statistical role. This
also means that background knowledge about the problem at hand should be
injected into the learning process mainly by encoding it into the kernel function.
This activity is sometimes carried out in an ad-hoc manner by guessing interesting features. However, if domain knowledge has been encoded formally (e.g.
in a declarative fashion using first-order logic, or by means of ontologies), then

L. De Raedt et al. (Eds.): Probabilistic ILP 2007, LNAI 4911, pp. 56–91, 2008.

it makes sense to use these representations as a starting point for building the kernel. An example along these lines is the work by Cumby & Roth [16] that uses description logic to specify features and that has been subsequently extended to specify kernels [17].

Within the field of inductive logic programming (ILP), a related area of research is the definition of distances in relational domains [18,19,20]. For every kernel function (intuitively, a kernel corresponds to a notion of similarity) elementary geometry allows us to derive an associated distance function in the feature space. Turning distances into valid (positive semi-definite) kernels, however, is not possible in general as the axiomatic definition of distance imposes less constraints. Thus, work on distance-based relational learning cannot be immediately translated into equivalent kernel methods

In this chapter, we describe a number of methods based on the combination of logical representations, kernel machines, and statistical learning. There are several reasons why seeking links between kernel methods and probabilistic logical learning can be interesting. First, background knowledge about a domain may be already available and described in a logical representation language. As we noted above, kernels are usually the main entry point for plugging background knowledge in the learning process. Therefore, from an engineering point of view, developing a flexible and systematic approach to kernel design starting from logical representations seems to be a natural choice. Second, learning algorithms based on kernel machines are very efficient from a computational point of view. After the Gram matrix has been computed, learning often consists of finding the (unique) solution of a convex numerical optimization problem. Additional efficiency can be gained by exploiting the sparsity of the structure of the solution, as it happens for example with support vector machines [21]. This scenario contrasts with the computational requirements of many ILP schemes that need to search hypotheses in a complex discrete space of logical programs [22]. Third, several types of learning problems, besides classification, can be solved under a uniform framework, including regression [23], ranking (ordinal regression) [24], novelty detection (one-class classification) [25], clustering [26], and principal component analysis [27]. Logic-based learning, on the other hand, has mainly focused on classification while other tasks such as regression often need ad-hoc solutions, except perhaps in the case of decision trees [28,29]. Fourth, kernel based learning algorithms can be naturally linked to regularization theory, where the complexity of the function calculated by a learning algorithm can be controlled via its norm in the so-called reproducing kernel Hilbert space [30]. Regularization restores well-posedness in learning algorithms based on empirical risk minimization, i.e. it ensures that the solution to the learning problem is unique and stable (small perturbations in the data lead to small variations of the learned function). Of course, uncertainty can also be handled using other probabilistic logic learning schemes, like those extensively presented elsewhere in this book, but from a different and complementary angle. Kernel-based approaches can be seen as taking the discriminant direction of learning, i.e. they attempt to identify the optimal prediction function (i.e. the well known Bayes function in the case of binary classification). Theory shows that machines based on regularized empirical risk minimization, such as the support vector machine (SVM), do converge to the optimal function as the number of examples goes to infinity [31,32]. This is a major difference with respect to other probabilistic ILP approaches that take the generative direction of modeling. Generative models require more knowledge about the structural form of the probability densities than their discriminant counterparts. If the underlying assumptions are wrong, they may converge to a sub-optimal asymptotic error, although faster than discriminant models constructed on the same model space of probability distributions (a classic propositional example is the model pair formed by Naive Bayes and logistic regression [33]).

There are, on the other hand, disadvantages when embracing the above framework, compared to learning with other probabilistic logic representations. Since the learning process only focuses on the discriminant function, it does not discover any new portion of the theory explaining the phenomena that underlie the data. Additionally the learned function does not provide any easily understandable explanations as to why certain predictions are associated with the input.

This chapter is a detailed review of several approaches that have been developed within APrIL II for statistical learning with kernels in the ILP setting. We start in Section 2 explaining some basic concepts about the statistical and the logical learning settings, in order to clarify our assumptions and for the benefit of readers who are not familiar with both areas. In Section 3 we present kernels on Prolog ground terms [34], a specialization to first-order logic of kernels on logical individuals introduced by Gaertner et al. [35]. In Section 4, we describe declarative kernels, a general approach for describing knowledge-informed kernels based on relations related to decomposition into parts and connection between parts. In Section 5, we present kernels based on Prolog proof trees [36], an approach where first a program is ran over instances, to inspect interesting features, and then program traces are compared by means of a kernel on ground terms. In Section 6, we describe kFOIL [37], an algorithm that is especially interesting from the point of view of the understandability of the learned solution. It constructs a kernel from data using a simple inductive logic programming engine (FOIL [38]) to generate the clauses that define the kernel. Finally, in Section 7, we report about two real-world applications that have been tackled with these techniques: information extraction from scientific literature, and prediction of protein folds. For the latter application, we also report novel results and comparisons for the task of multiclass protein fold classification.

2   Notation and Background Concepts

2.1   Supervised Learning in the Statistical Setting

In the typical statistical learning framework, a supervised learning algorithm is given a training set of input-output pairs $\mathcal{D} = \{(x_1, y_1), \ldots, (x_m, y_m)\}$, with $x_i \in \mathcal{X}$ and $y_i \in \mathcal{Y}$, sampled identically and independently from a fixed but unknown probability distribution ρ. The set $\mathcal{X}$ is called the input (or instance) space and can be any set. The set $\mathcal{Y}$ is called the output (or target) space; in the case of binary classification $\mathcal{Y} = \{-1, 1\}$ while the case of regression $\mathcal{Y}$ is the set of real numbers. The learning algorithm outputs a function $f : \mathcal{X} \mapsto \mathcal{Y}$ that approximates the probabilistic relation ρ between inputs and outputs. The class of functions that is searched is called the *hypothesis space*.

2.2 Supervised Learning with Kernel Machines

A kernel is a positive semi-definite (psd) symmetric function $K : \mathcal{X} \times \mathcal{X} \mapsto R$ that generalizes the notion of inner product to arbitrary domains [2]. Positive semi-definite here means that for all m and all finite data sets of size m, the Gram matrix with entries $K(x_i, x_j), i, j = 1, \ldots m$ has nonnegative eigenvalues. Each instance x is mapped to a corresponding element $\phi(x)$ in a Hilbert space commonly called the *feature space*. For example, a feature of a graph may be associated with the existence of a path with certain node labels; in this way, a graph is mapped to a sequence of booleans, each associated with a string over the node labels alphabet. Given this mapping, the kernel function is, by definition, the inner product $K(x, x') = \langle \phi(x), \phi(x') \rangle$. Mercer's theorem ensures that for any symmetric and psd function $K : \mathcal{X} \times \mathcal{X} \mapsto R$ there exists a mapping in a Hilbert space where K is the inner product.

When using kernel methods in supervised learning, the hypothesis space, denoted $\mathcal{F}_K$, is the so-called reproducing kernel Hilbert space (RKHS) associated with K [30]. Learning consists of solving the following Tikhonov regularized problem:

$$f = \arg\min_{h \in \mathcal{F}_K} C \sum_{i=1}^{m} V(y_i, h(x_i)) + \|h\|_K \tag{1}$$

where $V(y, h(x))$ is a positive function measuring the loss incurred in predicting $h(x)$ when the target is y, C is a positive regularization constant, and $\|\cdot\|_K$ is the norm in the RKHS. Popular algorithms in this framework include support vector machines [21], obtained using the "hinge" loss $V(y, a) = \max\{1 - ya, 0\}$, kernel ridge regression [39,40], obtained using the quadratic loss $V(y, a) = (v - a)^2$, and support vector regression [23], obtained using the ϵ-insensitive loss $V(y, a) = \max\{|y - a| - \epsilon, 0\}$. The representer theorem [41] shows that the solution to the above problem can be expressed as a linear combination of the kernel basis functions evaluated at the training examples:

$$f(x) = \sum_{i=1}^{m} c_i K(x, x_i) \tag{2}$$

where c_i are real coefficients expressing the solution of Eq. (1). The above form also encompasses the solution found by other algorithms not based on Eq. (1), such as the kernel perceptron [42].

2.3 Convolution Kernels for Discrete Structures

Suppose the instance space $\mathcal{X}$ is a set of composite structures and for $x \in \mathcal{X}$ let $\vec{x} = x_1, \ldots, x_D$ denote a tuple of "parts" of x, with $x_d \in \mathcal{X}_d$ (the d-th part type) for all $i \in [1, D]$. This decomposition can be formally represented by a relation R on $\mathcal{X}_1 \times \cdots \times \mathcal{X}_D \times \mathcal{X}$. For each $x \in \mathcal{X}$, $R^{-1}(x) = \{\vec{x} \in \vec{\mathcal{X}} : R(\vec{x}, x)\}$ denotes the multiset of all possible decompositions of x.

In order to complete the definition of convolution kernels, we assume that a kernel function $K_d : \mathcal{X}_d \times \mathcal{X}_d \to \mathbb{R}$ is given for each part type $\mathcal{X}_d, d = 1, \ldots, D$. The R-convolution kernel [43] is then defined as follows:

$$K_{R,\otimes}(x, z) = \sum_{(x_1,\ldots,x_D) \in R^{-1}(x)} \sum_{(z_1,\ldots,z_D) \in R^{-1}(z)} \prod_{d=1}^{D} K_d(x_d, z_d). \tag{3}$$

In the above formulation, a tensor product has been used to combine kernels between different part types. Haussler [43] showed that the tensor product is closed under positive definiteness and, therefore, R-convolution kernels that use tensor product as a combination operator are positive definite, provided that all K_d are. The result also holds for combinations based on other closed operators, such as direct sum, yielding

$$K_{R,\oplus}(x, z) = \sum_{(x_1,\ldots,x_D) \in R^{-1}(x)} \sum_{(z_1,\ldots,z_D) \in R^{-1}(z)} \sum_{d=1}^{D} K_d(x_d, z_d). \tag{4}$$

Convolution or decomposition kernels form a vast class of functions and need to be specialized to capture the correct notion of similarity required by the task at hand. For example, several kernels on discrete structures have been designed using $D = 1$ and defining a simple concept of part. These "all-substructures kernels" basically count the number of co-occurrences of substructures in two decomposable objects. Plain counting can be easily achieved by using the exact match kernel

$$\delta(x, z) = \begin{cases} 1 \text{ if } x = z \\ 0 \text{ otherwise.} \end{cases} \tag{5}$$

Interesting discrete data types that have been thoroughly studied in the literature include sequences [44,5,6], trees [45,8], and graphs [11,9]. The *set kernel* [2] is a special case of convolution kernel that will prove useful in defining logical kernels presented in this chapter and that has been also used in the context of multi-instance learning [46]. Suppose instances are sets and let us define the part-of relation as the usual set-membership. The kernel over sets K_{set} is then obtained from kernels between set members K_{member} as follows:

$$K_{set}(x, z) = \sum_{\xi \in x} \sum_{\zeta \in z} K_{member}(\xi, \zeta). \tag{6}$$

2.4   Normalization and Composition

In order to reduce the dependence on the dimension of the objects, kernels over discrete structures are often normalized. A common choice is that of using normalization in feature space, i.e., given a convolution kernel K_R:

$$K_{norm}(x, z) = \frac{K_R(x, z)}{\sqrt{K_R(x, x)}\sqrt{K_R(z, z)}}.$$ (7)

In the case of set kernels, an alternative is that of dividing by the cardinalities of the two sets, thus computing the mean value between pairwise comparisons[1]:

$$K_{mean}(x, z) = \frac{K_{set}(x, z)}{|x||z|}.$$ (8)

Richer families of kernels on data structures can be formed by applying composition to the feature mapping induced by a convolution kernel. For example, a convolution kernel K_R can be combined with a Gaussian kernel as follows:

$$K(x, z) = \exp\left(-\gamma\Big(K_R(x, x) - 2K_R(x, z) + K_R(z, z)\Big)\right).$$ (9)

2.5   A Framework for Statistical Logical Learning

One of the standard ILP frameworks is that of learning from entailment. In this setting, the learner is given a set of positive and negative examples, $\mathcal{D}^+$ and $\mathcal{D}^-$, respectively (in the form of ground facts), and a background theory $\mathcal{B}$ (as a set of definite clauses) and has to induce a hypothesis $\mathcal{H}$ (also a set of definite clauses) such that $\mathcal{B} \cup \mathcal{H}$ covers all positive examples and none of the negative ones. More formally, $\forall p(x) \in \mathcal{D}^+ : \mathcal{B} \cup \mathcal{H} \models p(x)$ and $\forall p(x) \in \mathcal{D}^- : \mathcal{B} \cup \mathcal{H} \not\models p(x)$. Note that the meaning of term *hypothesis* in this context is related but not coincident with its meaning in statistical learning, where the hypothesis space is a class of functions mapping instances to targets.

We now develop a framework aiming to combine some of the advantages of the statistical and the ILP settings, in particular: efficiency, stability, generality, and the possibility of describing background knowledge in a flexible declarative language. As in the ILP setting, we assume that a background theory $\mathcal{B}$ is available as a set of definite clauses. This background theory is divided into *intensional* predicates, $\mathcal{B}_I$, and *extensional* predicates, $\mathcal{B}_E$, the former relevant to all examples, and the latter that specify facts about specific examples. As in [48], examples will simply be individuals, i.e., first-order logic objects, syntactically denoted by a unique identifier. This means that we shall effectively refer to the examples by their identifier x rather than use the associated set of extensional clauses, $p(x) \subset \mathcal{B}_E$. The instance space $\mathcal{X}$ is therefore a set of individuals contained in the overall universe of discourse $\mathcal{U}$. As in the statistical setting, we assume that a fixed and unknown distribution ρ is defined on $\mathcal{X} \times \mathcal{Y}$ and that training data $\mathcal{D}$ consist of input-output pairs (x_i, y_i) sampled identically and independently from ρ. Note that the latter assumption is reasonable in the case of relational domains with independent examples (such as mutagenesis) but not,

---

[1] Note that normalizations such as those of Equations (7) and (8) can give indefinite results iff one of the two arguments (say x) is the null vector of the feature space associated to the original kernel (i.e., K_R or K_{set}). In such a case, we will define $K_{norm}(x, z) = K_{mean}(x, z) = 0 \; \forall z \in \mathcal{X}, z \neq x$.

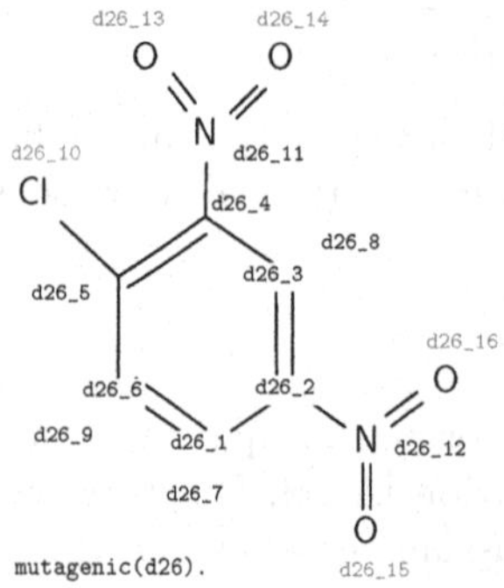

mutagenic(d26).

Extensional predicates

```
atm(d26,d26_1,c,22,-0.093).        bond(d26,d26_1,d26_2,7).
atm(d26,d26_2,c,22,-0.093).        bond(d26,d26_2,d26_3,7).
atm(d26,d26_3,c,22,-0.093).        bond(d26,d26_3,d26_4,7).
atm(d26,d26_4,c,22,-0.093).        bond(d26,d26_4,d26_5,7).
atm(d26,d26_5,c,22,-0.093).        bond(d26,d26_5,d26_6,7).
atm(d26,d26_6,c,22,-0.093).        bond(d26,d26_6,d26_1,7).
atm(d26,d26_7,h,3,0.167).          bond(d26,d26_1,d26_7,1).
atm(d26,d26_8,h,3,0.167).          bond(d26,d26_3,d26_8,1).
atm(d26,d26_9,h,3,0.167).          bond(d26,d26_6,d26_9,1).
atm(d26,d26_10,cl,93,-0.163).      bond(d26,d26_10,d26_5,1).
atm(d26,d26_11,n,38,0.836).        bond(d26,d26_4,d26_11,1).
atm(d26,d26_12,n,38,0.836).        bond(d26,d26_2,d26_12,1).
atm(d26,d26_13,o,40,-0.363).       bond(d26,d26_13,d26_11,2).
atm(d26,d26_14,o,40,-0.363).       bond(d26,d26_11,d26_14,2).
atm(d26,d26_15,o,40,-0.363).       bond(d26,d26_15,d26_12,2).
atm(d26,d26_16,o,40,-0.363).       bond(d26,d26_12,d26_16,2).
```

Intensional predicates

```
nitro(Drug,[Atom0,Atom1,Atom2,Atom3]) :-
   atm(Drug,Atom1,n,38,_),
   bondd(Drug,Atom0,Atom1,1),
   bondd(Drug,Atom1,Atom2,2),
   atm(Drug,Atom2,o,40,_),
   bondd(Drug,Atom1,Atom3,2),
   Atom3 @> Atom2,
   atm(Drug,Atom3,o,40,_).

bondd(Drug,Atom1,Atom2,Type) :-
   bond(Drug,Atom1,Atom2,Type).
bondd(Drug,Atom1,Atom2,Type) :-
   bond(Drug,Atom2,Atom1,Type).

benzene(Drug,Ring_list) :-
   atoms(Drug,6,Atom_list,[c,c,c,c,c,c]),
   ring6(Drug,Atom_list,Ring_list,[7,7,7,7,7,7]).

ring6(Drug,[Atom1|List],[Atom1,Atom2,Atom4,Atom6,Atom5,Atom3],
    [Type1,Type2,Type3,Type4,Type5,Type6]) :-
   bondd(Drug,Atom1,Atom2,Type1),   memberchk(Atom2,[Atom1|List]),
   bondd(Drug,Atom1,Atom3,Type2),   memberchk(Atom3,[Atom1|List]),
   Atom3 @> Atom2,
   bondd(Drug,Atom2,Atom4,Type3),   Atom4 \== Atom1,
   memberchk(Atom4,[Atom1|List]),
   bondd(Drug,Atom3,Atom5,Type4),   Atom5 \== Atom1,
   memberchk(Atom5,[Atom1|List]),
   bondd(Drug,Atom4,Atom6,Type5),   Atom6 \== Atom2,
   memberchk(Atom6,[Atom1|List]),
   bondd(Drug,Atom5,Atom6,Type6),   Atom6 \== Atom3.
```

Fig. 1. Example from the mutagenesis domain [47] illustrating the framework for statistical logic learning we use in this chapter

in general, when examples are linked by extensional predicates and collective prediction schemes are required (e.g. [49,50]).

In Figure 1 we exemplify our framework in the well known mutagenesis domain [47]. The extensional predicates are in this case `atm/5` and `bond/4`, describing the input portion of the data. The predicate `mutagenic/1` is also extensional. It describes the target class y and is not included in $\mathcal{B}$. The instance identifier in this case is `d26`, while $p($`d26`$)$, the extensional clauses associated with example `d26` $\in \mathcal{X}$, are listed in the upper right box of Figure 1. Intensional predicates include, among others, `nitro/2` and `benzene/2`, listed in the bottom box of Figure 1.

The output produced by statistical and ILP-based learning algorithms is also typically different. Rather than having to find a set of clauses that, added to the background theory, covers the examples, the main goal of a statistical learning algorithm is to find a function f that maps instances into their targets and whose general form is given by the representer theorem as in Eq. (2). Concerning the methods reviewed in this chapter, when the kernel function is fixed before learning, (as it happens in the methods presented in Sections 3, 4, and 5), predictions on new instances will be essentially opaque. However, when the kernel is learned together with the target function (see Section 6), the learning process also produces a collection of clauses, like an hypothesis in the ILP setting.

2.6 Types

A finer level of granularity in the definition of some of the logic-based kernels presented in this chapter can be gained from the use of typed terms. This extra flexibility may be necessary to specify different kernel functions associated with constants (e.g. to distinguish between numerical and categorical constants) or to different arguments of compound terms.

Following [51], we use a ranked set of type constructors $\mathcal{T}$, that contains at least the nullary constructor $\perp$. We allow polymorphism through type parameters. For example $list\alpha$ is a unary type constructor for the type of lists whose elements have type α. The arity of a type constructor is the number of type parameters it accepts. The set $\mathcal{T}$ is closed with respect to type variable substitution. Thus if $\tau\alpha_1, \ldots, \alpha_m \in \mathcal{T}$ is an m-ary type constructor (with type variables $\alpha_1, \ldots, \alpha_m$) and $\tau_1, \ldots, \tau_m \in \mathcal{T}$ then $\tau\tau_1, \ldots, \tau_m \in \mathcal{T}$.

The type signature of a function of arity n has the form $\tau_1 \times, \ldots, \times \tau_n \mapsto \tau'$ where $n \geq 0$ is the number of arguments, $\tau_1, \ldots, \tau_k \in \mathcal{T}$ their types, and $\tau' \in \mathcal{T}$ the type of the result. Functions of arity 0 have signature $\perp \mapsto \tau'$ and can be therefore interpreted as constants of type τ'. The type signature of a predicate of arity n has the form $\tau_1 \times, \ldots, \times \tau_n \mapsto \Omega$ where $\Omega \in \mathcal{T}$ is the type of booleans. We write $t : \tau$ to assert that t is a term of type τ.

A special case is when $\mathcal{T} = \{\tau_1, \ldots, \tau_n\}$ is a partition of $\mathcal{U}$. In this case $\mathcal{T}$ can be viewed as an equivalence relation $=_T$ as follows: $\forall x, y \in \mathcal{U}$ $x =_T y$ iff $\exists \tau_i \in \mathcal{T} s.t.(x : \tau_i \Leftrightarrow y : \tau_i)$. Another interesting situation is when type names are hierarchically organized in a partial order $\prec_T \subset \mathcal{T} \times \mathcal{T}$, with $\sigma \prec_T \tau$ meaning that σ *is a* τ (e.g. dog $\prec_T$ animal).

3 Kernels on Prolog Ground Terms

3.1 Motivations

We begin linking statistical and logic learning by introducing a family of kernels for Prolog terms. Convolution kernels over complex individuals have been recently defined using higher order logic abstractions [52]. The functions defined in this section can be seen as a specialization of such kernels to the case of Prolog and are motivated by the following considerations. First, Prolog and first-order logic representations provide a *simpler* representational framework than higher order logics. Second, Prolog expressiveness is *sufficient* for most application domains (for example, higher order structures such as sets can be simulated and types can also be introduced). Third, Prolog is a widespread and well supported language and many inductive logic programming systems and knowledge bases are actually based on (fragments of) first order logic. Finally, no probabilistic logic representations (like those thoroughly discussed elsewhere in this book) are yet available for higher-order logics.

The kernels introduced here have of course interesting connections to relational distances such as those described in [53,54]. It should be noted, however, that a distance function can trivially obtained from a kernel just by taking the

Euclidean distance in feature space, while a metric does not necessarily map into a Mercer kernel.

3.2   Untyped Terms

We begin with kernels on untyped terms. Let $\mathcal{C}$ be a set of constants and $\mathcal{F}$ a set of functors, and denote by $\mathcal{U}$ the corresponding Herbrand universe (the set of all ground terms that can be formed from constants in $\mathcal{C}$ and functors in $\mathcal{F}$). Let $f^{/n} \in \mathcal{F}$ denote a functor having name f and arity n. The kernel between two terms t and s is a function $K : \mathcal{U} \times \mathcal{U} \mapsto R$ defined inductively as follows:

– if $s \in \mathcal{C}$ and $t \in \mathcal{C}$ then

$$K(s,t) = \kappa(s,t) \tag{10}$$

where $\kappa : \mathcal{C} \times \mathcal{C} \mapsto R$ is a valid kernel on constants;
– else if s and t are compound terms and have different functors, i.e., $s = f(s_1, \ldots, s_n)$ and $t = g(t_1, \ldots, t_m)$, then

$$K(s,t) = \iota(f^{/n}, g^{/m}) \tag{11}$$

where $\iota : \mathcal{F} \times \mathcal{F} \mapsto R$ is a valid kernel on functors;
– else if s and t are compound terms and have the same functor, i.e., $s = f(s_1, \ldots, s_n)$ and $t = f(t_1, \ldots, t_n)$, then

$$K(s,t) = \iota(f^{/n}, f^{/n}) + \sum_{i=1}^{n} K(s_i, t_i) \tag{12}$$

– in all other cases $K(s,t) = 0$.

Functions κ and ι are *atomic* kernels that operate on non-structured symbols. A special but useful case is the atomic exact match kernel δ defined in Eq. (5).

3.3   Typed Terms

The kernel between two typed terms t and s (see Section 2.6) is defined inductively as follows:

– if $s \in \mathcal{C}$, $t \in \mathcal{C}$, $s : \tau$, $t : \tau$ then

$$K(s,t) = \kappa_\tau(s,t) \tag{13}$$

where $\kappa_\tau : \mathcal{C} \times \mathcal{C} \mapsto R$ is a valid kernel on constants of type τ;
– else if s and t are compound terms that have the same type but different functors or signatures, i.e., $s = f(s_1, \ldots, s_n)$ and $t = g(t_1, \ldots, t_m)$, $s : \sigma_1 \times, \ldots, \times \sigma_n \mapsto \tau'$, $t : \tau_1 \times, \ldots, \times \tau_m \mapsto \tau'$, then

$$K(s,t) = \iota_{\tau'}(f^{/n}, g^{/m}) \tag{14}$$

where $\iota_{\tau'} : \mathcal{F} \times \mathcal{F} \mapsto R$ is a valid kernel on functors that construct terms of type τ'

– else if s and t are compound terms and have the same functor and type signature, i.e., $s = f(s_1, \ldots, s_n)$, $t = f(t_1, \ldots, t_n)$, and $s, t : \tau_1 \times, \ldots, \times \tau_n \mapsto \tau'$, then

$$K(s,t) = \begin{cases} \kappa_{\tau_1 \times, \ldots, \times \tau_n \mapsto \tau'}(s,t) \\ \qquad \text{if } (\tau_1 \times, \ldots, \times \tau_n \mapsto \tau') \in \overline{T} \\ \iota_{\tau'}(f^{/n}, f^{/n}) + \sum_{i=1}^{n} K(s_i, t_i) \quad \text{otherwise} \end{cases} \qquad (15)$$

where $\overline{T} \subset T$ denotes a (possibly empty) set of *distinguished* type signatures that can be useful to specify ad-hoc kernel functions on certain compound terms, and $\kappa_{\tau_1 \times, \ldots, \times \tau_n \mapsto \tau'} : \mathcal{U} \times \mathcal{U} \mapsto \mathbb{R}$ is a valid kernel on terms having distinguished type signature $\tau_1 \times, \ldots, \times \tau_n \mapsto \tau' \in \overline{T}$.
– in all other cases $K(s,t) = 0$.

Positive semi-definiteness of these kernels follows from their being special cases of decomposition kernels (see [55] for details). Variants where direct summations over sub-terms are replaced by tensor products are also possible.

3.4 A Guided Example: Alkanes

We demonstrate here the use of kernels over logical terms in a simple application of quantitative structure-property relationship (QSPR) consisting in the prediction of boiling point of alkanes [56]. Alkanes (except cycloalkanes, which are not considered here) are naturally represented as trees and a root can be chosen using a very simple procedure. The resulting rooted trees are encoded as Prolog ground terms. Figure 2 shows an example of molecule encoding, where we actually employed a reversed ordering of the children of each node with respect to the procedure described in [56], in order to have the backbone of the molecule on the right hand side of the tree.

We designed a kernel on untyped terms by using exact match for comparing functors (carbon atoms), and the null function for comparing constants (hydrogen atoms). The resulting kernel counts the number of carbon atoms in corresponding positions of two alkanes. As an additional source of information, we

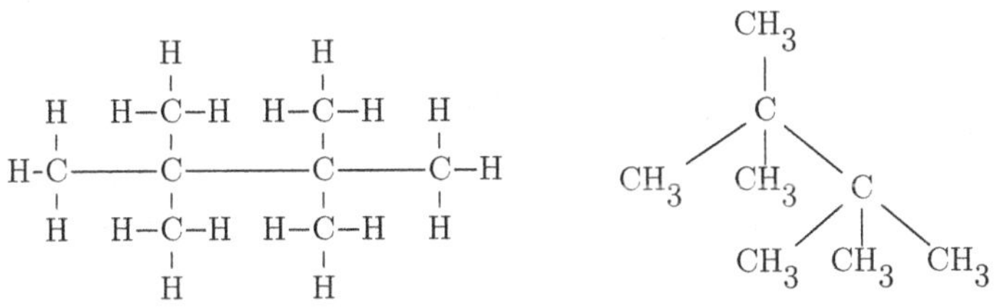

c(h,h,h,c(c(h,h,h),c(h,h,h),c(c(h,h,h),c(h,h,h),c(h,h,h))))

Fig. 2. An alkane, its canonical representation as a rooted tree, and the corresponding Prolog ground term

extracted the depths of the trees representing the molecules, and summed their product to the term kernel, obtaining a more informed kernel K'. The resulting function was composed with a Gaussian kernel.

The above kernel was used in conjunction with ridge regression to solve the boiling point prediction problem. Performance was evaluated by a ten fold cross validation procedure, removing the methane compound from the test results as suggested in [56], being it an outlier with basically no structure. Hyperparameters (namely, the Gaussian width and the regularization parameter), were chosen by a hold-out procedure on the training set of the first fold, and kept fixed for the successive 10 fold cross validation procedure. When using kernel K we obtained an average mean square error of 4.6 Celsius degrees while using K' the error can be reduced to 3.8 degrees. These results are comparable to those produced by the highly tuned neural networks developed in [56].

4   Declarative Kernels

We present in this section a logical framework for kernel specification that provides a simple interface for the incorporation of background knowledge. The relational feature generation process is controlled by an additional set of facts and axioms, developed on the basis of the available background theory $\mathcal{B}$. Although, in general, any set of relational features could be used, we start from a specific setting in which these additional facts and axioms refer to special and germane relations for reasoning about *parts* and *places*.

4.1   Mereotopology

The parthood relation has been formally investigated by logicians and philosophers for almost a century since the early work of Leśniewski [57] followed by Leonard & Goodman's calculus of individuals [58]. The axiomatic theory of parts is referred to as *mereology* (from the Greek μερος, "part"). It has obvious connections to decomposition of data structures in convolution kernels (see Section 2.3). The theory can be enriched with additional topological predicates and axioms aiming to describe wholeness. As pointed out by Varzi [59], topology is much needed because "mereological reasoning by itself cannot do justice to the notion of a whole (a one-piece, self-connected whole, such as a stone or a whistle, as opposed to a scattered entity made up of several disconnected parts, such as a broken glass, an archipelago, or the sum of two distinct cats)." These ideas can be also leveraged in machine learning to increase the kernel expressiveness with respect to pure decompositional approaches like the all-substructures kernels discussed in Section 2.3 that are only based on the notion of parts.

We formally introduce two special predicates: $\preceq_P$ and Connected, with the following intended meaning. For any two objects x and y, $x \preceq_P y$ declares x to be a part of y and Connected(x, y) declares x to be connected to y. Well-behaved definitions of parthood and connection should satisfy some given axiomatic structure [59]. In the context of knowledge representation, it is widely accepted that $\preceq_P$ should be a partial order, i.e. $\forall x, y, z \in \mathcal{U}$

$$x \preceq_P x \tag{P1}$$

$$x \preceq_P y \wedge y \preceq_P x \Rightarrow y =_P x \tag{P2}$$

$$x \preceq_P y \wedge y \preceq_P z \Rightarrow x \preceq_P z \tag{P3}$$

The theory defined by the above axioms is referred to as *ground mereology*. Interestingly, the above theory immediately provides us with a natural identity predicate $=_P$ that may be used as a basic elementary operator for comparing parts. Additional useful relations are supported by the theory, in particular

$$x \prec_P y \quad \text{iff} \quad x \preceq_P y \wedge \neg y \preceq_P x \qquad \text{proper part} \tag{16}$$

$$\text{Overlap}(x, y) \quad \text{iff} \quad \exists z.(z \preceq_P x \wedge z \preceq_P y) \qquad \text{overlap} \tag{17}$$

$$\text{Underlap}(x, y) \quad \text{iff} \quad \exists z.(x \preceq_P z \wedge y \preceq_P z) \qquad \text{underlap} \tag{18}$$

The *supplementation* axiom, if added to the theory, supports the notion of extensionality:

$$\forall z.(z \preceq_P x \Rightarrow \text{Overlap}(z, y)) \Rightarrow x \preceq_P y. \tag{P4}$$

Following [59], the following axioms characterize topology and its link to mereology:

$$\text{Connected}(x, x) \tag{C1}$$

$$\text{Connected}(x, y) \Rightarrow \text{Connected}(y, x) \tag{C2}$$

$$x \preceq_P y \Rightarrow \forall z.(\text{Connected}(z, x) \Rightarrow \text{Connected}(z, y)) \tag{C3}$$

Additional useful relations are supported by the theory, in particular

$$\text{Externally\_Connected}(x, y) \quad \text{iff} \quad \text{Connected}(x, y) \wedge \neg \text{Overlap}(x, y) \tag{19}$$

Mereotopology can be used to enrich the given background theory in the hope that it will generate further instances of the parthood and connection relations that will be *useful* for learning. It may also serve the purpose of checking the correctness of the declared parts and connections. When used for generating new instances of mereotopological relations, axioms should be used wisely to avoid an explosion of uninteresting parts and connections. Thus, depending on the application domain, axioms can be selectively omitted — for example (P4) will be typically avoided.

4.2 Mereotopological Relations

Several mereotopological relations (MR) can be introduced to characterize an instance x, for example:

i) The proper parts of x: $\mathcal{R}_P(x) = \{y : y \prec_P x\}$;
ii) The connected proper parts of x: $\mathcal{R}_C(x) = \{(y, z) : y \prec_P x \wedge z \prec_P x \wedge \text{Connected}(y, z)\}$;
iii) The overlapping parts in x, along with their common proper parts:
$\mathcal{R}_I(x) = \{(y, z, w) : y \neq z \wedge y \prec_P x \wedge z \prec_P x \wedge w \prec_P y \wedge w \prec_P z\}$;

iv) The externally connected parts in x along with the associated linking terminals:

$$\mathcal{R}_L(x) = \{(y, z, u, v) : z \prec_P x \wedge y \prec_P x \wedge \neg \text{Overlap}(z, y) \wedge u \prec_P z \wedge v \prec_P y$$
$$\wedge \text{Connected}(u, v)\}.$$

Additional MRs can be defined if necessary. We denote by $\mathcal{M}$ the set of declared MRs. As detailed below, a declarative kernel compares two instances by comparing the corresponding MRs, so adding relations to $\mathcal{M}$ plays a crucial role in shaping the feature space.

4.3   The Contribution of Parts

The kernel on parts, denoted K_P, is naturally defined as the *set kernel* between the sets of proper parts:

$$K_P(x, x') = \sum_{y \in \mathcal{P}(x)} \sum_{y' \in \mathcal{P}(x')} k_P(y, y') \qquad (20)$$

where k_P denotes a kernel function on parts, defined recursively using K_P. Types (see Section 2.6) can be used to fine-tune the definition of k_P. It types can be viewed as an equivalence relation ($\mathcal{T}$ is a partition of $\mathcal{U}$), then

$$k_P(y, y') = \begin{cases} \iota(y, y') & \text{if } y =_T y' \text{ and } y, y' \text{ are atomic objects;} \\ K_P(y, y') + \iota(y, y') & \text{if } y =_T y' \text{ and } y, y' \text{ are non atomic objects;} \\ 0 & \text{otherwise (i.e. } y \neq_T y'). \end{cases}$$
$$(21)$$

In the above definition, $\iota(y, y')$ is a kernel function that depends on properties or attributes of y and y' (not on their parts).

If types are hierarchically organized, then the test for type equality in Eq. (21) can be replaced by a more relaxed test on type compatibility. In particular, if $y : \tau$, $y' : \tau'$ and there exists a least general supertype $\sigma : \tau \prec_T \sigma, \tau' \prec_T \sigma$, then we may type cast y and y' to σ and evaluate κ on the generalized objects $\sigma(y)$ and $\sigma(y')$. In this case the function $\iota(y, y')$ depends only on properties or attributes that are common to y and y' (i.e. those that characterize the type σ).

4.4   The Contribution of Other MRs

The kernel on connected parts compares the sets of objects $\mathcal{R}_C(x)$ and $\mathcal{R}_C(x')$ as follows:

$$K_C(x, x') = \sum_{(y, z) \in \mathcal{R}_C(x)} \sum_{(y', z') \in \mathcal{R}_C(x')} K_P(y, y') \cdot K_P(z, z'). \qquad (22)$$

The kernel on overlapping parts compares the sets of objects $\mathcal{R}_I(x)$ and $\mathcal{R}_I(x')$ as follows:

$$K_I(x, x') = \sum_{(y, z, w) \in \mathcal{R}_I(x)} \sum_{(y', z', w') \in \mathcal{R}_L(x')} K_P(w, w') \delta(y, y') \delta(z, z') \qquad (23)$$

where $\delta(x, y) = 1$ if x and y have the same type and 0 otherwise. The kernel $K_L(x, x')$ on externally connected parts is defined in a similar way:

$$K_L(x, x') = \sum_{(y,z,u,v)\in\mathcal{R}_L(x)} \sum_{(y',z',u',v')\in\mathcal{R}_L(x')} K_P(u, u')K_P(v, v')\delta(y, y')\delta(z, z').$$

(24)

4.5   The General Case

Given a set $\mathcal{M}$ of MRs (such as those defined above), the final form of the kernel is

$$K(x, x') = \sum_{M\in\mathcal{M}} K_M(x, x').$$

(25)

Alternatively, a convolution-type form of the kernel can be defined as

$$K(x, x') = \prod_{M\in\mathcal{M}} K_M(x, x').$$

(26)

To equalize the contributions due to different MRs, the kernels K_M can be normalized before combining them with sum or product. Positive semi-definiteness follows, as in the case of convolution kernels, from the closeness with respect to direct sum and tensor product operators [43].

4.6   Remarks

The kernel of Eq. (25) could have been obtained also without the support of logic programming. However, deductive reasoning greatly simplifies the task of recognizing parts and connected parts and at the same time, the declarative style of programming makes it easy and natural to define the features that are implicitly defined by the kernel.

Declarative kernels and Haussler's convolution kernels [43] are intimately related. However the concept of *parts* in [43] is very broad and does not necessarily satisfy mereological assumptions.

4.7   A Guided Example: Mutagenesis

Defining and applying declarative kernels involves a three-step process: (1) collect data and background knowledge; (2) interface mereotopology to the available data and knowledge; (3) calculate the kernel on pairs of examples. We illustrate the process in the mutagenesis domain. The first step in this case simply consists of acquiring the atom-bond data and the ring theory developed by Srinivasan et al. [47], that comes in the usual form described in Figure 1. The second step consists of interfacing the available data and knowledge to the kernel. For this purpose, we first need to provide a set of declarations for types, objects, and basic instances of mereotopological relations. Objects are declared using the predicate obj(X,T) meaning that X is an object of type T. For example types include atoms and functional groups (see Figure 3a).

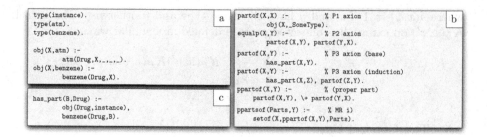

Fig. 3. Code fragments for the guided example (see text)

Then we declare basic proper parts via the predicate **has_part(X,Y)** that is true when **Y** is known to be a proper part of **X** . For example if an instance D (a molecule in this case) contains a benzene ring B, then $B \prec_P D$ (Figure 3b).

Note that the use of a predicate called **has_part** (rather than **partof**) is necessary to avoid calling a recursive predicate in the Prolog implementation. The third step is independent of the domain. To calculate the kernel, we first make use of mereotopology to construct the MRs associated with each instance (for example, the code for computing proper parts is shown in Figure 3c). The resulting sets of ground facts are then passed to a modified version of SVM^{light} [60] for fast kernel calculation.

We can construct here an example where connected parts may produce interesting features. Let us denote by x the molecule, by y the benzene ring, by v the nitro group consisting of atoms **d26_11**, **d26_13**, and **d26_14**, and by w the nitro group consisting of atoms **d26_12**, **d26_15**, and **d26_16**. Then $(y, v, \text{d26\_4}, \text{d26\_11}) \in \mathcal{R}_L(x)$ and $(y, w, \text{d26\_2}, \text{d26\_12}) \in \mathcal{R}_L(x)$.

To show how learning takes place in this domain, we run a series of 10-fold cross-validation experiments on the regression friendly data set of 188 compounds. First, we applied a mature ILP technique constructing an ensemble of

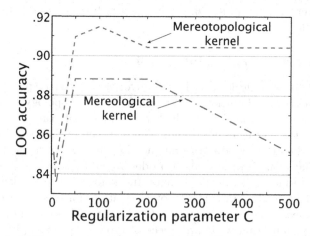

Fig. 4. LOO accuracy on the regression friendly mutagenesis data set

25 Aleph theories [61]. Aleph parameters *search, evalfn, clauselength* and *nodes* were set to be *bf, coverage*, 4 and 20000 respectively. The two tunable parameters *minacc* and *voting threshold* were selected by applying 3-fold cross validation in the training set of the first fold. Voting threshold ranges from 1 to the size of the ensemble and the set of values for minacc are given by $\{0.75, 0.9\}$. We obtained accuracy .88 ± .07 using atom-bond data and .89 ± .05 by adding the background ring theory. Next we applied declarative kernels with support vector machines (SVM), obtaining accuracy .90 ± .07. CPU time was of the order of minutes for the declarative kernel and days for the Aleph ensemble. Finally, we compared the expressive power of ground mereological relations with that of the full mereotopological theory. Figure 4 reports LOO accuracy for different values of the regularization parameter C, for both mereological and mereotopological kernels, showing the latter achieves both better optimal accuracy and more stable performances.

5   Kernels on Prolog Proof Trees

The main idea behind this family of kernels is the exploitation of program traces to define the kernel function. Traces have been extensively used in ILP and program synthesis (e.g. [62,63,64,65,66]). Kernels on Prolog proof trees are based on a new framework for learning from example-traces. The main assumption is that we are given a target program (called the *visitor*), that reflects background knowledge and that takes single examples as its input. The task consists of learning from the training set of traces obtained by executing the visitor program on each example. Hence, the statistical learning algorithm will employ a kernel on program traces rather than using directly a kernel on examples. The visitor acts therefore as a knowledge-based mediator between the data and the statistical learning algorithm. The bottom line is that similar instances should produce similar traces when probed with programs that express background knowledge and examine characteristics they have in common. These characteristics can be more general than parts. Hence, trace kernels can be introduced with the aim of achieving a greater generality and flexibility with respect to various decomposition kernels (including declarative kernels). These ideas will be developed in detail for logic programs, although nothing prevents, in principle, to use them in the context of different programming paradigms and in conjunction with alternative models of computation such as finite state automata or Turing machines.

Formally, a *visitor* program for a background theory $\mathcal{B}$ and domain $\mathcal{X}$ is a set $\mathcal{V}$ of definite clauses that contains at least one special clause (called a *visitor*) of the form $V \leftarrow B_1, \ldots, B_N$ and such that

- V is a predicate of arity 1
- for each $j = 1, \ldots, N$, B_j is declared in $\mathcal{B} \cup \mathcal{V}$;

Intuitively, if visit/1 is a visitor in $\mathcal{V}$, by answering the query visit(ex)? we explore the features of the instance whose constant identifier ex is passed to the visitor. Having multiple visitors in the program $\mathcal{V}$ allows us to explore different aspects of the examples and include multiple sources of information.

The visitor clauses should be designed to "inspect" examples using other predicates declared in $\mathcal{B}$, keeping in mind that the similarity between two examples is the similarity between the execution traces of visitors. Thus, we are not only simply interested in determining whether certain clauses succeed or fail on a particular example, but rather to ensure that visitors will construct useful features during their execution. This is a major difference with respect to other approaches in which features are explicitly constructed by computing the truth value for predicates [67].

The learning setting can be briefly sketched as follows. The learner is given a data set $\mathcal{D} = \{(x_1, y_1), \ldots, (x_m, y_m)\}$, background knowledge $\mathcal{B}$, and a visitor program $\mathcal{V}$. For each instance x_i, a trace T_{x_i} (see Eq. 28) is obtained by running the visitor program. A kernel machine (e.g., an SVM) is then trained to form the function $f : \mathcal{X} \mapsto \mathcal{Y}$ defined as

$$f(x) = \sum_{i=1}^{m} c_i K(T_{x_i}, T_x).$$

In the following, we give some details about the definition of traces and kernels between traces.

5.1   Traces and Proof Trees

In order to record a trace, we should store all steps in the proofs of a given visitor goal called on a given example. We may think that SLD-trees are a rather obvious representation of proofs when using Prolog. A path in an SLD-tree is indeed an execution sequence of the Prolog interpreter. Unfortunately, SLD-trees are too complex for our purposes, containing too many details and prone to generate irrelevant features such as those associated with failed paths. In order to obtain simple and still useful traces we prefer proof trees (see e.g. [68]). Given a program $\mathcal{P}$ and a goal G, the proof tree for G is empty if $\mathcal{P} \not\models G$ or, otherwise, it is a tree t recursively defined as follows:

- if there is a fact f in $\mathcal{P}$ and a substitution θ such that $G\theta = f\theta$, then $G\theta$ is a leaf of t.
- otherwise there must be a clause $H \leftarrow B_1, \ldots, B_n \in \mathcal{P}$ and a substitution θ' such that $H\theta' = G\theta'$ and $\mathcal{P} \models B_j\theta' \; \forall j$, $G\theta'$ is the root of t and there is a subtree of t for each $B_j\theta'$ that is a proof tree for $B_j\theta'$.

A second aspect is that we would like to deal with ground traces in order to simplify the definition of the kernel. On the other hand, proof trees may contain free variables. There are at least three ways of ensuring that proof trees are ground: first, we can use skolemization (naming existentially quantified variables with a specific constant symbol). A second option is to require that all clauses be range-restricted. Finally, we can make specific assumptions about the mode of head variables not occurring in the body, ensuring that these variables will be instantiated when proving the goal.

Goals can be satisfied in multiple ways, thus each query generates a (possibly empty) forest of proof trees. Since multiple visitors may be available, the trace of an instance is actually a *tuple of proof forests*. Formally, let N be the number of visitors in $\mathcal{V}$ and for each $l = 1, \ldots, N$ let $T_{lj,x}$ denote the proof tree that represents the j-th proof of the goal $V_l(x)$, i.e., a proof that $\mathcal{B} \cup \mathcal{V} \models V_l(x)$. Let

$$T_{l,x} = \{T_{l1,x}, \ldots, T_{ls_{l,x},x}\} \tag{27}$$

where $s_{l,x} \geq 0$ is the number of alternative proofs of goal $V_l(x)$. The trace of an instance x is then defined as the tuple

$$T_x = [T_{1,x}, \ldots, T_{N,x}]. \tag{28}$$

A proof tree can be pruned to remove unnecessary details and reduce the complexity of the feature space. Let us explain this concept with an example based on mutagenesis (see Figure 1). In this domain, it may be useful to define visitors that explore groups such as benzene rings:

```
atoms(X,[]).                    visit_benzene(X):-
atoms(X,[H|T]):-                    benzene(X,Atoms),
    atm(X,H,_,_,_),                 atoms(X,Atoms).
    atoms(X,T).
```

If we believe that the presence of the ring and the nature of the involved atoms represent a sufficient set of features, we may want to ignore details about the proof of the predicate **benzene** by pruning the corresponding proof subtree. This can be accomplished by including the following fact in the visitor program:

```
leaf(benzene(_,_)).
```

5.2   Kernels on Traces

A kernel over program traces can be defined in a top-down fashion. First, let us decompose traces into parts associated with different visitors (i.e., the elements of the tuple in Eq. (28)). The direct sum decomposition kernel of Eq. (4) applied to these parts yields:

$$K(T_x, T_z) = \sum_{l=1}^{N} K_l(T_{l,x}, T_{l,z}). \tag{29}$$

We always compare proofs of the same visitor since there is a unique decomposition of T_x and T_z. By definition of trace (see Eq. (28)), $T_{l,x}$ and $T_{l,z}$, $l = 1, \ldots, N$, are proof forests. Hence, the set kernel of Eq. (6) yields:

$$K_l(T_{l,x}, T_{l,z}) = \sum_{p=1}^{s_{l,x}} \sum_{q=1}^{s_{l,z}} K_{tree}(T_{lp,x}, T_{lq,z}). \tag{30}$$

We now need to introduce a kernel K_{tree} over individual proof trees. In principle, existing tree kernels (e.g. [7,8]) could be used for this purpose. However, we suggest here representing proof trees as typed Prolog ground terms. This option allows us to provide a fine-grained definition of kernel by exploiting type information on constants and functors (so that each object type can be compared by its own sub-kernel). Moreover, the kernel on ground terms introduced in Section 3 is able to compares sub-proofs only if they are reached as a result of similar inference steps. This distinction would be difficult to implement with traditional tree kernels. A ground term can be readily constructed from a proof tree as follows:

- Base step: if a node contains a fact, this is already a ground term.
- Induction: if a node contains a clause, then let n be the number of arguments in the head and m the number of atoms in the body (corresponding to the m children of the node). A ground compound term t having $n+1$ arguments is then formed as follows:
  - the functor name of t is the functor name of the head of the clause;
  - the first n arguments of t are the arguments of the clause head;
  - the last argument of t is a compound term whose functor name is a Prolog constant obtained from the clause number[2], and whose m arguments are the ground term representations of the m children of the node.

At the highest level of kernel between visitor programs (Eq. (29)), it is advisable to employ a feature space normalization using Eq. (7). In some cases it may also be useful to normalize lower-level kernels, in order to rebalance contributions of individual parts. In particular, the mean normalization of Eq. (8) can be applied to the kernel over individual visitors (Eq. (30)) and it is also possible to normalize kernels between individual proof trees, in order to reduce the influence of the proof size. Of course it is easy to gain additional expressiveness by defining specific kernels on proof trees that originate from different visitors.

In order to employ kernels on typed terms (see Section 3), we need a typed syntax for representing proof trees as ground terms. Constants can be of two main types: num (numerical) and cat (categorical). Types for compounds terms include fact (leaves) clause (internal nodes), and body (containing the body of a clause).

A number of special cases of kernels can be implemented in this framework. The simplest kernel is based on proof equivalence (two proofs being equivalent if the same sequence of clauses is proven in the two cases, and the head arguments in corresponding clauses satisfy a given equivalence relation): $K_{equiv}(s,t) = 1$ iff $s \equiv t$.

The *functor equality* kernel can be used when we want to ignore the arguments in the head of a clause. Given two ground terms $s = f(s_1, \ldots, s_n)$ and $t = g(t_1, \ldots, t_m)$, it is defined as:

$$K_f(s,t) = \begin{cases} 0 & \text{if } type(s) \neq type(t) \\ \delta(f^{/n}, g^{/m}) & \text{if } s, t : \texttt{fact} \\ \delta(f^{/n}, g^{/m}) \star K(s_n, t_m) & \text{if } s, t : \texttt{clause} \\ K(s,t) & \text{if } s, t : \texttt{body} \end{cases} \tag{31}$$

---

[2] Since numbers cannot be used as functor names, this constant can be simply obtained by prefixing the clause number by 'cbody'.

where K is a kernel on ground terms and the operator $\star$ can be either sum or product. Note that if s and t represent clauses (i.e., internal nodes of the proof tree), the comparison skips clause head arguments, represented by the first $n - 1$ (resp. $m - 1$) arguments of the terms, and compares the bodies (the last argument) thus proceeding on the children of the nodes. This kernel allows to define a non trivial equivalence between proofs (or parts of them) checking which clauses are proved in sequence and ignoring the specific values of their head arguments.

5.3   A Guided Example: Bongard Problems

One nice example showing the potential of learning from program traces is a very simple Bongard problem [69] in which the goal is to classify two-dimensional scenes consisting of sets of nested polygons (triangles, rectangles, and circles). In particular, we focus on the target concept defined by the pattern *triangle-X^n-triangle* for a given n, meaning that a positive example is a scene containing two triangles nested into one another with exactly n objects (possibly triangles) in between. Figure 5 shows a pair of examples of such scenes with their representation as Prolog facts and their classification according to the pattern for $n = 1$.

A possible example of background knowledge introduces the concepts of *nesting* in containment and *polygon* as a generic object, and can be represented by the following intensional predicates:

```
inside(X,A,B):- in(X,A,B).              % clause nr 1
inside(X,A,B):-                          % clause nr 2
    in(X,A,C),
    inside(X,C,B).
polygon(X,A) :-  triangle(X,A).          % clause nr 3
polygon(X,A) :-  rectangle(X,A).         % clause nr 4
polygon(X,A) :-  circle(X,A).            % clause nr 5
```

A visitor exploiting such background knowledge, and having hints on the target concept, could be looking for two polygons contained one into the other. This can be represented as:

```
visit(X):-                               % clause nr 6
    inside(X,A,B),polygon(X,A),polygon(X,B).
```

Figure 6 shows the proofs trees obtained running such a visitor on the first Bongard problem in Figure 5. A very simple kernel can be employed to solve such a task, namely an equivalence kernel with functor equality for nodewise comparison. For any value of n, such a kernel maps the examples into a feature space where there is a single feature discriminating between positive and negative examples, while the simple use of ground facts without intensional background knowledge would not provide sufficient information for the task.

The data set was generated by creating m scenes each containing a series of ℓ randomly chosen objects nested one into the other, and repeating the procedure

76   P. Frasconi and A. Passerini

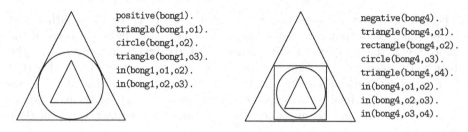

Fig. 5. Graphical and Prolog facts representation of two Bongard scenes. The left and right examples are positive and negative, respectively, according to the pattern *triangle-X-triangle*.

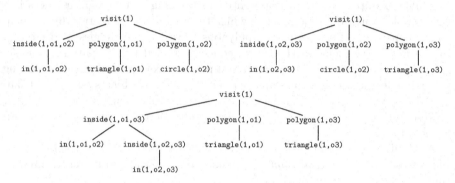

Fig. 6. Proof trees obtained by running the visitor on the first Bongard problem in Figure 5

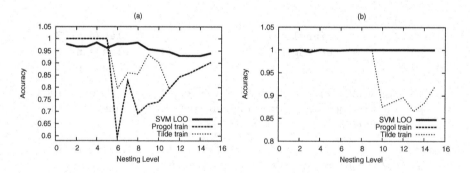

Fig. 7. Comparison between SVM leave-one-out error, Progol and Tilde empirical error in learning the *triangle-X^n-triangle* for different values of n, for data sets corresponding to $m = 10$ (a) and $m = 50$ (b)

for ℓ varying from 2 to 20. Moreover, we generated two different data sets by choosing $m = 10$ and $m = 50$ respectively. Finally, for each data set we obtained 15 experimental settings denoted by $n \in [0, 14]$. In each setting, positive examples were scenes containing the pattern *triangle-X^n-triangle*. We run an SVM with the above mentioned proof tree kernel and a fixed value $C = 10$ for the regularization parameter, on the basis that the data set is noise free. We evaluated its performance with a leave-one-out (LOO) procedure, and compared it to the empirical error of Tilde and Progol trained on the same data and background knowledge (including the visitor). Here we focus on showing that ILP algorithms have troubles finding a consistent hypothesis for this problem, hence we did not measure their generalization.

Figure 7(a) plots results for $m = 10$. Both Tilde and Progol stopped learning the concept for $n > 4$. Progol found the trivial empty hypothesis for all $n > 4$ apart from $n = 6$, and Tilde for all $n > 9$. While never learning the concept with 100% generalization accuracy, the SVM performance was much more stable when increasing the nesting level corresponding to positive examples. Figure 7(b) plots results for $m = 50$. Progol was extremely expensive to train with respect to the other methods. It successfully learned the concept for $n \leq 2$, but we stopped training for $n = 3$ after more than one week training time on a 3.20 GHz PENTIUM IV. Tilde stopped learning the concept for $n > 8$, and found the trivial empty hypothesis for $n > 12$. Conversely, the SVM was almost always able to learn the concept with 100% generalization accuracy, regardless of its complexity level.

Note that in order for the ILP algorithms to learn the target concept regardless of the nesting level, it would be necessary to provide a more informed inside predicate, which explicitly contains such nesting level as one of its arguments. The ability of the kernel to extract information from the predicate proof, on the other hand, allows our method to be employed when only partial background knowledge is available, which is typically the case in real world applications.

6   kFOIL

The above approaches for combining ILP and kernel can be expected to be highly effective from several points of view, in particular stability (i.e. robustness to noise), uniformity (i.e. classification and regression tasks can be handled in a uniform way) and expressivity (a rich hypothesis space is explored in domains consisting of independent relational objects). However, the function determined by these methods as a solution to the supervised learning problem is opaque, i.e. does not provide human-readable insights. In addition, although the feature space is rich, its definition must be specified *before* learning takes place. The idea behind kFOIL is radically different from this point of view. Unlike previous approaches, the feature space in kFOIL is *dynamically* constructed during learning (using a FOIL-like [38] covering algorithm) and can be effectively seen as an additional output of the learning problem (besides the prediction function). In this sense, kFOIL is similar to a recently introduced probabilistic ILP algorithm, nFOIL, that combines Naive Bayes and FOIL [70]. While nFOIL takes the generative direction of modeling, kFOIL is based on regularized empirical risk minimization (e.g. support vector machine learning). kFOIL preserves all the advantages of previously introduced kernels, in particular uniformity of representation across different supervised learning tasks, stability and robustness with respect to noise, expressivity of the representation language, and ability to reuse declarative background knowledge. The strength of kFOIL is its ability to provide additional explanations about the domain that can be read in the set of constructed clauses. However, since FOIL is used as an internal subroutine, the efficiency of other kernel based learning approaches cannot be preserved.

6.1   The Feature Space of kFOIL

In the kFOIL setting, the output of the learning process consists of both a prediction function f (as in Eq. (2)) and a kernel function K between examples. Each example $x \in \mathcal{X}$ is a first-order individual and $p(x)$ denotes the associated extensional clauses, as explained in Section 2.5. The function K is represented by means of a collection of clauses

$$\mathcal{H} = \{c_1, \ldots, c_n\}$$

that play the same role of a hypothesis in the learning from entailment ILP setting. In particular, the feature space associated with K consists of Boolean vectors, indexed by clauses in the current hypothesis $\mathcal{H}$. Formally, the feature space representation can be written as $\phi_\mathcal{H}(x) = \phi_{\mathcal{H},1}(x), \ldots, \phi_{\mathcal{H},n}(x)$ where

$$\phi_{\mathcal{H},i}(x) = \begin{cases} 1 \text{ if } \mathcal{B}_I \cup \{c_i\} \models p(x) \\ 0 \text{ otherwise} \end{cases}$$

The feature space representation is defined by the clauses in the current hypothesis and each feature simply check whether $p(x)$ is logically entailed by background knowledge and one given clause.

In this way, the kernel between two examples x and x' is simply the number of clauses firing on both examples, in the context of the given background knowledge:

$$K_\mathcal{H}(x, x') = \#ent_\mathcal{H}(p(x) \wedge p(x')) \tag{32}$$

where $\#ent_\mathcal{H}(a) = |\{c \in \mathcal{H} | \mathcal{B}_I \cup \{c\} \models a\}|$. The prediction function f has the standard form of Eq. (2), using $K_\mathcal{H}$ as kernel.

6.2   The kFOIL Learning Algorithm

The hypothesis $\mathcal{H}$ is induced by a modified version of the well-known FOIL algorithm [38], which essentially implements a *separate-and-conquer* rule learning algorithm in a relational setting.

KFOIL$(\mathcal{D}, \mathcal{B}, \epsilon)$

1   $\mathcal{H} := \emptyset$
2   **repeat**
3              $c := \text{"}pos(x) \leftarrow\text{"}$
4   **repeat**
5                      $c := \arg\max_{c' \in \rho(c)} \text{SCORE}(\mathcal{D}, \mathcal{H} \cup \{c'\}, \mathcal{B})$
6   **until** stopping criterion
7              $\mathcal{H} := \mathcal{H} \cup \{c\}$
8   **until** score improvement is smaller than ϵ
9   **return** $\mathcal{H}$

The kFOIL algorithm, sketched in the above pseudo-code, is similar to the general FOIL algorithm. It repeatedly searches for clauses that score well with respect to the data set $\mathcal{D}$ and the current hypothesis $\mathcal{H}$ and adds them to the current hypothesis. The most general clause which succeeds on all examples is "$pos(x) \leftarrow$" where pos is the predicate being learned. The "best" clause c is found in the inner loop according to a general-to-specific hill-climbing search strategy, using a refinement operator $\rho(c)$ that generates the set of all possible refinements of clause c. In the case of kFOIL, each refinement is obtained by simply adding a literal to the right-hand side of c. Different choices for the scoring function SCORE have been used with FOIL. The scoring function of kFOIL is computed by wrapping around a kernel machine (such as an SVM). Specifically, SCORE$(\mathcal{D}, \mathcal{H}, \mathcal{B})$ is computed by training a kernel machine on $\mathcal{D}$ and measuring the empirical risk

$$\text{SCORE}(\mathcal{D}, \mathcal{H}, \mathcal{B}) = \sum_{(x_i, y_i) \in \mathcal{D}} V(y_i, f(x_i))$$

being V a suitable loss function (that depends on the specific kernel machine, see Eq. (1) and following). kFOIL is stopped when the score improvement between two consecutive iterations falls below a given threshold ϵ. This a smoothed version of FOIL's criterion which is stopped when no clause can be found that cover additional positive examples. Finally, note that the data set size is reduced at each iteration of FOIL by removing examples that are already covered. However, this step is omitted from kFOIL as the kernel machine needs to be retrained (with a different kernel) on the entire data set.

In the case of kFOIL, a significant speedup can be obtained by working explicitly in a sparse feature space, rather than evaluating the kernel function according to its definition. This is because, especially at the early iterations, many examples are mapped to the same point in feature space and can be merged in a single point (multiplying the corresponding contribution to the loss function by the number of collisions).

6.3   A Guided Example: Biodegradability

In order to apply kFOIL to a certain learning task, three steps have to be accomplished: (1) collect data and background knowledge; (2) write the inductive bias

```
nitro(mol30, [atom1,atom2,atom3], [atom4]).                                          a
methyl(mol32, [atom1,atom2,atom3.atom4], [atom5]).bond(mol1,atom1,atom2,1).
atm(mol1,atom1,h,0,0).
logP(mol1, 0.35).
mweight(mol1,0.258).
```

```
gt(X,Y):- X > Y.                                           b
lt(X,Y):- X < Y.

num(N):-
   member(N,[0.0,0.1,0.2,0.3,0.4,0.5,0.6,0.7,0.8,0.9,1.0]).
sbond(Mol,Atom1,Atom2,Bondtype):-
   bond(Mol,Atom1,Atom2,Bondtype);bond(Mol,Atom2,Atom1,Bondtype).
```

```
                                                           c
rmode(lt(+,N)):-numrmode(lt(+,N)):-num(N).
type(nitro(compound,struct1,struct2)).
type(lt(number,number)).
rmode(atom(+,+-,c)).
rmode(nitro(+,-,-)).
```

Fig. 8. Code fragments for the kFOIL guided example on biodegradability

that will determine all possible refinements of clauses; (3) run kFOIL. We will show an example of such process on a real world task concerning biodegradability of molecules. Degradation is the process by which chemicals are transformed into components which are not considered pollutants. A number of different pathways are responsible for such process, depending on environmental conditions. Blockeel et al. [71] conducted a study focused on aqueous biodegradation under aerobic conditions. Low and high estimates of half life time degradation rate were collected for 328 commercial chemical compounds. In this application domain, one is interested in the half-life time of the biodegradation process. The regression task consists in predicting the natural logarithm of the arithmetic mean of the low and high estimate for a given molecule. Available data include the atom/bond representation of molecules as well as global physico-chemical properties such as weight and logP. Rings and functional groups within a molecule are also represented as facts[3], where each groups is described by its constituent atoms as well as the atoms connecting it to the rest of the molecule. Figure 8(a) shows extracts of such data[4]. Additional background knowledge (see Figure 8(b) for an extract) includes comparison operators between numbers (lt, gt) and the set of allowed numerical values (num) as well as a predicate (sbond) defining symmetric bonds. The second step consists of writing the configuration file for the FOIL part of the algorithm, as a combination of type and mode declarations. Figure 8(c) contains an extract of such configuration. The final step consists of running kFOIL in regression mode providing as inputs data, background knowledge and configuration files. Note that the first two steps are independent of the type of task to be learned (e.g. binary classification or regression), which will only influence the type of kernel machine to be employed in computing the score of a given clause and in producing the output for a test example (e.g. SVM or Support Vector Regression).

---

[3] Intensional predicates representing functional groups were saturated on the examples in this dataset, thus generating extensional predicates.

[4] Note that we are not considering facts representing counts of groups and small substructures, which were also included in [71], as they slightly degrade performances for all methods in almost all cases.

Table 1. Result on the Biodegradability dataset. The results for Tilde and S-CART have been taken from [71]. 5 runs of 10 fold cross-validation have been performed, on the same splits into training and test set as used in [71]. We report both Pearson correlation and RMSE as evaluation measures. • indicates that the result for kFOIL is significantly better than for other method (unpaired two-sided t-test, p = 0.05).

Evaluation measure	kFOIL	Tilde	S-CART
Correlation	0.609 ± 0.047	0.616 ± 0.021	0.605 ± 0.023
Root Mean Squared Error	1.196 ± 0.023	1.265 ± 0.033•	1.290 ± 0.038•

Table 1 shows the regression performance of kFOIL on the Biodegradability dataset, as compared to the results reported in [71] for Tilde and S-CART. As our aim here is showing that kFOIL is competitive to other state-of-the-art techniques, and not to boost performance, we did not try to specifically optimize any parameter. We thus used default settings for the FOIL parameters: maximum number of clauses in a hypothesis was set to 25, maximum number of literals in a clause to 10 and the threshold for the stopping criterion to 0.1%. However, we performed a beam search with beam size 5 instead of simple greedy search. The kernel machine employed was support vector regression, with regularization constant $C = 0.01$ and ϵ tube parameter set to 0.001. A polynomial kernel of degree 2 was used on top of the kernel induced by the learned clauses. The results obtained show that kFOIL is competitive with the first-order decision tree systems S-CART and Tilde at maximizing correlation, and slightly superior at minimizing RMSE.

7 Applications

7.1 Declarative Kernels for Information Extraction

In these experiments we apply declarative kernels to the extraction of relational information from free text. Specifically, we focus on multi-slot extraction of binary relations between candidate named entities. Our experiments were carried out on the yeast protein localization data set described in [72] and subsequently used as a testbed for state-of-the-art methods based on ILP [73]. The task consists of learning the relation `protein_location` between two named entities representing *candidate* protein names and cell locations. Instances are ordered pairs of noun phrases (NP) extracted from MEDLINE abstracts and with stemmed words. An instance is positive iff the first NP is a protein and the second NP is a location, for example:

```
protein_location("the mud2 gene product","earli spliceosom assembl",pos).
protein_location("sco1", "the inner mitochondri membran",pos).
protein_location("the ept1 gene product","membran topographi",pos).
protein_location("a reductas activ", "the cell", neg).
protein_location("the ace2 gene", "multipl copi", neg).
```

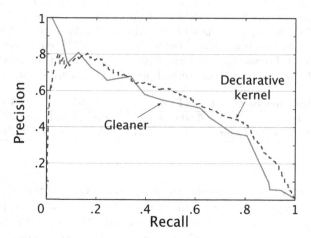

Fig. 9. Comparing Gleaner and the declarative kernel on the information extraction task (fold 5)

The data set is a collection of 7, 245 sentences from 871 abstracts, yielding 1, 773 positive and 279, 154 negative instances. The data is enriched by a large body of domain knowledge, including relations about the structure of sentences and abstracts, lexical knowledge, and biological knowledge derived from several specialized vocabularies and ontologies such as MeSH and Gene Ontology. For simplicity, only a fraction of the available knowledge has been used in our experiments. The main data types in this domain are: `instance` (pairs of candidate NP's); `cp_NP` (candidate protein NP); `cl_NP` (candidate location NP); `word_p` (word in a protein NP); `word_l` (word in a location NP). Basic parthood rules in the ontology declare that phrases (`cp_NP` and `cl_NP`) are parts of instances and words are parts of phrases. For this task we used a minimal mereological kernel with no connections and no axiomatic theory to avoid explosion of features due to words appearing both as part of NP's and instances. We compared declarative kernels to state-of-the-art ILP-based system for this domain: Aleph and Gleaner [73]. We used the same setting as in [73], performing a five-fold cross validation, with approximately 250 positive and 120, 000 negative examples in each fold (split at the level of abstracts), and measuring the quality of the predictor by means of the *area under the recall-precision curve* (AURPC). As reported in [73], Aleph attains its best performance (area .45) by learning on the order of 10^8 rules, while Gleaner attains similar performance (.43 ± .6) but using several orders of magnitude less rules [74]. We trained five SVMs using the declarative kernel composed with a Gaussian kernel. Gaussian width and the regularization parameter were selected by reserving a validation set inside each fold. The obtained AURPC was .47 ± .7. Figure 9 compares the recall precision curve reported in [73], which is produced by Gleaner using 1, 000, 000 candidate clauses on fold five, with that obtained by the declarative kernel. The result is very encouraging given that only a fraction of the available knowledge has been used. Training took less than three hours on a single 3.00GHz Pentium while Aleph and Gleaner run for several days on a large PC cluster on the same task [74].

7.2   Proof Tree Kernels for Protein Fold Classification

Binary Classification. In our first experiment, we tested our methodology on the protein fold classification problem studied by Turcotte et al. [75]. The task consists of classifying proteins into SCOP folds, given their high-level logical descriptions about secondary structure and amino acid sequence. SCOP is a manually curated database of proteins hierarchically organized according to their structural properties. At the top level SCOP groups proteins into four main classes (all-α, all-β, α/β, and $\alpha + \beta$). Each class is then divided into folds that group together proteins with similar secondary structures and three-dimensional arrangements. We used the data set made available as a supplement to the paper by Turcotte et al. [75][5] that consists of the five most populated folds from each of the four main SCOP classes. This setting yields 20 binary classification problems. The data sets for each of the 20 problems are relatively small (from about 30 to about 160 examples per fold, totaling 1143 examples).

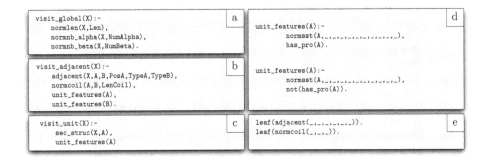

Fig. 10. Visitors for the protein fold classification problem

We relied on the background knowledge provided in [75], to design a set of visitors managing increasingly complex information. Visitors are shown in Figure 10. The "global" visitor `visit_global/1` is meant to extract protein level information, such as its length and the number of its α or β secondary structure segments. A "local" visitor `visit_unit/1` explores the details of each of these segments. In particular, after determining the secondary structure element, it explores the general features of the element using `normsst/11` and checks for the presence of proline (an amino acid that is known to have important effects on the secondary structure). Note that since traces are recorded as proof trees, the first clause of the predicate `unit_features/1` above produces information only in the case a proline is present. Finally, the "connection" visitor `visit_adjacent/1` inspects pairs of adjacent segments within the protein.

Numerical values were normalized within each top level fold class. The kernel configuration mainly consisted of type signatures aiming to ignore identifiers

---

[5] Available at `http://www.bmm.icnet.uk/ilp/data/ml_2000.tar.gz`

Table 2. Protein fold classification: 10-fold cross validation accuracy (%) for Tilde, Progol and SVM for the different classification tasks, and micro averaged accuracies with 95% confidence intervals. Results for Progol are taken from [75].

	Tilde	Progol	SVM
All-α:			
Globin-like	97.4	95.1	94.9
DNA-binding 3-helical bundle	81.1	83.0	88.9
4-helical cytokines	83.3	70.7	86.7
lambda repressor-like DNA-binding domains	70.0	73.4	83.3
EF Hand-like	71.4	77.6	85.7
All-β:			
Immunoglobulin-like beta-sandwich	74.1	76.3	85.2
SH3-like barrel	91.7	91.4	93.8
OB-fold	65.0	78.4	83.3
Trypsin-like serine proteases	95.2	93.1	93.7
Lipocalins	83.3	88.3	92.9
α/β:			
beta/alpha (TIM)-barrel	69.7	70.7	73.3
NAD(P)-binding Rossmann-fold domains	79.4	71.6	84.1
P-loop containing nucleotide triphosphate hydrolases	64.3	76.0	76.2
alpha/beta-Hydrolases	58.3	72.2	86.1
Periplasmic binding protein-like II	79.5	68.9	79.5
$\alpha + \beta$:			
Interleukin 8-like chemokines	92.6	92.9	96.3
beta-Grasp	52.8	71.7	88.9
Ferredoxin-like	69.2	83.1	76.9
Zincin-like	51.3	64.3	79.5
SH2-like	82.1	76.8	66.7
Micro average:	75.2	78.3	83.6
	±2.5	±2.4	±2.2

and treat some of the numerical features as categorical ones. A functor equality kernel was employed for those nodes of the proofs which did not contain valuable information in their arguments.

Following [75], we measured prediction accuracy by 10-fold cross-validation, micro-averaging the results over the 20 experiments by summing contingency tables. The proof-tree kernel was combined with a Gaussian kernel (see Eq. (9)) in order to model nonlinear interactions between the features extracted by the visitor program. Model selection (i.e., choice of the Gaussian width γ and the SVM regularization parameter C) was performed for each binary problem with a LOO procedure before running the 10-fold cross validation. Table 2 shows comparisons between the best setting for Progol (as reported by [75]), which uses both propositional and relational background knowledge, results for Tilde using the same setting, and SVM with our kernel over proof trees. The difference between Tilde and Progol is not significant, while our SVM achieves significantly higher overall accuracy with respect to both methods. The only task where our predictor performed worse than both ILP methods was the SH2-like one (the last one in Table 2). It is interesting to note that a simple global visitor would achieve 84.6% accuracy on this task, while in most other tasks full relational features produce consistently better results. This can suggest that even if SVMs are capable of effectively dealing with huge feature spaces, great amounts of uninformative or noisy features can also degrade performance, especially if only few examples are available.

Multiclass Classification. We additionally evaluated our proof tree kernels on the multiclass setting of the protein fold prediction task as described in [76]. The problem is based on the same 20 SCOP folds previously used for binary classification, but the data set contains only the chains considered as positive examples for one of the SCOP folds in the binary classification problems. Four independent multiclass classification problems are defined, one for each of the main fold classes in SCOP. A single multiclass problem consists of discriminating between chains belonging to the same fold class, by assigning each of them to one of the five main folds in the fold class. The statistics of the dataset are reported in Table 3, and show the unbalancing of the distribution of examples between folds. We employed a one-vs-all strategy to address each multiclass classification task: we trained a number of binary classifiers equal to the number of classes, each trained to discriminate between examples of one class and examples of all other classes; during testing, we presented each example to all binary classifiers, and assigned it to the class for which the corresponding binary classifier was the most confident, as measured by the margin of the prediction for the example. We employed the same 5-fold cross validation procedure as reported in [76] and

Table 3. Number of examples for each multiclass problem (fold class) both divided by single class (fold) and overall

fold class	fold$_1$	fold$_2$	fold$_3$	fold$_4$	fold$_5$	overall
all-α	13	30	10	10	14	77
all-β	90	32	40	42	28	116
α/β	55	21	14	12	13	115
$\alpha + \beta$	9	12	26	13	13	73

Table 4. Microaveraged accuracies with standard errors for the four multiclass problems and overall accuracy microaveraged over problems: comparison between SVM, SLP and ILP with majority class

fold class	SVM	SLP	ILP + majority class
all-α	80.5±4.5 (62/77)	76.6±4.8 (59/77)	71.4±5.2 (55/77)
all-β	87.1±3.1(101/116)	81.0±3.6 (94/116)	69.8±4.3 (81/116)
α/β	61.7±4.5 (71/115)	51.3±4.7 (59/115)	44.4±4.6 (51/115)
$\alpha + \beta$	60.3±5.7 (44/73)	82.2±4.5 (60/73)	80.8±4.6 (59/73)
overall	73.0±2.3 (278/381)	71.4±2.3 (272/381)	64.6±2.5 (246/381)

used exactly the same CV folds. Model selection (Gaussian width and regularization parameter) was conducted by a preliminary LOO procedure as in the case of binary classification, but using the F_1 measure (the harmonic mean of precision and recall) as a guiding criterion. We kept the same visitors developed for the binary setting. Table 4 reports accuracies with standard errors for the four multiclass problems, microaveraged on the CV folds, and overall accuracy microaveraged on CV folds and multiclass problems. Reported results include our SVM with proof tree kernels together to the results obtained by a stochastic logic program (SLP) and ILP with majority class prediction as reported in [76].

Results show that both the SLP and the kernel machine outperform the non-probabilistic ILP approach. In three out of four SCOP folds the SVM obtained a higher microaveraged accuracy than the SLP although the data sets have small size and the standard deviation is rather high. Interestingly, the SLP seems to perform better on smaller data set, which might indicate a faster rate of convergence of the SLP to its asymptotic error.

8   Conclusions

In this chapter we have pursued the construction of a bridge between two very different paradigms of machine learning: statistical learning with kernels, and inductive logic programming. In particular, we have shown in several ways that the use of stable (robust to noise) machine learning techniques are applicable in the ILP setting without resorting to propositionalization. This is especially interesting in cases where the feature space needed for representing the solution has a dimension that is not known in advance. The artificial Bongard data set shows this clearly. Of course one could have solved the Bongard problem even with traditional ILP techniques by adding to the background theory a predicate counting the number of polygons nested one inside another, but kernels on program traces can effectively discover this concept without the additional hint.

The algorithmic stability achieved by combining ILP with regularization can be seen as an interesting alternative to fully fledged probabilistic ILP where structure and parameters of a stochastic program are learned from data. Empirical evidence on real-world problems such as the protein fold classification task demonstrates that proof tree kernels can achieve better accuracy than non probabilistic ILP and similar accuracy as learning stochastic logic programs. Of course the solution found by the kernel machine in this case lacks interpretability. However, computational efficiency is another factor that in some cases needs to be taken into account. For example, problems like the information extraction task presented in Section 7.1 can be solved in a fraction of the time required by a state-of-the-art ILP system.

kFOIL is perhaps the less developed and tested method but at the same time very promising. Its approach to propositionalization is effectively dynamic and can be interpreted in close connection to methods that attempt to learn the kernel matrix from data [77,78,79]. Moreover, the solution found by kFOIL combines the advantages of kernel machines and ILP systems. It consists of both a robust decision function and a kernel function defined by interpretable first-order clauses. This is a direction of research that certainly deserves more investigation since interpretability is one of the weak aspects of many statistical learning methods.

Finally, some of the ideas seeded here may deserve more study from an engineering perspective, as the availability of languages for feature description (like declarative kernels) can help leveraging machine learning into routine software development. One of the main obstacles towards this goal is maybe the difficulty of integrating machine learning capabilities in a typical programming environment. The recent growth of interest in knowledge representation and ontologies suggests that logic-based representations may be much more widespread in the future, and that the attention to learning modules that can take advantage of data and existing knowledge with minimal programmer intervention could increase.

Acknowledgments

We would like to thank Luc De Raedt, Niels Landwehr, and Stephen Muggleton, with whom we collaborated in the development of the methods presented in this chapter (in particular, we collaborated with LDR on kernels on Prolog proof trees, with SM on declarative kernels, with LDR and NL on kFOIL; NL also carried out the experiments reported in Section 6.3). A special thanks goes to Alessio Ceroni, Fabrizio Costa, Kristian Kersting, Sauro Menchetti, and Jan Ramon, who helped us in numerous occasions with fruitful discussions. This work was supported by the European Union, contract number FP6-508861, Applications of Probabilistic Inductive Logic Programming II.

References

1. Schölkopf, B., Smola, A.: Learning with Kernels. The MIT Press, Cambridge (2002)
2. Shawe-Taylor, J., Cristianini, N.: Kernel Methods for Pattern Analysis. Cambridge University Press, Cambridge (2004)
3. Lodhi, H., Saunders, C., Shawe-Taylor, J., Cristianini, N., Watkins, C.: Text classification using string kernels. J. Mach. Learn. Res. 2, 419–444 (2002)
4. Jaakkola, T., Haussler, D.: Exploiting generative models in discriminative classifiers. In: Advances in Neural Information Processing Systems 11, pp. 487–493. MIT Press, Cambridge (1999)
5. Leslie, C.S., Eskin, E., Noble, W.S.: The spectrum kernel: A string kernel for svm protein classification. In: Pacific Symposium on Biocomputing, pp. 566–575 (2002)
6. Cortes, C., Haffner, P., Mohri, M.: Rational kernels: Theory and algorithms. Journal of Machine Learning Research 5, 1035–1062 (2004)
7. Collins, M., Duffy, N.: New ranking algorithms for parsing and tagging: Kernels over discrete structures, and the voted perceptron. In: Proceedings of the Fortieth Annual Meeting on Association for Computational Linguistics, Philadelphia, PA, USA, pp. 263–270 (2002)

8. Viswanathan, S., Smola, A.J.: Fast kernels for string and tree matching. In: Becker, S.T., S., Obermayer, K. (eds.) Advances in Neural Information Processing Systems, vol. 15, pp. 569–576. MIT Press, Cambridge (2003)
9. Gärtner, T.: A survey of kernels for structured data. SIGKDD Explorations Newsletter 5(1), 49–58 (2003)
10. Smola, A.J., Kondor, R.: Kernels and Regularization on Graphs. In: Schölkopf, B., Warmuth, M.K. (eds.) COLT/Kernel 2003. LNCS (LNAI), vol. 2777, pp. 144–158. Springer, Heidelberg (2003)
11. Kashima, H., Tsuda, K., Inokuchi, A.: Marginalized kernels between labeled graphs. In: Proceedings of ICML 2003 (2003)
12. Mahé, P., Ueda, N., Akutsu, T., Perret, J.L., Vert, J.P.: Extensions of marginalized graph kernels. In: Greiner, R., D. Schuurmans, A.P. (eds.) Proceedings of the Twenty-first International Conference on Machine Learning, Banff, Alberta, Canada, pp. 552–559 (2004)
13. Horváth, T., Gärtner, T., Wrobel, S.: Cyclic pattern kernels for predictive graph mining. In: Proceedings of the Tenth ACM SIGKDD International Conference on Knowledge Discovery and Data Mining, pp. 158–167. ACM Press, New York (2004)
14. Menchetti, S., Costa, F., Frasconi, P.: Weighted decomposition kernels. In: Proceedings of the Twenty-second International Conference on Machine Learning, pp. 585–592. ACM Press, New York (2005)
15. Kramer, S., Lavrac, N., Flach, P.: Propositionalization approaches to relational data mining. In: Relational Data Mining, pp. 262–286. Springer, Heidelberg (2000)
16. Cumby, C.M., Roth, D.: Learning with feature description logics. In: Matwin, S., Sammut, C. (eds.) ILP 2002. LNCS (LNAI), vol. 2583, pp. 32–47. Springer, Heidelberg (2003)
17. Cumby, C.M., Roth, D.: On kernel methods for relational learning. In: Proceedings of ICML 2003 (2003)
18. Ramon, J., Bruynooghe, M.: A Framework for Defining Distances Between First-Order Logic Objects. In: Proc. of the 8th International Conf. on Inductive Logic Programming, pp. 271–280 (1998)
19. Kirsten, M., Wrobel, S., Horváth, T.: Distance based approaches to relational learning and clustering. In: Relational Data Mining, pp. 213–230. Springer, Heidelberg (2001)
20. Ramon, J.: Clustering and instance based learning in first order logic. AI Communications 15(4), 217–218 (2002)
21. Cortes, C., Vapnik, V.N.: Support vector networks. Machine Learning 20, 1–25 (1995)
22. De Raedt, L.: Logical and Relational Learning: From ILP to MRDM. Springer, Heidelberg (2006)
23. Vapnik, V.N.: The Nature of Statistical Learning Theory. Springer, New York (1995)
24. Herbrich, R., Graepel, T., Obermayer, K.: Support vector learning for ordinal regression. In: Artificial Neural Networks, 1999. ICANN 1999. Ninth International Conference on (Conf. Publ. No. 470), vol. 1 (1999)
25. Tax, D., Duin, R.: Support vector domain description. Pattern Recognition Letters 20, 1991–1999 (1999)
26. Ben-Hur, A., Horn, D., Siegelmann, H., Vapnik, V.: Support vector clustering. Journal of Machine Learning Research 2, 125–137 (2001)
27. Schölkopf, B., Smola, A., Müller, K.: Nonlinear component analysis as a kernel eigenvalue problem. Neural computation 10(5), 1299–1319 (1998)

28. Kramer, S.: Structural regression trees. In: Proceedings of the Thirteenth National Conference on Artificial Intelligence, pp. 812–819 (1996)
29. Kramer, S.: Prediction of Ordinal Classes Using Regression Trees. Fundamenta Informaticae 47(1), 1–13 (2001)
30. Cucker, F., Smale, S.: On the mathematical foundations of learning. Bulletin (New Series) of the American Mathematical Society 39(1), 1–49 (2002)
31. Lin, Y.: Support Vector Machines and the Bayes Rule in Classification. Data Mining and Knowledge Discovery 6(3), 259–275 (2002)
32. Bartlett, P., Jordan, M., McAuliffe, J.: Large margin classifiers: Convex loss, low noise, and convergence rates. Advances in Neural Information Processing Systems 16 (2003)
33. Ng, A., Jordan, M.: On Discriminative vs. Generative classifiers: A comparison of logistic regression and naive Bayes. Neural Information Processing Systems (2001)
34. Passerini, A., Frasconi, P.: Kernels on prolog ground terms. In: Proceedings of the Nineteenth International Joint Conference on Artificial Intelligence, Edinburgh, Scotland, UK, pp. 1626–1627 (2005)
35. Gärtner, T., Lloyd, J., Flach, P.: Kernels for structured data. In: Matwin, S., Sammut, C. (eds.) ILP 2002. LNCS (LNAI), vol. 2583, pp. 66–83. Springer, Heidelberg (2003)
36. Passerini, A., Frasconi, P., De Raedt, L.: Kernels on prolog proof trees: Statistical learning in the ILP setting. Journal of Machine Learning Research 7, 307–342 (2006)
37. Landwehr, N., Passerini, A., Raedt, L.D., Frasconi, P.: kFOIL: Learning simple relational kernels. In: Gil, Y., Mooney, R. (eds.) Proc. Twenty-First National Conference on Artificial Intelligence (AAAI 2006), AAAI Press, Menlo Park (2006)
38. Quinlan, J.R.: Learning Logical Definitions from Relations. Machine Learning 5, 239–266 (1990)
39. Saunders, G., Gammerman, A., Vovk, V.: Ridge regression learning algorithm in dual variables. In: Proc. 15th International Conf. on Machine Learning, pp. 515–521 (1998)
40. Poggio, T., Smale, S.: The mathematics of learning: Dealing with data. Notices of the American Mathematical Society 50(5), 537–544 (2003)
41. Kimeldorf, G.S., Wahba, G.: A correspondence between Bayesian estimation on stochastic processes and smoothing by splines. The Annals of Mathematical Statistics 41, 495–502 (1970)
42. Freund, Y., Schapire, R.E.: Large margin classification using the perceptron algorithm. Machine Learning 37(3), 277–296 (1999)
43. Haussler, D.: Convolution kernels on discrete structures. Technical Report UCSC-CRL-99-10, University of California, Santa Cruz (1999)
44. Lodhi, H., Shawe-Taylor, J., Cristianini, N., Watkins, C.: Text classification using string kernels. Advances in Neural Information Processing Systems, 563–569 (2000)
45. Collins, M., Duffy, N.: Convolution kernels for natural language. In: NIPS 14, pp. 625–632 (2001)
46. Gärtner, T., Flach, P., Kowalczyk, A., Smola, A.: Multi-instance kernels. In: Sammut, C., Hoffmann, A. (eds.) Proceedings of the 19th International Conference on Machine Learning, pp. 179–186. Morgan Kaufmann, San Francisco (2002)
47. Srinivasan, A., Muggleton, S., Sternberg, M.J.E., King, R.D.: Theories for mutagenicity: A study in first-order and feature-based induction. Artificial Intelligence 85(1-2), 277–299 (1996)
48. Lloyd, J.W.: Logic for learning: Learning comprehensible theories from structured data. Springer, Heidelberg (2003)

49. Taskar, B., Abbeel, P., Koller, D.: Discriminative probabilistic models for relational data. In: Proceedings of the Eighteenth Conference on Uncertainty in Artificial Intelligence, Morgan Kaufmann, San Francisco (2002)
50. Neville, J., Jensen, D.: Collective classification with relational dependency networks. In: Proceedings of the Second International Workshop on Multi-Relational Data Mining, pp. 77–91 (2003)
51. Lakshman, T.K., Reddy, U.S.: Typed prolog: A semantic reconstruction of the mycroft-O'keefe type system. In: Saraswat, Vijay, Ueda, K. (eds.) Proceedings of the 1991 International Symposium on Logic Programming (ISLP 1991), pp. 202–220. MIT Press, San Diego (1991)
52. Gärtner, T., Lloyd, J., Flach, P.: Kernels and distances for structured data. Machine Learning 57(3), 205–232 (2004)
53. Ramon, J., Bruynooghe, M.: A polynomial time computable metric between point sets. Acta Informatica 37(10), 765–780 (2001)
54. Horváth, T., Wrobel, S., Bohnebeck, U.: Relational instance-based learning with lists and terms. Machine Learning 43(1/2), 53–80 (2001)
55. Passerini, A., Frasconi, P., De Raedt, L.: Kernels on prolog proof trees: Statistical learning in the ILP setting. Journal of Machine Learning Research 7, 307–342 (2006)
56. Bianucci, A., Micheli, A., Sperduti, A., Starita, A.: Application of cascade correlation networks for structures to chemistry. Appl. Intell. 12, 117–146 (2000)
57. Leśniewski, S.: Podstawy ogólnej teorii mnogości. Moscow (1916)
58. Leonard, H.S., Goodman, N.: The calculus of individuals and its uses. Journal of Symbolic Logic 5(2), 45–55 (1940)
59. Casati, R., Varzi, A.: Parts and places: The structures of spatial representation. MIT Press, Cambridge, MA and London (1999)
60. Joachims, T.: Making large-scale SVM learning practical. In: Schölkopf, B., Burges, C., Smola, A. (eds.) Advances in Kernel Methods – Support Vector Learning, pp. 169–185. MIT Press, Cambridge (1998)
61. Srinivasan, A.: The Aleph Manual. Oxford University Computing Laboratory (2001)
62. Biermann, A., Krishnaswamy, R.: Constructing programs from example computations. IEEE Transactions on Software Engineering 2(3), 141–153 (1976)
63. Mitchell, T.M., Utgoff, P.E., Banerji, R.: Learning by experimentation: Acquiring and refining problem-solving heuristics. In: Machine learning: An artificial intelligence approach, vol. 1, pp. 163–190. Morgan Kaufmann, San Francisco (1983)
64. Shapiro, E.Y.: Algorithmic program debugging. MIT Press, Cambridge (1983)
65. Zelle, J.M., Mooney, R.J.: Combining FOIL and EBG to speed-up logic programs. In: Proceedings of the Thirteenth International Joint Conference on Artificial Intelligence, Chambéry, France, pp. 1106–1111 (1993)
66. De Raedt, L., Kersting, K., Torge, S.: Towards learning stochastic logic programs from proof-banks. In: Proceedings of the Twentieth National Conference on Artificial Intelligence (AAAI 2005), pp. 752–757 (2005)
67. Muggleton, S., Lodhi, H., Amini, A., Sternberg, M.: Support vector inductive logic programming. In: Hoffmann, A., Motoda, H., Scheffer, T. (eds.) DS 2005. LNCS (LNAI), vol. 3735, pp. 163–175. Springer, Heidelberg (2005)
68. Russell, S., Norvig, P.: Artifical Intelligence: A Modern Approach, 2nd edn. Prentice-Hall, Englewood Cliffs (2002)
69. Bongard, M.: Pattern Recognition. Spartan Books (1970)

70. Landwehr, N., Kersting, K., De Raedt, L.: nFOIL: Integrating Naïve Bayes and FOIL. In: Proc. of the 20th National Conf. on Artificial Intelligence, pp. 795–800 (2005)
71. Blockeel, H., Dzeroski, S., Kompare, B., Kramer, S., Pfahringer, B., Laer, W.: Experiments in Predicting Biodegradability. Applied Artificial Intelligence 18(2), 157–181 (2004)
72. Ray, S., Craven, M.: Representing sentence structure in hidden Markov models for information extraction. In: Proceedings of IJCAI 2001, pp. 1273–1279 (2001)
73. Goadrich, M., Oliphant, L., Shavlik, J.W.: Learning ensembles of first-order clauses for recall-precision curves: A case study in biomedical information extraction. In: Camacho, R., King, R., Srinivasan, A. (eds.) ILP 2004. LNCS (LNAI), vol. 3194, pp. 98–115. Springer, Heidelberg (2004)
74. Goadrich, M.: Personal communication (2005)
75. Turcotte, M., Muggleton, S., Sternberg, M.: The effect of relational background knowledge on learning of protein three-dimensional fold signatures. Machine Learning 43(1-2), 81–96 (2001)
76. Chen, J., Kelley, L., Muggleton, S., Sternberg, M.: Multi-class prediction using stochastic logic programs. In: Muggleton, S., Otero, R., Tamaddoni-Nezhad, A. (eds.) ILP 2006. LNCS (LNAI), vol. 4455, Springer, Heidelberg (2007)
77. Lanckriet, G.R.G., Cristianini, N., Bartlett, P., Ghaoui, L.E., Jordan, M.I.: Learning the kernel matrix with semidefinite programming. J. Mach. Learn. Res. 5, 27–72 (2004)
78. Ong, C.S., Smola, A.J., Williamson, R.C.: Hyperkernels. In: Adv. in Neural Inf. Proc. Systems (2002)
79. Micchelli, C.A., Pontil, M.: Learning the Kernel Function via Regularization. Journal of Machine Learning Research 6, 1099–1125 (2005)

Markov Logic

Pedro Domingos[1], Stanley Kok[1], Daniel Lowd[1], Hoifung Poon[1],
Matthew Richardson[2], and Parag Singla[1]

[1] Department of Computer Science and Engineering
University of Washington
Seattle, WA 98195-2350, USA
{pedrod,koks,lowd,hoifung,parag}@cs.washington.edu
[2] Microsoft Research
Redmond, WA 98052
mattri@microsoft.com

Abstract. Most real-world machine learning problems have both statistical and relational aspects. Thus learners need representations that combine probability and relational logic. Markov logic accomplishes this by attaching weights to first-order formulas and viewing them as templates for features of Markov networks. Inference algorithms for Markov logic draw on ideas from satisfiability, Markov chain Monte Carlo and knowledge-based model construction. Learning algorithms are based on the conjugate gradient algorithm, pseudo-likelihood and inductive logic programming. Markov logic has been successfully applied to problems in entity resolution, link prediction, information extraction and others, and is the basis of the open-source Alchemy system.

1 Introduction

Two key challenges in most machine learning applications are uncertainty and complexity. The standard framework for handling uncertainty is probability; for complexity, it is first-order logic. Thus we would like to be able to learn and perform inference in representation languages that combine the two. This is the focus of the burgeoning field of statistical relational learning [11]. Many approaches have been proposed in recent years, including stochastic logic programs [33], probabilistic relational models [9], Bayesian logic programs [17], relational dependency networks [34], and others. These approaches typically combine probabilistic graphical models with a subset of first-order logic (e.g., Horn clauses), and can be quite complex. Recently, we introduced Markov logic, a language that is conceptually simple, yet provides the full expressiveness of graphical models and first-order logic in finite domains, and remains well-defined in many infinite domains [44,53]. Markov logic extends first-order logic by attaching weights to formulas. Semantically, weighted formulas are viewed as templates for constructing Markov networks. In the infinite-weight limit, Markov logic reduces to standard first-order logic. Markov logic avoids the assumption of i.i.d. (independent and identically distributed) data made by most statistical learners

L. De Raedt et al. (Eds.): Probabilistic ILP 2007, LNAI 4911, pp. 92–117, 2008.

by leveraging the power of first-order logic to compactly represent dependencies among objects and relations. In this chapter, we describe the Markov logic representation and give an overview of current inference and learning algorithms for it. We begin with some background on Markov networks and first-order logic.

2   Markov Networks

A *Markov network* (also known as *Markov random field*) is a model for the joint distribution of a set of variables $X = (X_1, X_2, \ldots, X_n) \in \mathcal{X}$ [37]. It is composed of an undirected graph G and a set of potential functions ϕ_k. The graph has a node for each variable, and the model has a potential function for each clique in the graph. A potential function is a non-negative real-valued function of the state of the corresponding clique. The joint distribution represented by a Markov network is given by

$$P(X = x) = \frac{1}{Z} \prod_k \phi_k(x_{\{k\}}) \tag{1}$$

where $x_{\{k\}}$ is the state of the kth clique (i.e., the state of the variables that appear in that clique). Z, known as the *partition function*, is given by $Z = \sum_{x \in \mathcal{X}} \prod_k \phi_k(x_{\{k\}})$. Markov networks are often conveniently represented as *log-linear models*, with each clique potential replaced by an exponentiated weighted sum of features of the state, leading to

$$P(X = x) = \frac{1}{Z} \exp\left(\sum_j w_j f_j(x)\right) \tag{2}$$

A feature may be any real-valued function of the state. This chapter will focus on binary features, $f_j(x) \in \{0, 1\}$. In the most direct translation from the potential-function form (Equation 1), there is one feature corresponding to each possible state $x_{\{k\}}$ of each clique, with its weight being $\log \phi_k(x_{\{k\}})$. This representation is exponential in the size of the cliques. However, we are free to specify a much smaller number of features (e.g., logical functions of the state of the clique), allowing for a more compact representation than the potential-function form, particularly when large cliques are present. Markov logic will take advantage of this.

Inference in Markov networks is #P-complete [47]. The most widely used method for approximate inference in Markov networks is Markov chain Monte Carlo (MCMC) [12], and in particular Gibbs sampling, which proceeds by sampling each variable in turn given its Markov blanket. (The Markov blanket of a node is the minimal set of nodes that renders it independent of the remaining network; in a Markov network, this is simply the node's neighbors in the graph.) Marginal probabilities are computed by counting over these samples; conditional probabilities are computed by running the Gibbs sampler with the conditioning variables clamped to their given values. Another popular method for inference in Markov networks is belief propagation [59].

Maximum-likelihood or MAP estimates of Markov network weights cannot be computed in closed form but, because the log-likelihood is a concave function of the weights, they can be found efficiently (modulo inference) using standard gradient-based or quasi-Newton optimization methods [35]. Another alternative is iterative scaling [7]. Features can also be learned from data, for example by greedily constructing conjunctions of atomic features [7].

3   First-Order Logic

A *first-order knowledge base (KB)* is a set of sentences or formulas in first-order logic [10]. Formulas are constructed using four types of symbols: constants, variables, functions, and predicates. Constant symbols represent objects in the domain of interest (e.g., people: Anna, Bob, Chris, etc.). Variable symbols range over the objects in the domain. Function symbols (e.g., MotherOf) represent mappings from tuples of objects to objects. Predicate symbols represent relations among objects in the domain (e.g., Friends) or attributes of objects (e.g., Smokes). An *interpretation* specifies which objects, functions and relations in the domain are represented by which symbols. Variables and constants may be *typed*, in which case variables range only over objects of the corresponding type, and constants can only represent objects of the corresponding type. For example, the variable x might range over people (e.g., Anna, Bob, etc.), and the constant C might represent a city (e.g, Seattle, Tokyo, etc.).

A *term* is any expression representing an object in the domain. It can be a constant, a variable, or a function applied to a tuple of terms. For example, Anna, x, and GreatestCommonDivisor(x, y) are terms. An *atomic formula* or *atom* is a predicate symbol applied to a tuple of terms (e.g., Friends(x, MotherOf(Anna))). Formulas are recursively constructed from atomic formulas using logical connectives and quantifiers. If F_1 and F_2 are formulas, the following are also formulas: $\neg F_1$ (negation), which is true iff F_1 is false; $F_1 \wedge F_2$ (conjunction), which is true iff both F_1 and F_2 are true; $F_1 \vee F_2$ (disjunction), which is true iff F_1 or F_2 is true; $F_1 \Rightarrow F_2$ (implication), which is true iff F_1 is false or F_2 is true; $F_1 \Leftrightarrow F_2$ (equivalence), which is true iff F_1 and F_2 have the same truth value; $\forall x \, F_1$ (universal quantification), which is true iff F_1 is true for every object x in the domain; and $\exists x \, F_1$ (existential quantification), which is true iff F_1 is true for at least one object x in the domain. Parentheses may be used to enforce precedence. A *positive literal* is an atomic formula; a *negative literal* is a negated atomic formula. The formulas in a KB are implicitly conjoined, and thus a KB can be viewed as a single large formula. A *ground term* is a term containing no variables. A *ground atom* or *ground predicate* is an atomic formula all of whose arguments are ground terms. A *possible world* (along with an interpretation) assigns a truth value to each possible ground atom.

A formula is *satisfiable* iff there exists at least one world in which it is true. The basic inference problem in first-order logic is to determine whether a knowledge base KB *entails* a formula F, i.e., if F is true in all worlds where KB is true (denoted by $KB \models F$). This is often done by *refutation*: KB entails F iff $KB \cup \neg F$ is unsatisfiable. (Thus, if a KB contains a contradiction, all formulas trivially follow from it, which makes painstaking knowledge engineering a necessity.) For

Table 1. Example of a first-order knowledge base and MLN. Fr() is short for Friends(), Sm() for Smokes(), and Ca() for Cancer().

First-Order Logic	Clausal Form	Weight
"Friends of friends are friends."		
$\forall x \forall y \forall z \; Fr(x,y) \wedge Fr(y,z) \Rightarrow Fr(x,z)$	$\neg Fr(x,y) \vee \neg Fr(y,z) \vee Fr(x,z)$	0.7
"Friendless people smoke."		
$\forall x \; (\neg (\exists y \; Fr(x,y)) \Rightarrow Sm(x))$	$Fr(x,g(x)) \vee Sm(x)$	2.3
"Smoking causes cancer."		
$\forall x \; Sm(x) \Rightarrow Ca(x)$	$\neg Sm(x) \vee Ca(x)$	1.5
"If two people are friends, then either both smoke or neither does."	$\neg Fr(x,y) \vee Sm(x) \vee \neg Sm(y),$	1.1
$\forall x \forall y \; Fr(x,y) \Rightarrow (Sm(x) \Leftrightarrow Sm(y))$	$\neg Fr(x,y) \vee \neg Sm(x) \vee Sm(y)$	1.1

automated inference, it is often convenient to convert formulas to a more regular form, typically *clausal form* (also known as *conjunctive normal form (CNF)*). A KB in clausal form is a conjunction of *clauses*, a clause being a disjunction of literals. Every KB in first-order logic can be converted to clausal form using a mechanical sequence of steps.[1] Clausal form is used in resolution, a sound and refutation-complete inference procedure for first-order logic [46].

Inference in first-order logic is only semidecidable. Because of this, knowledge bases are often constructed using a restricted subset of first-order logic with more desirable properties. The most widely-used restriction is to *Horn clauses*, which are clauses containing at most one positive literal. The Prolog programming language is based on Horn clause logic [25]. Prolog programs can be learned from databases by searching for Horn clauses that (approximately) hold in the data; this is studied in the field of inductive logic programming (ILP) [22].

Table 1 shows a simple KB and its conversion to clausal form. Notice that, while these formulas may be *typically* true in the real world, they are not *always* true. In most domains it is very difficult to come up with non-trivial formulas that are always true, and such formulas capture only a fraction of the relevant knowledge. Thus, despite its expressiveness, pure first-order logic has limited applicability to practical AI problems. Many *ad hoc* extensions to address this have been proposed. In the more limited case of propositional logic, the problem is well solved by probabilistic graphical models. The next section describes a way to generalize these models to the first-order case.

4   Markov Logic

A first-order KB can be seen as a set of hard constraints on the set of possible worlds: if a world violates even one formula, it has zero probability. The basic idea in Markov logic is to soften these constraints: when a world violates one

---

[1] This conversion includes the removal of existential quantifiers by Skolemization, which is not sound in general. However, in finite domains an existentially quantified formula can simply be replaced by a disjunction of its groundings.

formula in the KB it is less probable, but not impossible. The fewer formulas a world violates, the more probable it is. Each formula has an associated weight (e.g., see Table 1) that reflects how strong a constraint it is: the higher the weight, the greater the difference in log probability between a world that satisfies the formula and one that does not, other things being equal.

Definition 1. [44] *A Markov logic network (MLN) L is a set of pairs (F_i, w_i), where F_i is a formula in first-order logic and w_i is a real number. Together with a finite set of constants $C = \{c_1, c_2, \ldots, c_{|C|}\}$, it defines a Markov network $M_{L,C}$ (Equations 1 and 2) as follows:*

1. *$M_{L,C}$ contains one binary node for each possible grounding of each atom appearing in L. The value of the node is 1 if the ground atom is true, and 0 otherwise.*
2. *$M_{L,C}$ contains one feature for each possible grounding of each formula F_i in L. The value of this feature is 1 if the ground formula is true, and 0 otherwise. The weight of the feature is the w_i associated with F_i in L.*

Thus there is an edge between two nodes of $M_{L,C}$ iff the corresponding ground atoms appear together in at least one grounding of one formula in L. For example, an MLN containing the formulas ∀x Smokes(x) ⇒ Cancer(x) (smoking causes cancer) and ∀x∀y Friends(x, y) ⇒ (Smokes(x) ⇔ Smokes(y)) (friends have similar smoking habits) applied to the constants Anna and Bob (or A and B for short) yields the ground Markov network in Figure 1. Its features include Smokes(Anna) ⇒ Cancer(Anna), etc. Notice that, although the two formulas above are false as universally quantified logical statements, as weighted features of an MLN they capture valid statistical regularities, and in fact represent a standard social network model [55].

An MLN can be viewed as a *template* for constructing Markov networks. From Definition 1 and Equations 1 and 2, the probability distribution over possible worlds x specified by the ground Markov network $M_{L,C}$ is given by

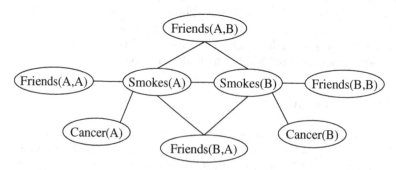

Fig. 1. Ground Markov network obtained by applying an MLN containing the formulas ∀x Smokes(x) ⇒ Cancer(x) and ∀x∀y Friends(x, y) ⇒ (Smokes(x) ⇔ Smokes(y)) to the constants Anna(A) and Bob(B)

$$P(X = x) = \frac{1}{Z} \exp \left(\sum_{i=1}^{F} w_i n_i(x) \right) \qquad (3)$$

where F is the number of formulas in the MLN and $n_i(x)$ is the number of true groundings of F_i in x. As formula weights increase, an MLN increasingly resembles a purely logical KB, becoming equivalent to one in the limit of all infinite weights. When the weights are positive and finite, and all formulas are simultaneously satisfiable, the satisfying solutions are the modes of the distribution represented by the ground Markov network. Most importantly, Markov logic allows contradictions between formulas, which it resolves simply by weighing the evidence on both sides. This makes it well suited for merging multiple KBs. Markov logic also provides a natural and powerful approach to the problem of merging knowledge and data in different representations that do not align perfectly, as will be illustrated in the application section.

It is interesting to see a simple example of how Markov logic generalizes first-order logic. Consider an MLN containing the single formula $\forall x \, R(x) \Rightarrow S(x)$ with weight w, and $C = \{A\}$. This leads to four possible worlds: $\{\neg R(A), \neg S(A)\}$, $\{\neg R(A), S(A)\}$, $\{R(A), \neg S(A)\}$, and $\{R(A), S(A)\}$. From Equation 3 we obtain that $P(\{R(A), \neg S(A)\}) = 1/(3e^w + 1)$ and the probability of each of the other three worlds is $e^w/(3e^w + 1)$. (The denominator is the partition function Z; see Section 2.) Thus, if $w > 0$, the effect of the MLN is to make the world that is inconsistent with $\forall x \, R(x) \Rightarrow S(x)$ less likely than the other three. From the probabilities above we obtain that $P(S(A)|R(A)) = 1/(1 + e^{-w})$. When $w \to \infty$, $P(S(A)|R(A)) \to 1$, recovering the logical entailment.

It is easily seen that all discrete probabilistic models expressible as products of potentials, including Markov networks and Bayesian networks, are expressible in Markov logic. In particular, many of the models frequently used in AI can be stated quite concisely as MLNs, and combined and extended simply by adding the corresponding formulas. Most significantly, Markov logic facilitates the construction of non-i.i.d. models (i.e., models where objects are not independent and identically distributed).

When working with Markov logic, we typically make three assumptions about the logical representation: different constants refer to different objects (unique names), the only objects in the domain are those representable using the constant and function symbols (domain closure), and the value of each function for each tuple of arguments is always a known constant (known functions). These assumptions ensure that the number of possible worlds is finite and that the Markov logic network will give a well-defined probability distribution. These assumptions are quite reasonable in most practical applications, and greatly simplify the use of MLNs. We will make these assumptions for the remainder of the chapter. See Richardson and Domingos [44] for further details on the Markov logic representation.

Markov logic can also be applied to a number of interesting infinite domains where some of these assumptions do not hold. See Singla and Domingos [53] for details on Markov logic in infinite domains.

5   Inference

5.1   MAP/MPE Inference

In the remainder of this chapter, we assume that the MLN is in function-free clausal form for convenience, but these methods can be applied to other MLNs as well. A basic inference task is finding the most probable state of the world given some evidence. (This is known as MAP inference in the Markov network literature, and MPE inference in the Bayesian network literature.) Because of the form of Equation 3, in Markov logic this reduces to finding the truth assignment that maximizes the sum of weights of satisfied clauses. This can be done using any weighted satisfiability solver, and (remarkably) need not be more expensive than standard logical inference by model checking. (In fact, it can be faster, if some hard constraints are softened.) We have successfully used MaxWalkSAT, a weighted variant of the WalkSAT local-search satisfiability solver, which can solve hard problems with hundreds of thousands of variables in minutes [16]. MaxWalkSAT performs this stochastic search by picking an unsatisfied clause at random and flipping the truth value of one of the atoms in it. With a certain probability, the atom is chosen randomly; otherwise, the atom is chosen to maximize the sum of satisfied clause weights when flipped. This combination of random and greedy steps allows MaxWalkSAT to avoid getting stuck in local optima while searching. Pseudocode for MaxWalkSAT is shown in Algorithm 1.

Algorithm 1. MaxWalkSAT(*weighted_clauses, max_flips, max_tries, target, p*)

$vars \leftarrow$ variables in *weighted_clauses*
for $i \leftarrow 1$ to *max_tries* **do**
  $soln \leftarrow$ a random truth assignment to *vars*
  $cost \leftarrow$ sum of weights of unsatisfied clauses in *soln*
  for $i \leftarrow 1$ to *max_flips* **do**
    if $cost \leq target$ **then**
      return "Success, solution is", *soln*
    end if
    $c \leftarrow$ a randomly chosen unsatisfied clause
    if Uniform(0,1) $< p$ **then**
      $v_f \leftarrow$ a randomly chosen variable from c
    else
      for each variable v in c **do**
        compute DeltaCost(v)
      end for
      $v_f \leftarrow v$ with lowest DeltaCost(v)
    end if
    $soln \leftarrow soln$ with v_f flipped
    $cost \leftarrow cost +$ DeltaCost(v_f)
  end for
end for
return "Failure, best assignment is", best *soln* found

DeltaCost(v) computes the change in the sum of weights of unsatisfied clauses that results from flipping variable v in the current solution. Uniform(0,1) returns a uniform deviate from the interval $[0,1]$.

One problem with this approach is that it requires propositionalizing the domain (i.e., grounding all atoms and clauses in all possible ways), which consumes memory exponential in the arity of the clauses. We have overcome this by developing LazySAT, a lazy version of MaxWalkSAT which grounds atoms and clauses only as needed [52]. This takes advantage of the sparseness of relational domains, where most atoms are false and most clauses are trivially satisfied. For example, in the domain of scientific research, most groundings of the atom Author(person, paper) are false, and most groundings of the clause Author(person1, paper)∧Author(person2, paper)⇒Coauthor(person1, person2) are satisfied. In LazySAT, the memory cost does not scale with the number of possible clause groundings, but only with the number of groundings that are potentially unsatisfied at some point in the search.

Algorithm 2. LazySAT(*weighted_KB, DB,* *max_flips, max_tries, target, p***)**

---

for $i \leftarrow 1$ to *max_tries* do

    active_atoms ← atoms in clauses not satisfied by *DB*

    active_clauses ← clauses activated by *active_atoms*

    soln ← a random truth assignment to *active_atoms*

    cost ← sum of weights of unsatisfied clauses in *soln*

    for $i \leftarrow 1$ to *max_flips* do

      if *cost* $\leq$ *target* then

        return "Success, solution is", *soln*

      end if

      $c \leftarrow$ a randomly chosen unsatisfied clause

      if Uniform(0,1) $< p$ then

        $v_f \leftarrow$ a randomly chosen variable from c

      else

        for each variable v in c do

          compute DeltaCost(v), using *weighted_KB* if $v \notin$ *active_atoms*

        end for

        $v_f \leftarrow v$ with lowest DeltaCost(v)

      end if

      if $v_f \notin$ *active_atoms* then

        add v_f to *active_atoms*

        add clauses activated by v_f to *active_clauses*

      end if

      soln ← *soln* with v_f flipped

      cost ← *cost* + DeltaCost(v_f)

    end for

end for

return "Failure, best assignment is", best *soln* found

Algorithm 2 gives pseudo-code for LazySAT, highlighting the places where it differs from MaxWalkSAT. LazySAT maintains a set of *active atoms* and a set of *active clauses*. A clause is active if it can be made unsatisfied by flipping zero or more of its active atoms. (Thus, by definition, an unsatisfied clause is always active.) An atom is active if it is in the initial set of active atoms, or if it was flipped at some point in the search. The initial active atoms are all those appearing in clauses that are unsatisfied if only the atoms in the database are true, and all others are false. The unsatisfied clauses are obtained by simply going through each possible grounding of all the first-order clauses and materializing the groundings that are unsatisfied; search is pruned as soon as the partial grounding of a clause is satisfied. Given the initial active atoms, the definition of active clause requires that some clauses become active, and these are found using a similar process (with the difference that, instead of checking whether a ground clause is unsatisfied, we check whether it should be active). Each run of LazySAT is initialized by assigning random truth values to the active atoms. This differs from MaxWalkSAT, which assigns random values to all atoms. However, the LazySAT initialization is a valid MaxWalkSAT initialization, and we have verified experimentally that the two give very similar results. Given the same initialization, the two algorithms will produce exactly the same results.

At each step in the search, the variable that is flipped is activated, as are any clauses that by definition should become active as a result. When evaluating the effect on cost of flipping a variable v, if v is active then all of the relevant clauses are already active, and DeltaCost(v) can be computed as in MaxWalkSAT. If v is inactive, DeltaCost(v) needs to be computed using the knowledge base. This is done by retrieving from the KB all first-order clauses containing the atom that v is a grounding of, and grounding each such clause with the constants in v and all possible groundings of the remaining variables. As before, we prune search as soon as a partial grounding is satisfied, and add the appropriate multiple of the clause weight to DeltaCost(v). (A similar process is used to activate clauses.) While this process is costlier than using pre-grounded clauses, it is amortized over many tests of active variables. In typical satisfiability problems, a small core of "problem" clauses is repeatedly tested, and when this is the case LazySAT will be quite efficient.

At each step, LazySAT flips the same variable that MaxWalkSAT would, and hence the result of the search is the same. The memory cost of LazySAT is on the order of the maximum number of clauses active at the end of a run of flips. (The memory required to store the active atoms is dominated by the memory required to store the active clauses, since each active atom appears in at least one active clause).

Experiments on entity resolution and planning problems show that this can yield very large memory reductions, and these reductions increase with domain size [52]. For domains whose full instantiations fit in memory, running time is comparable; as problems become larger, full instantiation for MaxWalkSAT becomes impossible.

5.2   Marginal and Conditional Probabilities

Another key inference task is computing the probability that a formula holds, given an MLN and set of constants, and possibly other formulas as evidence. By definition, the probability of a formula is the sum of the probabilities of the worlds where it holds, and computing it by brute force requires time exponential in the number of possible ground atoms. An approximate but more efficient alternative is to use Markov chain Monte Carlo (MCMC) inference [12], which samples a sequence of states according to their probabilities, and counting the fraction of sampled states where the formula holds. This can be extended to conditioning on other formulas by rejecting any state that violates one of them.

For the remainder of the chapter, we focus on the typical case where the evidence is a conjunction of ground atoms. In this scenario, further efficiency can be gained by applying a generalization of knowledge-based model construction [57]. This constructs only the minimal subset of the ground network required to answer the query, and runs MCMC (or any other probabilistic inference method) on it. The network is constructed by checking if the atoms that the query formula directly depends on are in the evidence. If they are, the construction is complete. Those that are not are added to the network, and we in turn check the atoms they depend on. This process is repeated until all relevant atoms have been retrieved. While in the worst case it yields no savings, in practice it can vastly reduce the time and memory required for inference. See Richardson and Domingos [44] for details.

One problem with applying MCMC to MLNs is that it breaks down in the presence of deterministic or near-deterministic dependencies (as do other probabilistic inference methods, e.g., belief propagation [59]). Deterministic dependencies break up the space of possible worlds into regions that are not reachable from each other, violating a basic requirement of MCMC. Near-deterministic dependencies greatly slow down inference, by creating regions of low probability that are very difficult to traverse. Running multiple chains with random starting points does not solve this problem, because it does not guarantee that different regions will be sampled with frequency proportional to their probability, and there may be a very large number of regions.

We have successfully addressed this problem by combining MCMC with satisfiability testing in the MC-SAT algorithm [40]. MC-SAT is a *slice sampling* MCMC algorithm. It uses a combination of satisfiability testing and simulated annealing to sample from the slice. The advantage of using a satisfiability solver (WalkSAT) is that it efficiently finds isolated modes in the distribution, and as a result the Markov chain mixes very rapidly. The slice sampling scheme ensures that detailed balance is (approximately) preserved.

MC-SAT is orders of magnitude faster than standard MCMC methods such as Gibbs sampling and simulated tempering, and is applicable to any model that can be expressed in Markov logic, including many standard models in statistical physics, vision, natural language processing, social network analysis, spatial statistics, etc.

Slice sampling [5] is an instance of a widely used approach in MCMC inference that introduces *auxiliary variables* to capture the dependencies between observed

---

Algorithm 3. MC-SAT(*clauses, weights, num_samples*)

$x^{(0)} \leftarrow$ Satisfy(hard *clauses*)
for $i \leftarrow 1$ to *num_samples* **do**
  $M \leftarrow \emptyset$
  for all $c_k \in$ *clauses* satisfied by $x^{(i-1)}$ **do**
    With probability $1 - e^{-w_k}$ add c_k to M
  end for
  Sample $x^{(i)} \sim \mathcal{U}_{SAT(M)}$
end for

---

variables. For example, to sample from $P(X = x) = (1/Z)\prod_k \phi_k(x_{\{k\}})$, we can define $P(X = x, U = u) = (1/Z)\prod_k I_{[0,\phi_k(x_{\{k\}})]}(u_k)$, where ϕ_k is the kth potential function, u_k is the kth auxiliary variable, $I_{[a,b]}(u_k) = 1$ if $a \leq u_k \leq b$, and $I_{[a,b]}(u_k) = 0$ otherwise. The marginal distribution of X under this joint is $P(X = x)$, so to sample from the original distribution it suffices to sample from $P(x, u)$ and ignore the u values. $P(u_k|x)$ is uniform in $[0, \phi_k(x_{\{k\}})]$, and thus easy to sample from. The main challenge is to sample x given u, which is uniform among all $\mathcal{X}$ that satisfies $\phi_k(x_{\{k\}}) \geq u_k$ for all k. MC-SAT uses SampleSAT [56] to do this. In each sampling step, MC-SAT takes the set of all ground clauses satisfied by the current state of the world and constructs a subset, M, that must be satisfied by the next sampled state of the world. (For the moment we will assume that all clauses have positive weight.) Specifically, a satisfied ground clause is included in M with probability $1 - e^{-w}$, where w is the clause's weight. We then take as the next state a uniform sample from the set of states $SAT(M)$ that satisfy M. (Notice that $SAT(M)$ is never empty, because it always contains at least the current state.) Algorithm 3 gives pseudo-code for MC-SAT. $\mathcal{U}_S$ is the uniform distribution over set S. At each step, all hard clauses are selected with probability 1, and thus all sampled states satisfy them. Negative weights are handled by noting that a clause with weight $w < 0$ is equivalent to its negation with weight $-w$, and a clause's negation is the conjunction of the negations of all of its literals. Thus, instead of checking whether the clause is satisfied, we check whether its negation is satisfied; if it is, with probability $1 - e^w$ we select all of its negated literals, and with probability e^w we select none.

It can be shown that MC-SAT satisfies the MCMC criteria of detailed balance and ergodicity [40], assuming a perfect uniform sampler. In general, uniform sampling is #P-hard and SampleSAT [56] only yields approximately uniform samples. However, experiments show that MC-SAT is still able to produce very accurate probability estimates, and its performance is not very sensitive to the parameter setting of SampleSAT.

We have applied the ideas of LazySAT to implement a lazy version of MC-SAT that avoids grounding unnecessary atoms and clauses. A working version of this algorithm is present in the open-source Alchemy system [20].

It is also possible to carry out lifted first-order probabilistic inference (akin to resolution) in Markov logic [3]. These methods speed up inference by reasoning at the first-order level about groups of indistinguishable objects rather than propositionalizing the entire domain. This is particularly applicable when the population size is given but little is known about most individual members.

6   Learning

6.1   Generative Weight Learning

MLN weights can be learned generatively by maximizing the likelihood of a relational database (Equation 3). This relational database consists of one or more "possible worlds" that form our training examples. Note that we can learn to generalize from even a single example because the clause weights are shared across their many respective groundings. We assume that the set of constants of each type is known. We also make a closed-world assumption: all ground atoms not in the database are false. This assumption can be removed by using an EM algorithm to learn from the resulting incomplete data. The gradient of the log-likelihood with respect to the weights is

$$\frac{\partial}{\partial w_i} \log P_w(X{=}x) = n_i(x) - \sum_{x'} P_w(X{=}x')\, n_i(x') \tag{4}$$

where the sum is over all possible databases x', and $P_w(X = x')$ is $P(X = x')$ computed using the current weight vector $w = (w_1, \ldots, w_i, \ldots)$. In other words, the ith component of the gradient is simply the difference between the number of true groundings of the ith formula in the data and its expectation according to the current model. Unfortunately, computing these expectations requires inference over the model, which can be very expensive. Most fast numeric optimization methods (e.g., conjugate gradient with line search, L-BFGS) also require computing the likelihood itself and hence the partition function Z, which is also intractable. Although inference can be done approximately using MCMC, we have found this to be too slow. Instead, we maximize the pseudo-likelihood of the data, a widely-used alternative [2]. If x is a possible world (relational database) and x_l is the lth ground atom's truth value, the pseudo-log-likelihood of x given weights w is

$$\log P_w^*(X{=}x) = \sum_{l=1}^{n} \log P_w(X_l{=}x_l | MB_x(X_l)) \tag{5}$$

where $MB_x(X_l)$ is the state of X_l's Markov blanket in the data (i.e., the truth values of the ground atoms it appears in some ground formula with). Computing the pseudo-likelihood and its gradient does not require inference, and is therefore much faster. Combined with the L-BFGS optimizer [24], pseudo-likelihood yields efficient learning of MLN weights even in domains with millions of ground atoms [44]. However, the pseudo-likelihood parameters may lead to poor results when long chains of inference are required.

   In order to reduce overfitting, we penalize each weight with a Gaussian prior. We apply this strategy not only to generative learning, but to all of our weight learning methods, even those embedded within structure learning.

6.2   Discriminative Weight Learning

Discriminative learning is an attractive alternative to pseudo-likelihood. In many applications, we know *a priori* which atoms will be evidence and which ones will be queried, and the goal is to correctly predict the latter given the former. If we partition the ground atoms in the domain into a set of evidence atoms X and a set of query atoms Y, the *conditional likelihood (CLL)* of Y given X is $P(y|x) = (1/Z_x) \exp \left(\sum_{i \in F_Y} w_i n_i(x, y) \right) = (1/Z_x) \exp \left(\sum_{j \in G_Y} w_j g_j(x, y) \right)$, where F_Y is the set of all MLN clauses with at least one grounding involving a query atom, $n_i(x, y)$ is the number of true groundings of the ith clause involving query atoms, G_Y is the set of ground clauses in $M_{L,C}$ involving query atoms, and $g_j(x, y) = 1$ if the jth ground clause is true in the data and 0 otherwise. The gradient of the CLL is

$$\frac{\partial}{\partial w_i} \log P_w(y|x) = n_i(x, y) - \sum_{y'} P_w(y'|x) n_i(x, y')$$

$$= n_i(x, y) - E_w[n_i(x, y)] \tag{6}$$

As before, computing the expected counts $E_w[n_i(x, y)]$ is intractable. However, they can be approximated by the counts $n_i(x, y_w^*)$ in the MAP state $y_w^*(x)$ (i.e., the most probable state of y given x). This will be a good approximation if most of the probability mass of $P_w(y|x)$ is concentrated around $y_w^*(x)$. Computing the gradient of the CLL now requires only MAP inference to find $y_w^*(x)$, which is much faster than the full conditional inference for $E_w[n_i(x, y)]$. This is the essence of the voted perceptron algorithm, initially proposed by Collins [4] for discriminatively learning hidden Markov models. Because HMMs have a very simple linear structure, their MAP states can be found in polynomial time using the Viterbi algorithm, a form of dynamic programming [43]. The voted perceptron initializes all weights to zero, performs T iterations of gradient ascent using the approximation above, and returns the parameters averaged over all iterations, $w_i = \sum_{t=1}^{T} w_{i,t}/T$. The parameter averaging helps to combat overfitting. T is chosen using a validation subset of the training data. We have extended the voted perceptron to Markov logic simply by replacing Viterbi with MaxWalkSAT to find the MAP state [50].

In practice, the voted perceptron algorithm can exhibit extremely slow convergence when applied to MLNs. One cause of this is that the gradient can easily vary by several orders of magnitude among the different clauses. For example, consider a transitivity rule such as `Friends(x, y)` $\land$ `Friends(y, z)` $\Rightarrow$ `Friends(x, z)` compared to a simple attribute relationship such as `Smokes(x)` $\Rightarrow$ `Cancer(x)`. In a social network domain of 1000 people, the former clause has one billion groundings while the latter has only 1000. Since each dimension of the gradient is a difference of clause counts and these can vary by orders of magnitude from one clause to another, a learning rate that is small enough to avoid divergence in some weights is too small for fast convergence in others.

This is an instance of the well-known problem of ill-conditioning in numerical optimization, and many candidate solutions for it exist [35]. However, the most common ones are not easily applicable to MLNs because of the nature of the function being optimized. As in Markov networks, computing the likelihood in MLNs requires computing the partition function, which is generally intractable. This makes it difficult to apply methods that require performing line searches, which involve computing the function as well as its gradient. These include most conjugate gradient and quasi-Newton methods (e.g., L-BFGS). Two exceptions to this are scaled conjugate gradient [32] and Newton's method with a diagonalized Hessian [1]. In the remainder of this subsection, we focus on scaled conjugate gradient, since we found it to be the best-performing method for discriminative weight learning.

In many optimization problems, gradient descent can be sped up by performing a line search to find the optimum along the chosen descent direction instead of taking a small step of constant size at each iteration. However, on ill-conditioned problems this is still inefficient, because line searches along successive directions tend to partly undo the effect of each other: each line search makes the gradient along its direction zero, but the next line search will generally make it non-zero again. In long narrow valleys, instead of moving quickly to the optimum, gradient descent zigzags.

A solution to this is to impose at each step the condition that the gradient along previous directions remain zero. The directions chosen in this way are called *conjugate*, and the method *conjugate gradient* [49]. Conjugate gradient methods are some of the most efficient available, on a par with quasi-Newton ones. While the standard conjugate gradient algorithm uses line searches to choose step sizes, we can use the Hessian (matrix of second derivatives of the function) instead. This method is known as *scaled conjugate gradient* (SCG), and was originally proposed by Møller [32] for training neural networks.

In a Markov logic network, the Hessian is simply the negative covariance matrix of the clause counts:

$$\frac{\partial}{\partial w_i \partial w_j} \log P(Y=y|X=x) = E_w[n_i]E_w[n_j] - E_w[n_i n_j]$$

Both the gradient and the Hessian matrix can be estimated using samples collected with the MC-SAT algorithm, described earlier. While full convergence could require many samples, we find that as few as five samples are often sufficient for estimating the gradient and Hessian. This is due in part to the efficiency of MC-SAT as a sampler, and in part to the tied weights: the many groundings of each clause can act to reduce the variance.

Given a conjugate gradient search direction $\mathbf{d}$ and Hessian matrix $\mathbf{H}$, we compute the step size α as follows:

$$\alpha = \frac{\mathbf{d}^T \mathbf{g}}{\mathbf{d}^T \mathbf{H} \mathbf{d} + \lambda \mathbf{d}^T \mathbf{d}}$$

For a quadratic function and $\lambda = 0$, this step size would move to the minimum function value along $\mathbf{d}$. Since our function is not quadratic, a non-zero λ term serves to limit the size of the step to a region in which our quadratic approximation is good. After each step, we adjust λ to increase or decrease the size of the so-called *model trust region* based on how well the approximation matched the function. We cannot evaluate the function directly, but the dot product of the step we just took and the gradient after taking it is a lower bound on the improvement in the actual log-likelihood. This works because the log-likelihood of an MLN is convex.

In models with thousands of weights or more, storing the entire Hessian matrix becomes impractical. However, when the Hessian appears only inside a quadratic form, as above, the value of this form can be computed simply as:

$$\mathbf{d}^T \mathbf{H} \mathbf{d} = (E_w[\textstyle\sum_i d_i n_i])^2 - E_w[(\textstyle\sum_i d_i n_i)^2]$$

The product of the Hessian by a vector can also be computed compactly [38].

Conjugate gradient is usually more effective with a preconditioner, a linear transformation that attempts to reduce the condition number of the problem (e.g., [48]). Good preconditioners approximate the inverse Hessian. We use the inverse diagonal Hessian as our preconditioner. Performance with the preconditioner is much better than without.

See Lowd and Domingos [26] for more details and results.

6.3   Structure Learning

The structure of a Markov logic network is the set of formulas or clauses to which we attach weights. In principle, this structure can be learned or revised using any inductive logic programming (ILP) technique. However, since an MLN represents a probability distribution, much better results are obtained by using an evaluation function based on pseudo-likelihood, rather than typical ILP ones like accuracy and coverage [18]. Log-likelihood or conditional log-likelihood are potentially better evaluation functions, but are vastly more expensive to compute. In experiments on two real-world datasets, our MLN structure learning algorithm found better MLN rules than CLAUDIEN [6], FOIL [42], Aleph [54], and even a hand-written knowledge base.

MLN structure learning can start from an empty network or from an existing KB. Either way, we have found it useful to start by adding all unit clauses (single atoms) to the MLN. The weights of these capture (roughly speaking) the marginal distributions of the atoms, allowing the longer clauses to focus on modeling atom dependencies. To extend this initial model, we either repeatedly find the best clause using beam search and add it to the MLN, or add all "good" clauses of length l before trying clauses of length $l + 1$. Candidate clauses are formed by adding each predicate (negated or otherwise) to each current clause, with all possible combinations of variables, subject to the constraint that at least one variable in the new predicate must appear in the current clause. Hand-coded clauses are also modified by removing predicates.

We now discuss the evaluation measure, clause construction operators, search strategy, and speedup methods in greater detail.

As an evaluation measure, pseudo-likelihood (Equation 5) tends to give undue weight to the largest-arity predicates, resulting in poor modeling of the rest. We thus define the weighted pseudo-log-likelihood (WPLL) as

$$\log P_w^\bullet(X\!=\!x) = \sum_{r \in R} c_r \sum_{k=1}^{g_r} \log P_w(X_{r,k}\!=\!x_{r,k}|MB_x(X_{r,k})) \qquad (7)$$

where R is the set of first-order atoms, g_r is the number of groundings of first-order atom r, and $x_{r,k}$ is the truth value (0 or 1) of the kth grounding of r. The choice of atom weights c_r depends on the user's goals. In our experiments, we simply set $c_r = 1/g_r$, which has the effect of weighting all first-order predicates equally. If modeling a predicate is not important (e.g., because it will always be part of the evidence), we set its weight to zero. To combat overfitting, we penalize the WPLL with a structure prior of $e^{-\alpha \sum_{i=1}^{F} d_i}$, where d_i is the number of literals that differ between the current version of the clause and the original one. (If the clause is new, this is simply its length.) This is similar to the approach used in learning Bayesian networks [14].

A potentially serious problem that arises when evaluating candidate clauses using WPLL is that the optimal (maximum WPLL) weights need to be computed for each candidate. Given that this involves numerical optimization, and may need to be done thousands or millions of times, it could easily make the algorithm too slow to be practical. We avoid this bottleneck by simply initializing L-BFGS with the current weights (and zero weight for a new clause). Second-order, quadratic-convergence methods like L-BFGS are known to be very fast if started near the optimum. This is what happens in our case; L-BFGS typically converges in just a few iterations, sometimes one. The time required to evaluate a clause is in fact dominated by the time required to compute the number of its true groundings in the data. This time can be greatly reduced using sampling and other techniques [18].

When learning an MLN from scratch (i.e., from a set of unit clauses), the natural operator to use is the addition of a literal to a clause. When refining a hand-coded KB, the goal is to correct the errors made by the human experts. These errors include omitting conditions from rules and including spurious ones, and can be corrected by operators that add and remove literals from a clause. These are the basic operators that we use. In addition, we have found that many common errors (wrong direction of implication, wrong use of connectives with quantifiers, etc.) can be corrected at the clause level by flipping the signs of atoms, and we also allow this. When adding a literal to a clause, we consider all possible ways in which the literal's variables can be shared with existing ones, subject to the constraint that the new literal must contain at least one variable that appears in an existing one. To control the size of the search space, we set a limit on the number of distinct variables in a clause. We only try removing literals from the original hand-coded clauses or their descendants, and we only consider removing a literal if it leaves at least one path of shared variables between each pair of remaining literals.

We have implemented two search strategies, one faster and one more complete. The first approach adds clauses to the MLN one at a time, using beam search to find the best clause to add: starting with the unit clauses and the expert-supplied ones, we apply each legal literal addition and deletion to each clause, keep the b best ones, apply the operators to those, and repeat until no new clause improves the WPLL. The chosen clause is the one with highest WPLL found in any iteration of the search. If the new clause is a refinement of a hand-coded one, it replaces it. (Notice that, even though we both add and delete literals, no loops can occur because each change must improve WPLL to be accepted.)

The second approach adds k clauses at a time to the MLN, and is similar to that of McCallum [30]. In contrast to beam search, which adds the best clause of any length found, this approach adds all "good" clauses of length l before attempting any of length $l + 1$. We call it *shortest-first search*.

The algorithms described in the previous section may be very slow, particularly in large domains. However, they can be greatly sped up using a combination of techniques described in Kok and Domingos [18]. These include looser convergence thresholds, subsampling atoms and clauses, caching results, and ordering clauses to avoid evaluating the same candidate clause twice.

Recently, Mihalkova and Mooney [31] introduced BUSL, an alternative, bottom-up structure learning algorithm for Markov logic. Instead of blindly constructing candidate clauses one literal at a time, they let the training data guide and constrain clause construction. First, they use a propositional Markov network structure learner to generate a graph of relationships among atoms. Then they generate clauses from paths in this graph. In this way, BUSL focuses on clauses that have support in the training data. In experiments on three datasets, BUSL evaluated many fewer candidate clauses than our top-down algorithm, ran more quickly, and learned more accurate models.

We are currently investigating further approaches to learning MLNs, including automatically inventing new predicates (or, in statistical terms, discovering hidden variables) [19].

7   Applications

Markov logic has been successfully applied in a variety of areas. A system based on it recently won a competition on information extraction for biology [45]. Cycorp has used it to make parts of the Cyc knowledge base probabilistic [29]. The CALO project is using it to integrate probabilistic predictions from many components [8]. We have applied it to link prediction, collective classification, entity resolution, information extraction, social network analysis and other problems [44,50,18,51,40,41]. Applications to Web mining, activity recognition, natural language processing, computational biology, robot mapping and navigation, game playing and others are under way.

7.1   Entity Resolution

The application to entity resolution illustrates well the power of Markov logic [51]. Entity resolution is the problem of determining which observations (e.g., database records, noun phrases, video regions, etc.) correspond to the same real-world objects, and is of crucial importance in many areas. Typically, it is solved by forming a vector of properties for each pair of observations, using a learned classifier (such as logistic regression) to predict whether they match, and applying transitive closure. Markov logic yields an improved solution simply by applying the standard logical approach of removing the unique names assumption and introducing the equality predicate and its axioms: equality is reflexive, symmetric and transitive; groundings of a predicate with equal constants have the same truth values; and constants appearing in a ground predicate with equal constants are equal. This last axiom is not valid in logic, but captures a useful statistical tendency. For example, if two papers are the same, their authors are the same; and if two authors are the same, papers by them are more likely to be the same. Weights for different instances of these axioms can be learned from data. Inference over the resulting MLN, with entity properties and relations as the evidence and equality atoms as the query, naturally combines logistic regression and transitive closure. Most importantly, it performs *collective* entity resolution, where resolving one pair of entities helps to resolve pairs of related entities.

As a concrete example, consider the task of deduplicating a citation database in which each citation has author, title, and venue fields. We can represent the domain structure with eight relations: Author(bib, author), Title(bib, title), and Venue(bib, venue) relate citations to their fields; HasWord(author/title/venue, word) indicates which words are present in each field; SameAuthor (author, author), SameTitle(title, title), and SameVenue(venue, venue) represent field equivalence; and SameBib(bib, bib) represents citation equivalence. The truth values of all relations except for the equivalence relations are provided as background theory. The objective is to predict the SameBib relation.

We begin with a logistic regression model to predict citation equivalence based on the words in the fields. This is easily expressed in Markov logic by rules such as the following:

$$\text{Title(b1, t1)} \wedge \text{Title(b2, t2)} \wedge \text{HasWord(t1, +word)}$$
$$\wedge \text{HasWord(t2, +word)} \Rightarrow \text{SameBib(b1, b2)}$$

The '+' operator here generates a separate rule (and with it, a separate learnable weight) for each constant of the appropriate type. When given a positive weight, each of these rules increases the probability that two citations with a particular title word in common are equivalent. We can construct similar rules for other fields. Note that we may learn negative weights for some of these rules, just as logistic regression may learn negative feature weights. Transitive closure consists of a single rule:

$$\text{SameBib(b1, b2)} \wedge \text{SameBib(b2, b3)} \Rightarrow \text{SameBib(b1, b3)}$$

This model is similar to the standard solution, but has the advantage that the classifier is learned in the context of the transitive closure operation.

We can construct similar rules to predict the equivalence of two fields as well. The usefulness of Markov logic is shown further when we link field equivalence to citation equivalence:

$$\text{Author(b1, a1)} \wedge \text{Author(b2, a2)} \wedge \text{SameBib(b1, b2)} \Rightarrow \text{SameAuthor(a1, a2)}$$
$$\text{Author(b1, a1)} \wedge \text{Author(b2, a2)} \wedge \text{SameAuthor(a1, a2)} \Rightarrow \text{SameBib(b1, b2)}$$

The above rules state that if two citations are the same, their authors should be the same, and that citations with the same author are more likely to be the same. The last rule is not valid in logic, but captures a useful statistical tendency.

Most importantly, the resulting model can now perform *collective* entity resolution, where resolving one pair of entities helps to resolve pairs of related entities. For example, inferring that a pair of citations are equivalent can provide evidence that the names *AAAI-06* and *21st Natl. Conf. on AI* refer to the same venue, even though they are superficially very different. This equivalence can then aid in resolving other entities.

Experiments on citation databases like Cora and BibServ.org show that these methods can greatly improve accuracy, particularly for entity types that are difficult to resolve in isolation as in the above example [51]. Due to the large number of words and the high arity of the transitive closure formula, these models have thousands of weights and ground millions of clauses during learning, even after using canopies to limit the number of comparisons considered. Learning at this scale is still reasonably efficient: preconditioned scaled conjugate gradient with MC-SAT for inference converges within a few hours [26].

7.2   Information Extraction

In this citation example, it was assumed that the fields were manually segmented in advance. The goal of information extraction is to extract database records starting from raw text or semi-structured data sources. Traditionally, information extraction proceeds by first segmenting each candidate record separately, and then merging records that refer to the same entities. Such a pipeline achitecture is adopted by many AI systems in natural language processing, speech recognition, vision, robotics, etc. Markov logic allows us to perform the two tasks jointly [41]. While computationally efficient, this approach is suboptimal, because it ignores the fact that segmenting one candidate record can help to segment similar ones. This allows us to use the segmentation of one candidate record to help segment similar ones. For example, resolving a well-segmented field with a less-clear one can disambiguate the latter's boundaries. We will continue with the example of citations, but similar ideas could be applied to other data sources, such as Web pages or emails.

The main evidence predicate in the information extraction MLN is $\text{Token}(t, i, c)$, which is true iff token t appears in the ith position of the cth citation. A token can be a word, date, number, etc. Punctuation marks are not treated as separate tokens; rather, the predicate $\text{HasPunc}(c, i)$ is true iff a punctuation mark appears immediately after the ith position in the cth citation. The query predicates are $\text{InField}(i, f, c)$ and $\text{SameCitation}(c, c')$. $\text{InField}(i, f, c)$ is true iff the ith position of the cth citation is part of field f, where $f \in \{\text{Title}, \text{Author}, \text{Venue}\}$, and inferring it performs segmentation. $\text{SameCitation}(c, c')$ is true iff citations c and c' represent the same publication, and inferring it performs entity resolution.

Our segmentation model is essentially a hidden Markov model (HMM) with enhanced ability to detect field boundaries. The observation matrix of the HMM correlates tokens with fields, and is represented by the simple rule

$$\texttt{Token}(+\texttt{t}, \texttt{i}, \texttt{c}) \Rightarrow \texttt{InField}(\texttt{i}, +\texttt{f}, \texttt{c})$$

If this rule was learned in isolation, the weight of the (t, f)th instance would be $\log(p_{tf}/(1 - p_{tf}))$, where p_{tf} is the corresponding entry in the HMM observation matrix. In general, the transition matrix of the HMM is represented by a rule of the form

$$\texttt{InField}(\texttt{i}, +\texttt{f}, \texttt{c}) \Rightarrow \texttt{InField}(\texttt{i} + 1, +\texttt{f}', \texttt{c})$$

However, we (and others, e.g., [13]) have found that for segmentation it suffices to capture the basic regularity that consecutive positions tend to be part of the same field. Thus we replace $\texttt{f}'$ by $\texttt{f}$ in the formula above. We also impose the condition that a position in a citation string can be part of at most one field; it may be part of none.

The main shortcoming of this model is that it has difficulty pinpointing field boundaries. Detecting these is key for information extraction, and a number of approaches use rules designed specifically for this purpose (e.g., [21]). In citation matching, boundaries are usually marked by punctuation symbols. This can be incorporated into the MLN by modifying the rule above to

$$\texttt{InField}(\texttt{i}, +\texttt{f}, \texttt{c}) \wedge \neg\texttt{HasPunc}(\texttt{c}, \texttt{i}) \Rightarrow \texttt{InField}(\texttt{i} + 1, +\texttt{f}, \texttt{c})$$

The $\neg\texttt{HasPunc}(\texttt{c}, \texttt{i})$ precondition prevents propagation of fields across punctuation marks. Because propagation can occur differentially to the left and right, the MLN also contains the reverse form of the rule. In addition, to account for commas being weaker separators than other punctuation, the MLN includes versions of these rules with $\texttt{HasComma}()$ instead of $\texttt{HasPunc}()$.

Finally, the MLN contains rules capturing a variety of knowledge about citations: the first two positions of a citation are usually in the author field, and the middle one in the title; initials (e.g., "J.") tend to appear in either the author or the venue field; positions preceding the last non-venue initial are usually not part of the title or venue; and positions after the first venue keyword (e.g., "Proceedings", "Journal") are usually not part of the author or title.

By combining this segmentation model with our entity resolution model from before, we can exploit relational information as part of the segmentation process. In practice, something a little more sophisticated is necessary to get good results on real data. In Poon and Domingos [41], we define predicates and rules specifically for passing information between the stages, as opposed to just using the existing $\texttt{InField}()$ outputs. This leads to a "higher bandwidth" of communication between segmentation and entity resolution, without letting excessive segmentation noise through. We also define an additional predicate and modify rules to better exploit information from similar citations during the segmentation process. See [41] for further details.

We evaluated this model on the CiteSeer and Cora datasets. For entity resolution in CiteSeer, we measured *cluster recall* for comparison with previously published results. Cluster recall is the fraction of clusters that are correctly output by the system after taking transitive closure from pairwise decisions. For entity resolution in Cora, we measured both cluster recall and pairwise recall/precision.

Table 2. CiteSeer entity resolution: cluster recall on each section

Approach	Constr.	Face	Reason.	Reinfor.
Fellegi-Sunter	84.3	81.4	71.3	50.6
Lawrence et al. (1999)	89	94	86	79
Pasula et al. (2002)	93	97	96	94
Wellner et al. (2004)	95.1	96.9	93.7	94.7
Joint MLN	96.0	97.1	95.1	96.7

Table 3. Cora entity resolution: pairwise recall/precision and cluster recall

Approach	Pairwise Rec./Prec.	Cluster Recall
Fellegi-Sunter	78.0 / 97.7	62.7
Joint MLN	94.3 / 97.0	78.1

In both datasets we also compared with a "standard" Fellegi-Sunter model (see [51]), learned using logistic regression, and with oracle segmentation as the input.

In both datasets, joint inference improved accuracy and our approach outperformed previous ones. Table 2 shows that our approach outperforms previous ones on CiteSeer entity resolution. (Results for Lawrence et al. (1999) [23], Pasula et al. (2002) [36] and Wellner et al. (2004) [58] are taken from the corresponding papers.) This is particularly notable given that the models of [36] and [58] involved considerably more knowledge engineering than ours, contained more learnable parameters, and used additional training data.

Table 3 shows that our entity resolution approach easily outperforms Fellegi-Sunter on Cora, and has very high pairwise recall/precision.

8   The Alchemy System

The inference and learning algorithms described in the previous sections are publicly available in the open-source Alchemy system [20]. Alchemy makes it possible to define sophisticated probabilistic models with a few formulas, and to add probability to a first-order knowledge base by learning weights from a relevant database. It can also be used for purely logical or purely statistical applications, and for teaching AI. From the user's point of view, Alchemy provides a full spectrum of AI tools in an easy-to-use, coherent form. From the researcher's point of view, Alchemy makes it possible to easily integrate a new inference or learning algorithm, logical or statistical, with a full complement of other algorithms that support it or make use of it.

Alchemy can be viewed as a declarative programming language akin to Prolog, but with a number of key differences: the underlying inference mechanism is model checking instead of theorem proving; the full syntax of first-order logic is allowed, rather than just Horn clauses; and, most importantly, the ability to handle uncertainty and learn from data is already built in. Table 4

Table 4. A comparison of Alchemy, Prolog and BUGS

Aspect	Alchemy	Prolog	BUGS
Representation	First-order logic + Markov nets	Horn clauses	Bayes nets
Inference	Model checking, MCMC	Theorem proving	MCMC
Learning	Parameters and structure	No	Parameters
Uncertainty	Yes	No	Yes
Relational	Yes	Yes	No

compares Alchemy with Prolog and BUGS [28], one of the most popular toolkits for Bayesian modeling and inference.

9   Current and Future Research Directions

We are actively researching better learning and inference methods for Markov logic, as well as extensions of the representation that increase its generality and power.

Exact methods for learning and inference are usually intractable in Markov logic, but we would like to see better, more efficient approximations along with the automatic application of exact methods when feasible.

One method of particular interest is lifted inference. In short, we would like to reason with clusters of nodes for which we have exactly the same amount of information. The inspiration is from lifted resolution in first order logic, but must be extended to handle uncertainty. Prior work on lifted inference such as [39] and [3] mainly focused on exact inference which can be quite slow. There has been some recent work on lifted belief propagation in a Markov logic like setting [15], but only for the case in which there is no evidence. We would like to extend this body of work for approximate inference in the case where arbitrary evidence is given, potentially speeding up inference in Markov logic by orders of magnitude.

Numerical attributes must be discretized to be used in Markov logic, but we are working on extending the representation to handle continuous random variables and features. This is particularly important in domains like robot navigation, where the coordinates of the robot and nearby obstacles are real-valued. Even domains that are handled well by Markov logic, such as entity resolution, could still benefit from this extension by incorporating numeric features into similarities.

Another extension of Markov logic is to support uncertainty at multiple levels in the logical structure. A formula in first-order logic can be viewed as a tree, with a logical connective at each node, and a knowledge base can be viewed as a tree whose root is a conjunction. Markov logic makes this conjunction probabilistic, as well as the universal quantifiers directly under it, but the rest of the tree remains purely logical. Recursive random fields [27] overcome this by allowing the features to be nested MLNs instead of clauses. Unfortunately, learning them suffers from the limitations of backpropagation.

Statistical predicate invention is the problem of discovering new concepts, properties, and relations in structured data, and generalizes hidden variable discovery in statistical models and predicate invention in ILP. Rather than extending the model directly, statistical predicate invention enables richer models by extending the domain with discovered predicates. Our initial work in this area uses second-order Markov logic to generate multiple cross-cutting clusterings of constants and predicates [19]. Formulas in second-order Markov logic could also be used to add declarative bias to our structure learning algorithms.

Current work also includes semi-supervised learning, and learning with incomplete data in general. The large amount of unlabeled data on the Web is an excellent resource that, properly exploited, could lead to many exciting applications.

Finally, we would like to develop a general framework for decision-making in relational domains. This can be accomplished in Markov logic by adding utility weights to formulas and finding the settings of all action predicates that jointly maximize expected utility.

10   Conclusion

Markov logic is a simple yet powerful approach to combining logic and probability in a single representation. We have developed a series of learning and inference algorithms for it, and successfully applied them in a number of domains. These algorithms are available in the open-source Alchemy system. We hope that Markov logic and its implementation in Alchemy will be of use to researchers and practitioners who wish to have the full spectrum of logical and statistical inference and learning techniques at their disposal, without having to develop every piece themselves.

Acknowledgements

This research was partly supported by DARPA grant FA8750-05-2-0283 (managed by AFRL), DARPA contract NBCH-D030010, NSF grant IIS-0534881, ONR grants N00014-02-1-0408 and N00014-05-1-0313, a Sloan Fellowship and NSF CAREER Award to the first author, and a Microsoft Research fellowship awarded to the third author. The views and conclusions contained in this document are those of the authors and should not be interpreted as necessarily representing the official policies, either expressed or implied, of DARPA, NSF, ONR, or the United States Government.

References

1. Becker, S., Le Cun, Y.: Improving the convergence of back-propagation learning with second order methods. In: Proceedings of the 1988 Connectionist Models Summer School, San Mateo, CA, pp. 29–37. Morgan Kaufmann, San Francisco (1989)
2. Besag, J.: Statistical analysis of non-lattice data. The Statistician 24, 179–195 (1975)
3. Braz, R., Amir, E., Roth, D.: Lifted first-order probabilistic inference. In: Proceedings of the Nineteenth International Joint Conference on Artificial Intelligence, Edinburgh, UK, pp. 1319–1325. Morgan Kaufmann, San Francisco (2005)
4. Collins, M.: Discriminative training methods for hidden Markov models: Theory and experiments with perceptron algorithms. In: Proceedings of the 2002 Conference on Empirical Methods in Natural Language Processing, Philadelphia, PA, pp. 1–8. ACL (2002)

5. Damien, P., Wakefield, J., Walker, S.: Gibbs sampling for Bayesian non-conjugate and hierarchical models by auxiliary variables. Journal of the Royal Statistical Society, Series B 61 (1999)
6. De Raedt, L., Dehaspe, L.: Clausal discovery. Machine Learning 26, 99–146 (1997)
7. Della Pietra, S., Della Pietra, V., Lafferty, J.: Inducing features of random fields. IEEE Transactions on Pattern Analysis and Machine Intelligence 19, 380–392 (1997)
8. Dietterich, T.: Experience with Markov logic networks in a large AI system. In: Probabilistic, Logical and Relational Learning - Towards a Synthesis, number 05051 in Dagstuhl Seminar Proceedings. Internationales Begegnungs- und Forschungszentrum für Informatik (IBFI), Dagstuhl, Germany (2007)
9. Friedman, N., Getoor, L., Koller, D., Pfeffer, A.: Learning probabilistic relational models. In: Proceedings of the Sixteenth International Joint Conference on Artificial Intelligence, Stockholm, Sweden, pp. 1300–1307. Morgan Kaufmann, San Francisco (1999)
10. Genesereth, M.R., Nilsson, N.J.: Logical Foundations of Artificial Intelligence. Morgan Kaufmann, San Mateo (1987)
11. Getoor, L., Taskar, B. (eds.): Introduction to Statistical Relational Learning. MIT Press, Cambridge (2007)
12. Gilks, W.R., Richardson, S., Spiegelhalter, D.J.: Markov Chain Monte Carlo in Practice. Chapman and Hall, London, UK (1996)
13. Grenager, T., Klein, D., Manning, C.D.: Unsupervised learning of field segmentation models for information extraction. In: Proceedings of the Forty-Third Annual Meeting on Association for Computational Linguistics, Ann Arbor, Michigan, pp. 371–378. Association for Computational Linguistics (2005)
14. Heckerman, D., Geiger, D., Chickering, D.M.: Learning Bayesian networks: The combination of knowledge and statistical data. Machine Learning 20, 197–243 (1995)
15. Jaimovich, A., Meshi, O., Friedman, N.: Template based inference in symmetric relational markov random fields. In: Proceedings of the Twenty-Third Conference on Uncertainty in Artificial Intelligence, Vancouver, Canada, AUAI Press (2007)
16. Kautz, H., Selman, B., Jiang, Y.: A general stochastic approach to solving problems with hard and soft constraints. In: Gu, D., Du, J., Pardalos, P. (eds.) The Satisfiability Problem: Theory and Applications, pp. 573–586. American Mathematical Society, New York (1997)
17. Kersting, K., De Raedt, L.: Towards combining inductive logic programming with Bayesian networks. In: Proceedings of the Eleventh International Conference on Inductive Logic Programming, Strasbourg, France, pp. 118–131. Springer, Heidelberg (2001)
18. Kok, S., Domingos, P.: Learning the structure of Markov logic networks. In: Proceedings of the Twenty-Second International Conference on Machine Learning, Bonn, Germany, pp. 441–448. ACM Press, New York (2005)
19. Kok, S., Domingos, P.: Statistical predicate invention. In: Proceedings of the Twenty-Fourth International Conference on Machine Learning, Corvallis, OR, pp. 433–440. ACM Press, New York (2007)
20. Kok, S., Sumner, M., Richardson, M., Singla, P., Poon, H., Lowd, D., Domingos, P.: The Alchemy system for statistical relational AI. Technical report, Department of Computer Science and Engineering, University of Washington, Seattle, WA (2007), http://alchemy.cs.washington.edu
21. Kushmerick, N.: Wrapper induction: Efficiency and expressiveness. Artificial Intelligence 118(1-2), 15–68 (2000)

22. Lavrač, N., Džeroski, S.: Inductive Logic Programming: Techniques and Applications. Ellis Horwood, Chichester (1994)
23. Lawrence, S., Bollacker, K., Giles, C.L.: Autonomous citation matching. In: Proceedings of the Third International Conference on Autonomous Agents, ACM Press, New York (1999)
24. Liu, D.C., Nocedal, J.: On the limited memory BFGS method for large scale optimization. Mathematical Programming 45(3), 503–528 (1989)
25. Lloyd, J.W.: Foundations of Logic Programming. Springer, Berlin, Germany (1987)
26. Lowd, D., Domingos, P.: Efficient weight learning for Markov logic networks. In: Proceedings of the Eleventh European Conference on Principles and Practice of Knowledge Discovery in Databases, Warsaw, Poland, pp. 200–211. Springer, Heidelberg (2007)
27. Lowd, D., Domingos, P.: Recursive random fields. In: Proceedings of the Twentieth International Joint Conference on Artificial Intelligence, Hyderabad, India, AAAI Press, Menlo Park (2007)
28. Lunn, D.J., Thomas, A., Best, N., Spiegelhalter, D.: WinBUGS – a Bayesian modeling framework: Concepts, structure, and extensibility. Statistics and Computing 10, 325–337 (2000)
29. Matuszek, C., Witbrock, M.: Personal communication (2006)
30. McCallum, A.: Efficiently inducing features of conditional random fields. In: Proceedings of the Nineteenth Conference on Uncertainty in Artificial Intelligence, Acapulco, Mexico, Morgan Kaufmann, San Francisco (2003)
31. Mihalkova, L., Mooney, R.: Bottom-up learning of Markov logic network structure. In: Proceedings of the Twenty-Fourth International Conference on Machine Learning, Corvallis, OR, pp. 625–632. ACM Press, New York (2007)
32. Møller, M.: A scaled conjugate gradient algorithm for fast supervised learning. Neural Networks 6, 525–533 (1993)
33. Muggleton, S.: Stochastic logic programs. In: De Raedt, L. (ed.) Advances in Inductive Logic Programming, pp. 254–264. IOS Press, Amsterdam, Netherlands (1996)
34. Neville, J., Jensen, D.: Dependency networks for relational data. In: Proceedings of the Fourth IEEE International Conference on Data Mining, Brighton, UK, pp. 170–177. IEEE Computer Society Press, Los Alamitos (2004)
35. Nocedal, J., Wright, S.: Numerical Optimization. Springer, New York (2006)
36. Pasula, H., Marthi, B., Milch, B., Russell, S., Shpitser, I.: Identity uncertainty and citation matching. In: Advances in Neural Information Processing Systems 14, MIT Press, Cambridge (2002)
37. Pearl, J.: Probabilistic Reasoning in Intelligent Systems: Networks of Plausible Inference. Morgan Kaufmann, San Francisco (1988)
38. Pearlmutter, B.: Fast exact multiplication by the Hessian. Neural Computation 6(1), 147–160 (1994)
39. Poole, D.: First-order probabilistic inference. In: Proceedings of the Eighteenth International Joint Conference on Artificial Intelligence, Acapulco, Mexico, pp. 985–991. Morgan Kaufmann, San Francisco (2003)
40. Poon, H., Domingos, P.: Sound and efficient inference with probabilistic and deterministic dependencies. In: Proceedings of the Twenty-First National Conference on Artificial Intelligence, Boston, MA, pp. 458–463. AAAI Press, Menlo Park (2006)
41. Poon, H., Domingos, P.: Joint inference in information extraction. In: Proceedings of the Twenty-Second National Conference on Artificial Intelligence, Vancouver, Canada, pp. 913–918. AAAI Press, Menlo Park (2007)

42. Quinlan, J.R.: Learning logical definitions from relations. Machine Learning 5, 239–266 (1990)

43. Rabiner, L.R.: A tutorial on hidden Markov models and selected applications in speech recognition. Proceedings of the IEEE 77, 257–286 (1989)

44. Richardson, M., Domingos, P.: Markov logic networks. Machine Learning 62, 107–136 (2006)

45. Riedel, S., Klein, E.: Genic interaction extraction with semantic and syntactic chains. In: Proceedings of the Fourth Workshop on Learning Language in Logic, Bonn, Germany, pp. 69–74. IMLS (2005)

46. Robinson, J.A.: A machine-oriented logic based on the resolution principle. Journal of the ACM 12, 23–41 (1965)

47. Roth, D.: On the hardness of approximate reasoning. Artificial Intelligence 82, 273–302 (1996)

48. Sha, F., Pereira, F.: Shallow parsing with conditional random fields. In: Proceedings of the 2003 Human Language Technology Conference and North American Chapter of the Association for Computational Linguistics, Association for Computational Linguistics (2003)

49. Shewchuck, J.: An introduction to the conjugate gradient method without the agonizing pain. Technical Report CMU-CS-94-125, School of Computer Science, Carnegie Mellon University (1994)

50. Singla, P., Domingos, P.: Discriminative training of Markov logic networks. In: aaai05, Pittsburgh, PA, pp. 868–873. AAAI Press, Menlo Park (2005)

51. Singla, P., Domingos, P.: Entity resolution with Markov logic. In: Proceedings of the Sixth IEEE International Conference on Data Mining, Hong Kong, pp. 572–582. IEEE Computer Society Press, Los Alamitos (2006)

52. Singla, P., Domingos, P.: Memory-efficient inference in relational domains. In: Proceedings of the Twenty-First National Conference on Artificial Intelligence, Boston, MA, AAAI Press, Menlo Park (2006)

53. Singla, P., Domingos, P.: Markov logic in infinite domains. In: Proceedings of the Twenty-Third Conference on Uncertainty in Artificial Intelligence, Vancouver, Canada, pp. 368–375. AUAI Press (2007)

54. Srinivasan, A.: The Aleph manual. Technical report, Computing Laboratory, Oxford University (2000)

55. Wasserman, S., Faust, K.: Social Network Analysis: Methods and Applications. Cambridge University Press, Cambridge (1994)

56. Wei, W., Erenrich, J., Selman, B.: Towards efficient sampling: Exploiting random walk strategies. In: aaai04, San Jose, CA, AAAI Press, Menlo Park (2004)

57. Wellman, M., Breese, J.S., Goldman, R.P.: From knowledge bases to decision models. Knowledge Engineering Review 7 (1992)

58. Wellner, B., McCallum, A., Peng, F., Hay, M.: An integrated, conditional model of information extraction and coreference with application to citation matching. In: Proceedings of the Twentieth Conference on Uncertainty in Artificial Intelligence, Banff, Canada, pp. 593–601. AUAI Press (2004)

59. Yedidia, J.S., Freeman, W.T., Weiss, Y.: Generalized belief propagation. In: Leen, T., Dietterich, T., Tresp, V. (eds.) Advances in Neural Information Processing Systems 13, pp. 689–695. MIT Press, Cambridge (2001)

New Advances in Logic-Based Probabilistic Modeling by PRISM

Taisuke Sato and Yoshitaka Kameya

Tokyo Institute of Technology, Ookayama Meguro Tokyo, Japan
{sato,kameya}@mi.cs.titech.ac.jp

Abstract. We review a logic-based modeling language PRISM and report recent developments including belief propagation by the generalized inside-outside algorithm and generative modeling with constraints. The former implies PRISM subsumes belief propagation at the algorithmic level. We also compare the performance of PRISM with state-of-the-art systems in statistical natural language processing and probabilistic inference in Bayesian networks respectively, and show that PRISM is reasonably competitive.

1 Introduction

The objective of this chapter is to review PRISM,[1] a logic-based modeling language that has been developed since 1997, and report its current status.[2]

PRISM was born in 1997 as an experimental language for unifying logic programming and probabilistic modeling [1]. It is an embodiment of the *distribution semantics* proposed in 1995 [2] and the first programming language with the ability to perform EM (expectation-maximization) learning [3] of parameters in programs. Looking back, when it was born, it already subsumed BNs (Bayesian networks), HMMs (hidden Markov models) and PCFGs (probabilistic context free grammars) semantically and could compute their probabilities.[3] However there was a serious problem: most of probability computation was exponential. Later in 2001, we added a tabling mechanism [4,5] and "largely solved" this problem. Tabling enables both reuse of computed results and dynamic programming for probability computation which realizes standard polynomial time probability computations for singly connected BNs, HMMs and PCFGs [6].

Two problems remained though. One is the *no-failure condition* that dictates that failure must not occur in a probabilistic model. It is placed for mathematical consistency of defined distributions but obviously an obstacle against the use of constraints in probabilistic modeling. This is because constraints may be

---

[1] http://sato-www.cs.titech.ac.jp/prism/

[2] This work is supported in part by the 21st Century COE Program 'Framework for Systematization and Application of Large-scale Knowledge Resources' and also in part by Ministry of Education, Science, Sports and Culture, Grant-in-Aid for Scientific Research (B), 2006, 17300043.

[3] We assume that BNs and HMMs in this chapter are discrete.

L. De Raedt et al. (Eds.): Probabilistic ILP 2007, LNAI 4911, pp. 118–155, 2008.

unsatisfiable thereby causing failure of computation and the failed computation means the loss of probability mass. In 2005, we succeeded in eliminating this condition by merging the FAM (failure-adjusted maximization) algorithm [7] with the idea of logic program synthesis [8].

The other problem is inference in multiply connected BNs. When a Bayesian network is singly connected, it is relatively easy to write a program that simulates $\pi\lambda$ message passing [9] and see the correctness of the program [6]. When, on the other hand, the network is not singly connected, it has been customarily to use the junction tree algorithm but how to realize BP (belief propagation)[4] on junction trees in PRISM has been unclear.[5] In 2006 however, it was found and proved that BP on junction trees is a special case of probability computation by the IO (inside-outside) algorithm generalized for logic programs used in PRISM [10].

As a result, we can now claim that PRISM *uniformly* subsumes BNs, HMMs and PCFGs at the algorithmic level as well as at the semantic level. All we need to do is to write appropriate programs for each model so that they denote intended distributions. PRISM's probability computation and EM learning for these programs exactly coincides with the standard algorithms for each model, i.e. the junction tree algorithm for BNs [11,12], the Baum-Welch (forward-backward) algorithm for HMMs [13] and the IO algorithm for PCFGs [14] respectively.

This is just a theoretical statement though, and the actual efficiency of probability computation and EM learning is another matter which depends on implementation and should be gauged against real data. Since our language is at an extremely high level (predicate calculus) and the data structure is very flexible (terms containing variables), we cannot expect the same speed as a C implementation of a specific model. However due to the continuing implementation efforts made in the past few years, PRISM's execution speed has greatly improved to the point of being usable for medium-sized machine learning experiments. We have conducted comparative experiments with Dyna [15] and ACE [16,17,18]. Dyna is a dynamic programming system for statistical natural language processing and ACE is a compiler that compiles a Bayesian network into an arithmetic circuit to perform probabilistic inference. Both represent the state-of-the-art approach in each field. Results are encouraging and demonstrate PRISM's competitiveness in probabilistic modeling.

That being said, we would like to emphasize that although the generality and descriptive power of PRISM enables us to treat existing probabilistic models uniformly, it should also be exploited for exploring new probabilistic models. One such model, *constrained HMMs* that combine HMMs with constraints, is explained in Section 5.

In what follows, we first look at the basics of PRISM [6] in Section 2. Then in Section 3, we explain how to realize BP in PRISM using logically described

---

[4] We use BP as a synonym of the part of the junction tree algorithm concerning message passing.

[5] Contrastingly it is straightforward to simulate variable elimination for multiply connected BNs [6].

junction trees. Section 4 deals with the system performance of PRISM and contains comparative data with Dyna and ACE. Section 5 contains generative modeling with constraints made possible by the elimination of the no-failure condition. Related work and future topics are discussed in Section 6. We assume the reader is familiar with logic programming [19], PCFGs [20] and BNs [9,21].

2   The Basic System

2.1   Programs as Distributions

One distinguished characteristic of PRISM is its declarative semantics. For self-containedness, in a slightly different way from that of [6], we quickly define the semantics of PRISM programs, the distribution semantics [2], which regards programs as defining infinite probability distributions.

Overview of the Distribution Semantics: In the distribution semantics, we consider a logic program DB which consists of a set F of facts (unit clauses) and a set R of rules (non-unit definite clauses). That is, we have $DB = F \cup R$. We assume the *disjoint condition* that there is no atom in F unifiable with the head of any clause in R. Semantically DB is treated as the set of all ground instances of the clauses in DB. So in what follows, F and R are equated with their ground instantiations. In particular F is a set of ground atoms. Since our language includes countably many predicate and function symbols, F and R are countably infinite.

We construct an infinite distribution, or to be more exact, a probability measure P_{DB}[6] on the set of possible Herbrand interpretations [19] of DB as the denotation of DB in two steps.

Let a sample space Ω_F (resp. Ω_{DB}) be all interpretations (truth value assignments) for the atoms appearing in F (resp. DB). They are so called the "possible worlds" for F (resp. DB). We construct a probability space on Ω_F and then extend it to a larger probability space on Ω_{DB} where the probability mass is distributed only over the least Herbrand models made from DB. Note that Ω_F and Ω_{DB} are uncountably infinite. We construct their probability measures, P_F and P_{DB} respectively, from a family of finite probability measures using Kolmogorov's extension theorem.[7]

Constructing P_F: Let $A_1, A_2, \ldots$ be an enumeration of the atoms in F. A truth value assignment for the atoms in F is represented by an infinite vector

---

[6] A probability space is a triplet $(\Omega, \mathcal{F}, P)$ where Ω is a sample space (the set of possible outcomes), $\mathcal{F}$ a σ-algebra which consists of subsets of Ω and is closed under complementation and countable union, and P a probability measure which is a function from sets $\mathcal{F}$ to real numbers in $[0, 1]$. Every set S in $\mathcal{F}$ is said to be measurable by P and assigned probability $P(S)$.

[7] Given denumerably many, for instance, discrete joint distributions satisfying a certain condition, Kolmogorov's extension theorem guarantees the existence of an infinite distribution (probability measure) which is an extension of each component distribution [22].

of 0s and 1s in such way that i-th value is 1 when A_i is true and 0 otherwise. Thus the sample space, F's all truth value assignments, is represented by a set of infinite vectors $\Omega_F = \prod_{i=1}^{\infty} \{0,1\}_i$.

We next introduce finite probability measures $P_F^{(n)}$ on $\Omega_F^{(n)} = \prod_{i=1}^{n} \{0,1\}_i$ ($n = 1, 2, \ldots$). We choose 2^n real numbers $p_\nu^{(n)}$ for each n such that $0 \leq p_\nu^{(n)} \leq 1$ and $\sum_{\nu \in \Omega_F^{(n)}} p_\nu^{(n)} = 1$. We put $P_F^{(n)}(\{\nu\}) = p_\nu^{(n)}$ for $\nu = (x_1, \ldots, x_n) \in \Omega_F^{(n)}$, which defines a finite probability measure $P_F^{(n)}$ on $\Omega_F^{(n)}$ in an obvious way. $p_\nu^{(n)} = p_{(x_1, \ldots, x_n)}^{(n)}$ is a probability that $A_1^{x_1} \wedge \cdots \wedge A_n^{x_n}$ is true where $A^x = A$ (when $x = 1$) and $A^x = \neg A$ (when $x = 0$).

We here require that the *compatibility condition* below hold for the $p_\nu^{(n)}$s:

$$\sum_{x_{n+1} \in \{0,1\}} p_{(x_1, \ldots, x_n, x_{n+1})}^{(n+1)} = p_{(x_1, \ldots, x_n)}^{(n)}.$$

It follows from the compatibility condition and Kolmogorov's extension theorem [22] that we can construct a probability measure on Ω_F by merging the family of finite probability measures $P_F^{(n)}$ ($n = 1, 2, \ldots$). That is, there uniquely exists a probability measure P_F on the minimum σ-algebra $\mathcal{F}$ that contains subsets of Ω_F of the form

$$[A_1^{x_1} \wedge \cdots \wedge A_n^{x_n}]_F \stackrel{\text{def}}{=} \{\nu \mid \nu = (x_1, x_2, \ldots, x_n, *, *, \ldots) \in \Omega_F, * \text{ is either 1 or 0}\}$$

such that P_F is an extension of each $P_F^{(n)}$ ($n = 1, 2, \ldots$):

$$P_F([A_1^{x_1} \wedge \cdots \wedge A_n^{x_n}]_F) = p_{(x_1, \ldots, x_n)}^{(n)} .$$

In this construction of P_F, it must be emphasized that the choice of $P_F^{(n)}$ is arbitrary as long as they satisfy the compatibility condition.

Having constructed a probability space $(\Omega_F, \mathcal{F}, P_F)$, we can now consider each ground atom A_i in F as a random variable that takes 1 (true) or 0 (false). We introduce a probability function $P_F(A_1 = x_1, \ldots, A_n = x_n) = P_F([A_1^{x_1} \wedge \cdots \wedge A_n^{x_n}]_F)$. We here use, for notational convenience, P_F both as the probability function and as the corresponding probability measure. We call P_F a *base distribution* for DB.

Extending P_F to P_{DB}: Next, let us consider an enumeration $B_1, B_2, \ldots$ of atoms appearing in DB. Note that it necessarily includes some enumeration of F. Also let $\Omega_{DB} = \prod_{j=1}^{\infty} \{0,1\}_j$ be the set of all interpretations (truth assignments) for the B_is and $M_{DB}(\nu)$ ($\nu \in \Omega_F$) the least Herbrand model [19] of a program $R \cup F_\nu$ where F_ν is the set of atoms in F made true by the truth assignment ν. We consider the following sets of interpretations for F and DB, respectively:

$$[B_1^{y_1} \wedge \cdots \wedge B_n^{y_n}]_F \stackrel{\text{def}}{=} \{\nu \in \Omega_F \mid M_{DB}(\nu) \models B_1^{y_1} \wedge \cdots \wedge B_n^{y_n}\},$$
$$[B_1^{y_1} \wedge \cdots \wedge B_n^{y_n}]_{DB} \stackrel{\text{def}}{=} \{\omega \in \Omega_{DB} \mid \omega \models B_1^{y_1} \wedge \cdots \wedge B_n^{y_n}\}.$$

We remark that $[\cdot]_F$ is measurable by P_F. So define $P_{DB}^{(n)}$ $(n > 0)$ by:

$$P_{DB}^{(n)}([B_1^{y_1} \wedge \cdots \wedge B_n^{y_n}]_{DB}) \overset{\text{def}}{=} P_F([B_1^{y_1} \wedge \cdots \wedge B_n^{y_n}]_F).$$

It is easily seen from the definition of $[\cdot]_F$ that $[B_1^{y_1} \wedge \cdots \wedge B_n^{y_n} \wedge B_{n+1}]_F$ and $[B_1^{y_1} \wedge \cdots \wedge B_n^{y_n} \wedge \neg B_{n+1}]_F$ form a partition of $[B_1^{y_1} \wedge \cdots \wedge B_n^{y_n}]_F$, and hence the following compatibility condition holds:

$$\sum_{y_{n+1} \in \{0,1\}} P_{DB}^{(n+1)}([B_1^{y_1} \wedge \cdots \wedge B_n^{y_n} \wedge B_{n+1}^{y_{n+1}}]_{DB}) = P_{DB}^{(n)}([B_1^{y_1} \wedge \cdots \wedge B_n^{y_n}]_{DB}).$$

Therefore, similarly to P_F, we can construct a unique probability measure P_{DB} on the minimum σ-algebra containing open sets of Ω_{DB}[8] which is an extension of $P_{DB}^{(n)}$ and P_F, and obtain a (-n infinite) joint distribution such that $P_{DB}(B_1 = y_1, \ldots, B_n = y_n) = P_F([B_1^{y_1} \wedge \cdots \wedge B_n^{y_n}]_F)$ for any $n > 0$. The *distribution semantics* is the semantics that considers P_{DB} as the denotation of *DB*. It is a probabilistic extension of the standard least model semantics in logic programming and gives a probability measure on the set of possible Herbrand interpretations of *DB* [2,6].

Since $[G] \overset{\text{def}}{=} \{\omega \in \Omega_{DB} \mid \omega \models G\}$ is P_{DB}-measurable for any closed formula G built from the symbols appearing in *DB*, we can define the probability of G as $P_{DB}([G])$. In particular, quantifiers are numerically approximated as we have

$$\lim_{n \to \infty} P_{DB}([G(t_1) \wedge \cdots \wedge G(t_n)]) = P_{DB}([\forall x G(x)]),$$
$$\lim_{n \to \infty} P_{DB}([G(t_1) \vee \cdots \vee G(t_n)]) = P_{DB}([\exists x G(x)]),$$

where $t_1, t_2, \ldots$ is an enumeration of all ground terms.

Note that properties of the distribution semantics described so far only assume the disjoint condition. In the rest of this section, we may use $P_{DB}(G)$ instead of $P_{DB}([G])$. Likewise $P_F(A_{i_1} = 1, A_{i_2} = 1, \ldots)$ is sometimes abbreviated to $P_F(A_{i_1}, A_{i_2}, \ldots)$.

PRISM Programs: The distribution semantics has freedom in the choice of $P_F^{(n)}$ $(n = 1, 2, \ldots)$ as long as they satisfy the compatibility condition. In PRISM which is an embodiment of the distribution semantics, the following requirements are imposed on F and P_F w.r.t. a program $DB = F \cup R$:

- Each (ground) atom in F takes the form $\mathtt{msw}(i, n, v)$,[9] which is interpreted that "a switch named i randomly takes a value v at the trial n." For each switch i, a finite set V_i of possible values it can take is given in advance.[10]
- Each switch i chooses a value exclusively from V_i, i.e. for any ground terms i and n, it holds that $P_F(\mathtt{msw}(i, n, v_1), \mathtt{msw}(i, n, v_2)) = 0$ for every $v_1 \neq v_2 \in V_i$ and $\sum_{v \in V_i} P_F(\mathtt{msw}(i, n, v)) = 1$.
- The choices of each switch made at different trials obey the same distribution, i.e. for each ground term i and any different ground terms $n_1 \neq n_2$,

---

[8] Each component space $\{0, 1\}$ of Ω_{DB} carries the discrete topology.

[9] $\mathtt{msw}$ is an abbreviation for *multi-ary random switch*.

[10] The intention is that $\{\mathtt{msw}(i, n, v) \mid v \in V_i\}$ jointly represents a random variable X_i whose range is V_i. In particular, $\mathtt{msw}(i, n, v)$ represents $X_i = v$.

$P_F(\mathtt{msw}(i, n_1, v)) = P_F(\mathtt{msw}(i, n_2, v))$ holds. Hence we denote $P_F(\mathtt{msw}(i, \cdot, v))$ by $\theta_{i,v}$ and call it as a *parameter* of switch i.

- The choices of different switches or different trials are independent. The joint distribution of atoms in F is decomposed as $\prod_{i,n} P_F(\mathtt{msw}(i, n, v_{i1}) = x_{i1}^n, \ldots, \mathtt{msw}(i, n, v_{iK_i}) = x_{iK_i}^n)$, where $V_i = \{v_{i1}, v_{i2}, \ldots, v_{iK_i}\}$.

The disjoint condition is automatically satisfied since $\mathtt{msw/3}$, the only predicate for F, is a built-in predicate and cannot be redefined by the user.

To construct P_F that satisfies the conditions listed above, let $A_1, A_2, \ldots$ be an enumeration of the $\mathtt{msw}$ atoms in F and put $N_{i,n}^{(m)} \stackrel{\text{def}}{=} \{k \mid \mathtt{msw}(i, n, v_{ik}) \in \{A_1, \ldots, A_m\}\}$. We have $\{A_1, \ldots, A_m\} = \bigcup_{i,n}\{\mathtt{msw}(i, n, v_{ik}) \mid k \in N_{i,n}^{(m)}\}$. We introduce a joint distribution $P_F^{(m)}$ over $\{A_1, \ldots, A_m\}$ for each $m > 0$ which is decomposed as

$$\prod_{i,n:N_{i,n}^{(m)} \neq \phi} P_{i,n}^{(m)} \left(\bigwedge_{k \in N_{i,n}^{(m)}} \mathtt{msw}(i, n, v_{ik}) = x_{ik}^n \right) \text{ where}$$

$$P_{i,n}^{(m)} \left(\bigwedge_{k \in N_{i,n}^{(m)}} \mathtt{msw}(i, n, v_{ik}) = x_{ik}^n \right)$$
$$\stackrel{\text{def}}{=} \begin{cases} \theta_{i,v_{ik}} & \text{if } x_{ik}^n = 1, x_{ik'}^n = 0 \ (k' \neq k) \\ 1 - \sum_{k \in N_{i,n}^{(m)}} \theta_{i,v_{ik}} & \text{if } x_{ik}^n = 0 \text{ for every } k \in N_{i,n}^{(m)} \\ 0 & \text{otherwise.} \end{cases}$$

$P_F^{(m)}$ obviously satisfies the second and the third conditions. Besides the compatibility condition holds for the family $P_F^{(m)}$ ($m = 1, 2, \ldots$). Hence P_{DB} is definable for every DB based on $\{P_F^{(m)} \mid m = 1, 2, \ldots\}$.

We remark that the $\mathtt{msw}$ atoms can be considered as a syntactic specialization of assumables in PHA (probabilistic Horn abduction) [23] or atomic choices in ICL (independent choice logic) [24] (see also Chapter 9 in this volume), but without imposing restrictions on modeling itself. We also point out that there are notable differences between PRISM and PHA/ICL. First unlike PRISM, PHA/ICL has no explicitly defined infinite probability space. Second the role of assumptions differs in PHA/ICL. While the assumptions in Subsection 2.2 are introduced just for the sake of computational efficiency and have no role in the definability of semantics, the assumptions made in PHA/ICL are indispensable for their language and semantics.

Program Example: As an example of a PRISM program, let us consider a left-to-right HMM described in Fig. 1. This HMM has four states $\{\mathtt{s0}, \mathtt{s1}, \mathtt{s2}, \mathtt{s3}\}$ where $\mathtt{s0}$ is the initial state and $\mathtt{s3}$ is the final state. In each state, the HMM outputs a symbol either 'a' or 'b'. The program for this HMM is shown in Fig. 2. The first four clauses in the program are called *declarations* where $\mathtt{target}(p/n)$ declares that the observable event is represented by the predicate p/n, and $\mathtt{values}(i, V_i)$ says that V_i is a list of possible values the switch i can take (a $\mathtt{values}$ declaration can

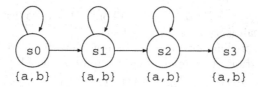

Fig. 1. Example of a left-to-right HMM with four states

```
target(hmm/1).
values(tr(s0),[s0,s1]).
values(tr(s1),[s1,s2]).
values(tr(s2),[s2,s3]).
values(out(_),[a,b]).

hmm(Cs):- hmm(0,s0,Cs).

hmm(T,s3,[C]):- msw(out(s3),T,C).  % If at the final state:
                                   %   output a symbol and then terminate.
hmm(T,S,[C|Cs]):- S\==s3,          % If not at the final state:
   msw(out(S),T,C),                %   choose a symbol to be output,
   msw(tr(S),T,Next),              %   choose the next state,
   T1 is T+1,                      %   Put the clock ahead,
   hmm(T1,Next,Cs).               %   and enter the next loop.
```

Fig. 2. PRISM program for the left-to-right HMM, which uses msw/3

be seen as a 'macro' notation for a set of facts in F). The remaining clauses define the probability distribution on the strings generated by the HMM. hmm(Cs) denotes a probabilistic event that the HMM generates a string Cs. hmm(T, S, Cs') denotes that the HMM, whose state is S at time T, generates a substring Cs' from that time on. The comments in Fig. 2 describe a procedural behavior of the HMM as a string generator. It is important to note here that this program has no limit on the string length, and therefore it implicitly contains countably infinite ground atoms. Nevertheless, thanks to the distribution semantics, their infinite joint distribution is defined with mathematical rigor.

One potentially confusing issue in our current implementation is the use of msw(i, v), where the second argument is omitted from the original definition for the sake of simplicity and efficiency. Thus, to run the program in Fig. 2 in practice, we need to delete the second argument in msw/3 and the first argument in hmm/3, i.e. the 'clock' variables T and T1 accordingly. When there are multiple occurrences of msw(i, v) in a proof tree, we assume by default that their original second arguments differ and their choices, though they happened to be the same, v, are made independently.[11] In the sequel, we will use msw/2 instead of msw/3.

---

[11] As a result $P(\text{msw}(i, v) \wedge \text{msw}(i, v))$ is computed as $\{P(\text{msw}(i, v))\}^2$ which makes the double occurrences of the same atom unequivalent to its single occurrence. In this sense, the current implementation of PRISM is not purely logical.

$$E_1 = \mathtt{m(out(s0),a)} \wedge \mathtt{m(tr(s0),s0)} \wedge \mathtt{m(out(s0),b)} \wedge \mathtt{m(tr(s0),s0)} \wedge \mathtt{m(out(s0),b)}$$
$$\wedge \mathtt{m(tr(s0),s1)} \wedge \mathtt{m(out(s1),b)} \wedge \mathtt{m(tr(s1),s2)} \wedge \mathtt{m(out(s2),b)} \wedge \mathtt{m(tr(s2),s3)}$$
$$\wedge \mathtt{m(out(s3),a)}$$
$$E_2 = \mathtt{m(out(s0),a)} \wedge \mathtt{m(tr(s0),s0)} \wedge \mathtt{m(out(s0),b)} \wedge \mathtt{m(tr(s0),s1)} \wedge \mathtt{m(out(s1),b)}$$
$$\wedge \mathtt{m(tr(s1),s1)} \wedge \mathtt{m(out(s1),b)} \wedge \mathtt{m(tr(s1),s2)} \wedge \mathtt{m(out(s2),b)} \wedge \mathtt{m(tr(s2),s3)}$$
$$\wedge \mathtt{m(out(s3),a)}$$

$$\vdots$$

$$E_6 = \mathtt{m(out(s0),a)} \wedge \mathtt{m(tr(s0),s1)} \wedge \mathtt{m(out(s1),b)} \wedge \mathtt{m(tr(s1),s2)} \wedge \mathtt{m(out(s2),b)}$$
$$\wedge \mathtt{m(tr(s2),s2)} \wedge \mathtt{m(out(s2),b)} \wedge \mathtt{m(tr(s2),s2)} \wedge \mathtt{m(out(s2),b)} \wedge \mathtt{m(tr(s2),s3)}$$
$$\wedge \mathtt{m(out(s3),a)}$$

Fig. 3. Six explanations for $\mathtt{hmm([a,b,b,b,b,a])}$. Due to the space limit, the predicate name $\mathtt{msw}$ is abbreviated to $\mathtt{m}$.

2.2   Realizing Generality with Efficiency

Explanations for Observations: So far we have been taking a model-theoretic approach to define the language semantics where an interpretation (a possible world) is a fundamental concept. From now on, to achieve efficient probabilistic inference, we take a proof-theoretic view and introduce the notion of *explanation*. Let us consider again a PRISM program $DB = F \cup R$. For a ground goal G, we can obtain $G \Leftrightarrow E_1 \vee E_2 \vee \cdots \vee E_K$ by logical inference from the completion [25] of R where each E_k $(k = 1, \ldots, K)$ is a conjunction of switches (ground atoms in F). We sometimes call G an *observation* and call each E_k an *explanation for* G. For instance, for the goal $G = \mathtt{hmm([a,b,b,b,b,a])}$ in the HMM program, we have six explanations shown in Fig. 3.

Intuitively finding explanations simulates the behavior of an HMM as a string analyzer, where each explanation corresponds to a state transition sequence. For example, E_1 indicates the transitions $\mathtt{s0} \to \mathtt{s0} \to \mathtt{s0} \to \mathtt{s1} \to \mathtt{s2} \to \mathtt{s3}$. It follows from the conditions on P_F of PRISM programs that the explanations $E_1, \ldots, E_6$ are all exclusive to each other (i.e. they cannot be true at the same time), so we can compute the probability of G by $P_{DB}(G) = P_{DB}(\bigvee_{k=1}^6 E_k) = \sum_{k=1}^6 P_{DB}(E_k)$. This way of probability computation would be satisfactory if the number of explanations is relatively small, but in general it is intractable. In fact, for left-to-right HMMs, the number of possible explanations (state transitions) is $_{T-2}C_{N-2}$, where N is the number of states and T is the length of the input string.[12]

---

[12] This is because in each transition sequence, there are $(N-2)$ state changes in $(T-2)$ time steps since there are two constraints — each sequence should start from the initial state, and the final state should appear only once at the last of the sequence. For fully-connected HMMs, on the other hand, it is easily seen that the number of possible state transitions is $O(N^T)$.

$$\mathtt{hmm}([a, b, b, b, b, a]) \Leftrightarrow \mathtt{hmm}(0, s0, [a, b, b, b, b, a])$$

$$\mathtt{hmm}(0, s0, [a, b, b, b, b, a]) \Leftrightarrow \mathtt{m}(\mathtt{out}(s0), a) \wedge \mathtt{m}(\mathtt{tr}(s0), s0) \wedge \mathtt{hmm}(1, s0, [b, b, b, b, a])$$
$$\vee \ \mathtt{m}(\mathtt{out}(s0), a) \wedge \mathtt{m}(\mathtt{tr}(s0), s1) \wedge \mathtt{hmm}(1, s1, [b, b, b, b, a])$$

$$\mathtt{hmm}(1, s0, [b, b, b, b, a]) \Leftrightarrow \mathtt{m}(\mathtt{out}(s0), b) \wedge \mathtt{m}(\mathtt{tr}(s0), s0) \wedge \mathtt{hmm}(2, s0, [b, b, b, a])$$
$$\vee \ \mathtt{m}(\mathtt{out}(s0), b) \wedge \mathtt{m}(\mathtt{tr}(s0), s1) \wedge \mathtt{hmm}(2, s1, [b, b, b, a])\ddagger$$

$$\mathtt{hmm}(2, s0, [b, b, b, a]) \Leftrightarrow \mathtt{m}(\mathtt{out}(s0), b) \wedge \mathtt{m}(\mathtt{tr}(s0), s1) \wedge \mathtt{hmm}(3, s1, [b, b, a])$$

$$\mathtt{hmm}(1, s1, [b, b, b, b, a]) \Leftrightarrow \mathtt{m}(\mathtt{out}(s1), b) \wedge \mathtt{m}(\mathtt{tr}(s1), s1) \wedge \mathtt{hmm}(2, s1, [b, b, b, a])\ddagger$$
$$\vee \ \mathtt{m}(\mathtt{out}(s1), b) \wedge \mathtt{m}(\mathtt{tr}(s1), s2) \wedge \mathtt{hmm}(2, s2, [b, b, b, a])$$

$$\mathtt{hmm}(2, s1, [b, b, b, a])\dagger \Leftrightarrow \mathtt{m}(\mathtt{out}(s1), b) \wedge \mathtt{m}(\mathtt{tr}(s1), s1) \wedge \mathtt{hmm}(3, s1, [b, b, a])$$
$$\vee \ \mathtt{m}(\mathtt{out}(s1), b) \wedge \mathtt{m}(\mathtt{tr}(s1), s2) \wedge \mathtt{hmm}(3, s2, [b, b, a])$$

$$\mathtt{hmm}(3, s1, [b, b, a]) \Leftrightarrow \mathtt{m}(\mathtt{out}(s1), b) \wedge \mathtt{m}(\mathtt{tr}(s1), s2) \wedge \mathtt{hmm}(4, s2, [b, a])$$

$$\mathtt{hmm}(2, s2, [b, b, b, a]) \Leftrightarrow \mathtt{m}(\mathtt{out}(s2), b) \wedge \mathtt{m}(\mathtt{tr}(s2), s2) \wedge \mathtt{hmm}(3, s2, [b, b, a])$$

$$\mathtt{hmm}(3, s2, [b, b, a]) \Leftrightarrow \mathtt{m}(\mathtt{out}(s2), b) \wedge \mathtt{m}(\mathtt{tr}(s2), s2) \wedge \mathtt{hmm}(4, s2, [b, a])$$

$$\mathtt{hmm}(4, s2, [b, a]) \Leftrightarrow \mathtt{m}(\mathtt{out}(s2), b) \wedge \mathtt{m}(\mathtt{tr}(s2), s3) \wedge \mathtt{hmm}(5, s3, [a])$$

$$\mathtt{hmm}(5, s3, [a]) \Leftrightarrow \mathtt{m}(\mathtt{out}(s3), a)$$

Fig. 4. Factorized explanations for $\mathtt{hmm}([a, b, b, b, b, a])$

Efficient Probabilistic Inference by Dynamic Programming: We know that there exist efficient algorithms for probabilistic inference for HMMs which run in $O(T)$ time — forward (backward) probability computation, the Viterbi algorithm, the Baum-Welch algorithm [13]. Their common computing strategy is dynamic programming. That is, we divide a problem into sub-problems recursively with memoizing and reusing the solutions of the sub-problems which appear repeatedly. To realize such dynamic programming for PRISM programs, we adopt a two-staged procedure. In the first stage, we run tabled search to find all explanations for an observation G, in which the solutions for a subgoal A are registered into a table so that they are reused for later occurrences of A. In the second stage, we compute probabilities while traversing an AND/OR graph called the *explanation graph for G*, extracted from the table constructed in the first stage.[13]

For instance from the HMM program, we can extract factorized iff formulas from the table as shown in Fig. 4 after the tabled search for $\mathtt{hmm}([a, b, b, b, b, a])$. Each iff formula takes the form $A \Leftrightarrow E_1' \vee \cdots \vee E_K'$ where A is a subgoal (also called a *tabled atom*) and E_k' (called a *sub-explanation*) is a conjunction of subgoals and switches. These iff formulas are graphically represented as the explanation graph for $\mathtt{hmm}([a, b, b, b, b, a])$ as shown in Fig. 5.

As illustrated in Fig. 4, in an explanation graph sub-structures are shared (e.g. a subgoal marked with $\dagger$ is referred to by two subgoals marked with $\ddagger$). Besides, it is reasonably expected that the iff formulas can be linearly ordered

---

[13] Our approach is an instance of a general scheme called *PPC (propositionalized probability computation)* which computes the sum-product of probabilities via propositional formulas often represented as a graph. Minimal AND/OR graphs proposed in [26] are another example of PPC specialized for BNs.

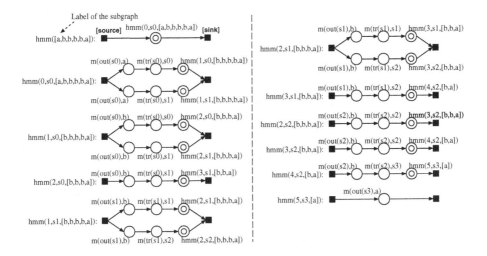

Fig. 5. Explanation graph

with respect to the caller-callee relationship in the program. These properties conjunctively enable us to compute probabilities in a dynamic programming fashion. For example, with an iff formula $A \Leftrightarrow E'_1 \vee \cdots \vee E'_K$ such that $E'_k = B_{k1} \wedge B_{k2} \wedge \cdots \wedge B_{kM_k}$, $P_{DB}(A)$ is computed as $\sum_{k=1}^{K} \prod_{j=1}^{M_k} P_{DB}(B_{kj})$ if $E'_1, \ldots, E'_6$ are exclusive and B_{kj} are independent. In the later section, we call $P_{DB}(A)$ the *inside probability* of A. The required time for computing $P_{DB}(A)$ is known to be linear in the size of the explanation graph, and in the case of the HMM program, we can see from Fig. 5 that it is $O(T)$, i.e. linear in the length of the input string. Recall that this is the same computation time as that of the forward (or backward) algorithm. Similar discussions can be made for the other types of probabilistic inference, and hence we can say that probabilistic inference with PRISM programs is as efficient as the ones by specific-purpose algorithms.

Assumptions for Efficient Probabilistic Inference: Of course, our efficiency depends on several assumptions. We first assume that $obs(DB)$, a countable subset of ground atoms appearing in the clause head, is given as a set of observable atoms. In the HMM program, we may consider that $obs(DB) = \{\texttt{hmm}([o_1, o_2, \ldots, o_T]) \mid o_t \in \{\texttt{a}, \texttt{b}\}, 1 \leq t \leq T, T \leq T_{\max}\}$ for some arbitrary finite $T_{\max}$. Then we roughly summarize the assumptions as follows (for the original definitions, please consult [6]):

Independence condition:
  For any observable atom $G \in obs(DB)$, the atoms appearing in the sub-explanations in the explanation graph for G are all independent. In the current implementation of $\texttt{msw}(i, v)$, this is unconditionally satisfied.

Exclusiveness condition:
  For any observable atom $G \in obs(DB)$, the sub-explanations for each sub-goal of G are exclusive to each other. The independence condition and the

exclusiveness condition jointly make the sum-product computation of probabilities possible.

Finiteness condition:

For any observable atom $G \in obs(DB)$, the number of explanations for G is finite. Without this condition, probability computation could be infinite.

Uniqueness condition:

Observable atoms are exclusive to each other, and the sum of probabilities of all observable atoms is equal to unity (i.e. $\sum_{G \in obs(DB)} P_{DB}(G) = 1$). The uniqueness condition is important especially for EM learning in which the training data is given as a bag of atoms from $obs(DB)$ which are observed as true. That is, once we find G_t as true at t-th observation, we immediately know from the uniqueness condition that the atoms in $obs(DB)$ other than G_t are false, and hence in EM learning, we can ignore the statistics on the explanations for these false atoms. This property underlies a dynamic-programming-based EM algorithm in PRISM [6].

Acyclic condition:

For any observable atom $G \in obs(DB)$, there is no cycle with respect to the caller-callee relationship among the subgoals for G. The acyclicity condition makes dynamic programming possible.

It may look difficult to satisfy all the conditions listed above. However, if we keep in mind to write a terminating program that generates the observations (by chains of choices made by switches), with care for the exclusiveness among disjunctive paths, these conditions are likely to be satisfied. In fact not only popular *generative models* such as HMMs, BNs and PCFGs but unfamiliar ones that have been little explored [27,28] can naturally be written in this style.

Further Issues: In spite of the general prospects of generative modeling, there are two cases where the uniqueness condition is violated and the PRISM's semantics is undefined. We call the first one *"probability-loss-to-infinity,"* in which an infinite generation process occurs with a non-zero probability.[14] The second one is called *"probability-loss-to-failure,"* in which there is a (finite) generation process with a non-zero probability that fails to yield any observable outcome. In Section 5, we discuss this issue, and describe a new learning algorithm that can deal with the second case.

Finally we address yet another view that takes explanation graphs as Boolean formulas consisting of ground atoms.[15] From this view, we can say that tabled search is a *propositionalization* procedure in the sense that it receives first-order expressions (a PRISM program) and an observation G as input, and generates as output a propositional AND/OR graph. In Section 6, we discuss advantages of such a propositionalization procedure.

---

[14] The HMM program in Fig. 2 satisfies the uniqueness condition provided the probability of looping state transitions is less than one, since in that case the probability of all infinite sequence becomes zero.

[15] Precisely speaking, while switches must be ground, subgoals can be an existentially quantified atom other than a ground atom.

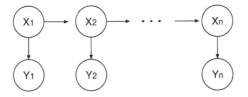

Fig. 6. BN for an HMM

3   Belief Propagation

3.1   Belief Propagation Beyond HMMs

In this section, we show that PRISM can subsume BP (belief propagation). What we actually show is that BP is nothing but a special case of generalized IO (inside-outside) probability computation[16] in PRISM applied to a junction tree expressed logically [6,10]. Symbolically we have

$$BP = \text{the generalized IO computation} + \text{junction tree.}$$

As far as we know this is the first link between BP and the IO algorithm. It looks like a mere variant of the well-known fact that BP applied to HMMs equals the forward-backward algorithm [31] but one should not miss the fundamental difference between HMMs and PCFGs.

Recall that an HMM deals with fixed-size sequences probabilistically generated from a finite state automaton, which is readily translated into a BN like the one in Fig. 6 where X_i's stand for hidden states and Y_i's stand for output symbols respectively. This translation is possible solely because an HMM has a finite number of states. Once HMMs are elevated to PCFGs however, a pushdown automaton for an underlying CFG has infinitely many stack states and it is by no means possible to represent them in terms of a BN which has only finitely many states. So any attempt to construct a BN that represents a PCFG and apply BP to it to upgrade the correspondence between BP and the forward-backward algorithm is doomed to fail. We cannot reach an algorithmic link between BNs and PCFGs this way.

We instead think of applying IO computation for PCFGs to BNs. Or more precisely we apply the *generalized IO computation*, a propositionally reformulated IO computation for PRISM programs [6], to a junction tree described logically as a PRISM program. Then we can prove that what the generalized IO computation does for the program is identical to what BP does on the junction tree [10], which we explain next.

---

[16] Wang et al. recently proposed the generalized inside-outside algorithm for a language model that combines a PCFG, n-gram and PLSA (probabilistic latent semantic analysis) [29]. It extends the standard IO algorithm but seems an instance of the gEM algorithm used in PRISM [30]. In fact such combination can be straightforwardly implemented in PRISM by using appropriate `msw` switches. For example an event of a preterminal node A's deriving a word w with two preceding words (trigram), u and v, under the semantic content h is represented by `msw([u,v,A,h],w)`.

3.2   Logical Junction Tree

Consider random variables $X_1, \ldots, X_N$ indexed by numbers $1, \ldots, N$. In the following we use Greek letters $\alpha, \beta, \ldots$ as a set of variable indices. Put $\alpha = \{n_1, \ldots, n_k\}$ ($\subseteq \{1, \ldots, N\}$). We denote by X_α the set (vector) of variables $\{X_{n_1}, \ldots, X_{n_k}\}$ and also by the lower case letter x_α the corresponding set (vector) of realizations of each variable in X_α.

A *junction tree* for a BN defining a distribution $P_{BN}(X_1 = x_1, \ldots, X_N = x_N) = \prod_{i=1}^{N} P_{BN}(X_i = x_i \mid X_{\pi(i)} = x_{\pi(i)})$, abbreviated to $P_{BN}(x_1, \ldots, x_N)$, where $\pi(i)$ denotes the indices of parent nodes of X_i, is a tree $T = (V, E)$ satisfying the following conditions [11,12,32].

- A node α ($\in V$) is a set of random variables X_α. An edge connecting X_α and X_β is labeled by $X_{\alpha \cap \beta}$. We use α instead of X_α etc to identify the node when the context is clear.
- A *potential* $\phi_\alpha(x_\alpha)$ is associated with each node α. It is a function consisting of a product of zero or more CPTs (conditional probability tables) like $\phi_\alpha(x_\alpha) = \prod_{\{j\} \cup \pi(j) \subseteq \alpha} P_{BN}(X_j = x_j \mid X_{\pi(j)} = x_{\pi(j)})$. It must hold that $\prod_{\alpha \in V} \phi_\alpha(x_\alpha) = P_{BN}(x_1, \ldots, x_N)$.
- RIP (*running intersection property*) holds which dictates that if nodes α and β have a common variable in the tree, it is contained in every node on the path between α and β.

After introducing junction trees, we show how to encode a junction tree T in a PRISM program. Suppose a node α has K ($K \geq 0$) child nodes $\beta_1, \ldots, \beta_K$ and a potential $\phi_\alpha(x_\alpha)$. We use a *node atom* $nd_\alpha(X_\alpha)$ to assert that the node α is in a state X_α and δ to denote the root node of T. Introduce for every $\alpha \in V$ W_α (*weight clause for α*) and C_α (*node clause for α*), together with the *top clause* C_{top} by

$$W_\alpha : weight_\alpha(X_\alpha) \Leftarrow \bigwedge_{P_{BN}(x_j \mid x_{\pi(j)}) \in \phi_\alpha} \texttt{msw}(\texttt{bn}(j, X_{\pi(j)}), X_j)$$
$$C_\alpha : \quad nd_\alpha(X_\alpha) \Leftarrow weight_\alpha(X_\alpha) \wedge nd_{\beta_1}(X_{\beta_1}) \wedge \cdots \wedge nd_{\beta_K}(X_{\beta_K})$$
$$C_{top} : \quad top \Leftarrow nd_\delta(X_\delta).$$

Here the X_j's denote logical variables, not random variables (we follow Prolog convention). W_α encodes ϕ_α as a conjunction of $\texttt{msw}(\texttt{bn}(j, X_{\pi(j)}), X_j)$s representing CPT $P_{BN}(X_j = x_j \mid X_{\pi(j)} = x_{\pi(j)})$. C_α is an encoding of the parent-child relation in T. C_{top} is a special clause to distinguish the root node δ in T.

Since we have $(\beta_i \setminus \alpha) \cap (\beta_j \setminus \alpha) = \phi$ if $i \neq j$ thanks to the RIP of T, we can rewrite $X_{(\bigcup_{i=1}^{K} \beta_i) \setminus \alpha}$, the set of variables appearing only in the right hand-side of C_α, to a disjoint union $\bigcup_{i=1}^{K} X_{\beta_i \setminus \alpha}$. Hence C_α is logically equivalent to

$$nd_\alpha(X_\alpha) \Leftarrow$$
$$weight_\alpha(X_\alpha) \wedge \exists X_{\beta_1 \setminus \alpha} nd_{\beta_1}(X_{\beta_1}) \wedge \cdots \wedge \exists X_{\beta_K \setminus \alpha} nd_{\beta_K}(X_{\beta_K}). \qquad (1)$$

Let F_T be the set of all ground $\texttt{msw}$ atoms of the form $\texttt{msw}(\texttt{bn}(i, x_{\pi(i)}), x_i)$. Give a joint distribution $P_{F_T}(\cdot)$ over F_T so that $P_{F_T}(\texttt{msw}(\texttt{bn}(i, x_{\pi(i)}), x_i))$, the probability of $\texttt{msw}(\texttt{bn}(i, x_{\pi(i)}), x_i)$ being true, is equal to $P_{BN}(X_i = x_i \mid X_{\pi(i)} = x_{\pi(i)})$.

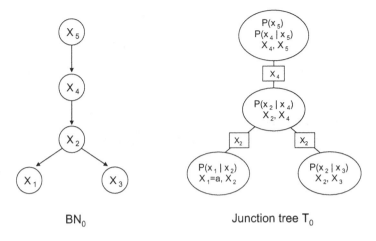

Fig. 7. Bayesian network BN_0 and a junction tree for BN_0

$$R_{T_0} \begin{cases} top \Leftarrow nd_{\{4,5\}}(X_4, X_5) \\ nd_{\{4,5\}}(X_4, X_5) \Leftarrow \mathtt{msw(bn}(5, []), X_5) \wedge \mathtt{msw(bn}(4, [X_5]), X_4) \wedge nd_{\{2,4\}}(X_2, X_4) \\ nd_{\{2,4\}}(X_2, X_4) \Leftarrow \mathtt{msw(bn}(2, [X_4]), X_2) \wedge nd_{\{1,2\}}(X_1, X_2) \wedge nd_{\{2,3\}}(X_2, X_3) \\ nd_{\{2,3\}}(X_2, X_3) \Leftarrow \mathtt{msw(bn}(3, [X_2]), X_3) \\ nd_{\{1,2\}}(X_1, X_2) \Leftarrow \mathtt{msw(bn}(1, [X_2]), X_1) \wedge X_1 = a \end{cases}$$

Fig. 8. Logical description of T_0

Sampling from $P_{F_T}(\cdot)$ is equivalent to simultaneous independent sampling from every $P_{BN}(X_i = x_i \mid X_{\pi(i)} = x_{\pi(i)})$ given i and $x_{\pi(i)}$. It yields a set S of true $\mathtt{msw}$ atoms. We can prove however that S uniquely includes a subset $S_{BN} = \{\mathtt{msw(bn}(i, x_{\pi(i)}), x_i) \mid 1 \leq i \leq N\}$ that corresponds to a draw from the joint distribution $P_{BN}(\cdot)$.

Finally define a program DB_T describing the junction tree T as a union $F_T \cup R_T$ having the base distribution $P_{F_T}(\cdot)$:

$$DB_T = F_T \cup R_T \quad \text{where } R_T = \{W_\alpha, C_\alpha \mid \alpha \in V\} \cup \{C_{top}\}. \tag{2}$$

Consider the Bayesian network BN_0 and a junction tree T_0 for BN_0 shown in Fig. 7. Suppose evidence $X_1 = a$ is given. Then R_{T_0} contains node clauses listed in Fig. 8.

3.3   Computing Generalized Inside-Outside Probabilities

After defining DB_T which is a logical encoding of a junction tree T, we demonstrate that the generalized IO probability computation for DB_T with a goal top by PRISM coincides with BP on T. Let $P_{DB_T}(\cdot)$ be the joint distribution defined by DB_T. According to the distribution semantics of PRISM, it holds that for any ground instantiations x_α of X_α,

$$P_{DB_T}(weight_\alpha(x_\alpha)) = P_{DB_T}\left(\bigwedge_{P_{BN}(x_j|x_{\pi(j)})\in\phi_\alpha} \mathtt{msw}(\mathtt{bn}(j,x_{\pi(j)}),x_j)\right) \quad (3)$$

$$= \prod_{P_{BN}(x_j|x_{\pi(j)})\in\phi_\alpha} P_F(\mathtt{msw}(\mathtt{bn}(j,x_{\pi(j)}),x_j)) \quad (4)$$

$$= \prod_{P_{BN}(x_j|x_{\pi(j)})\in\phi_\alpha} P_{BN}(x_j \mid x_{\pi(j)})$$

$$= \phi_\alpha(x_\alpha).$$

In the above transformation, (3) is rewritten to (4) using the fact that conditional distributions $\{P_{F_T}(x_j \mid x_{\pi(j)})\}$ contained in ϕ_α are pairwise different and $\mathtt{msw}(\mathtt{bn}(j,x_{\pi(j)}),x_j))$ and $\mathtt{msw}(\mathtt{bn}(j',x_{\pi(j')}),x_{j'}))$ are independent unless $j = j'$.

We now transform $P_{DB_T}(nd_\alpha(x_\alpha))$. Using (1), we see it holds that between the node α and its children $\beta_1,\ldots,\beta_K$

$$P_{DB_T}(nd_\alpha(x_\alpha))$$

$$= \phi_\alpha(x_\alpha)\cdot P_{DB_T}\left(\bigwedge_{i=1}^{K}\exists x_{\beta_i\setminus\alpha}nd_{\beta_i}(x_{\beta_i})\right)$$

$$= \phi_\alpha(x_\alpha)\cdot\prod_{i=1}^{K}P_{DB_T}\left(\exists x_{\beta_i\setminus\alpha}nd_{\beta_i}(x_{\beta_i})\right) \quad (5)$$

$$= \phi_\alpha(x_\alpha)\cdot\prod_{i=1}^{K}\sum_{x_{\beta_i\setminus\alpha}}P_{DB_T}(nd_{\beta_i}(x_{\beta_i})). \quad (6)$$

The passage from (5) to (6) is justified by Lemma 3.3 in [10]. We also have

$$P_{DB_T}(top) = 1 \quad (7)$$

because top is provable from $S \cup R_T$ where S is an arbitrary sample obtained from P_{F_T}.

Let us recall that in PRISM, two types of probability are defined for a ground atom A. The first one is $inside(A)$, the *generalized inside probability* of A, defined by $inside(A) \overset{\text{def}}{=} P_{DB}(A)$. It is just the probability of A being true. (6) and (7) give us a set of recursive equations to compute generalized inside probabilities of $nd_\alpha(x_\alpha)$ and top in a bottom-up manner [6].

The second one is the *generalized outside probability* of A with respect to a goal G, denoted by $outside(G\ ;\ A)$, which is more complicated than $inside(A)$, but can also be recursively computed in a top-down manner using generalized inside probabilities of other atoms [6]. We here simply list the set of equations to compute generalized outside probabilities with respect to top in DB_T. We describe the recursive equation between a child node β_j and its parent node α in (9).

$$outside(top\ ;\ top) = 1 \quad (8)$$

$$outside(top\ ;\ nd_{\beta_j}(x_{\beta_j}))$$

$$= \sum_{x_{\alpha\setminus\beta_j}}\left(\phi_\alpha(x_\alpha)\cdot outside(top\ ;\ nd_\alpha(x_\alpha))\prod_{i\neq j}^{K}\sum_{x_{\beta_i\setminus\alpha}}inside(nd_{\beta_i}(x_{\beta_i}))\right). \quad (9)$$

3.4 Marginal Distribution

The important property of generalized inside and outside probabilities is that their product, $inside(A) \cdot outside(G \ ; \ A)$, gives the expected number of occurrences of A in a(-n SLD) proof of G from msws drawn from P_F. In our case where $A = nd_\alpha(x_\alpha)$ and $G = top$, each node atom occurs at most once in an SLD proof of top. Hence $inside(A) \cdot outside(top \ ; \ A)$ is equal to the probability that top has a(-n SLD) proof containing A from $R_T \cup S$ where S is a sample from P_{F_T}. In this proof, not all members of S are relevant but only a subset $S_{BN}(x_1, \ldots, x_N) = \{\text{msw}(\text{bn}(i, x_{\pi(i)}), x_i) \mid 1 \le i \le N\}$ which corresponds to a sample $(x_1, \ldots, x_N)$ from P_{BN} (and vice versa) is relevant. By analyzing the proof, we know that sampled values x_α of X_α appearing in S_{BN} is identical to those in $A = nd_\alpha(x_\alpha)$ in the proof. We therefore have

$inside(A) \cdot outside(top \ ; \ A)$

$= $ Probability of sampling $S_{BN}(x_1, \ldots, x_N)$ from P_{F_T} such that
$\quad S_{BN}(x_1, \ldots, x_N) \cup R_T \vdash A$

$= $ Probability of sampling $(x_1, \ldots, x_N)$ from $P_{BN}(\cdot)$ such that
$\quad S_{BN}(x_1, \ldots, x_N) \cup R_T \vdash A$

$= $ Probability of sampling $(x_1, \ldots, x_N)$ from $P_{BN}(\cdot)$ such that
$\quad (x_1, \ldots, x_N)_\alpha = x_\alpha$ in $A = nd_\alpha(x_\alpha)$

$= $ Probability of sampling x_α from $P_{BN}(\cdot)$

$= P_{BN}(x_\alpha).$

When some of the X_i's in the BN are observed and fixed as evidence e, a slight modification of the above derivation gives $inside(A) \cdot outside(top \ ; \ A) = P_{BN}(x_\alpha, e)$. We summarize this result as a theorem.

Theorem 1. *Suppose DB_T is the PRISM program describing a junction tree T for a given BN. Let e be evidence. Also let α be a node in T and $A = nd_\alpha(x_\alpha)$ an atom describing the state of the node α. We then have*

$$inside(A) \cdot outside(top \ ; \ A) = P_{BN}(x_\alpha, e).$$

3.5 Deriving BP Messages

We now introduce *messages* which represent messages in BP value-wise in terms of inside-outside probabilities of ground node atoms. Let nodes $\beta_1, \ldots, \beta_K$ be the child nodes of α as before and γ be the parent node of α in the junction tree T. Define a *child-to-parent message* by

$$msg_{\alpha \rhd \gamma}(x_{\alpha \cap \gamma}) \overset{\text{def}}{=} \textstyle\sum_{x_{\alpha \setminus \gamma}} inside(nd_\alpha(x_\alpha))$$
$$= \textstyle\sum_{x_{\alpha \setminus \gamma}} P_{DB_T}(nd_\alpha(x_\alpha)).$$

The equation (6) for generalized inside probability is rewritten in terms of child-to-parent message as follows.

$$msg_{\alpha \triangleright \gamma}(x_{\alpha \cap \gamma}) = \sum_{x_{\alpha \setminus \gamma}} P_{DB_T}(nd_\alpha(x_\alpha))$$

$$= \sum_{x_{\alpha \setminus \gamma}} \left(\phi_\alpha(x_\alpha) \cdot \prod_{i=1}^{K} msg_{\beta_i \triangleright \alpha}(x_{\beta_i \cap \alpha}) \right). \qquad (10)$$

Next define a *parent-to-child message* from the parent node α to the j-th child node β_j by

$$msg_{\alpha \triangleright \beta_j}(x_{\alpha \cap \beta_j}) \stackrel{\text{def}}{=} outside(top \; ; \; nd_{\beta_j}(x_{\beta_j})).$$

Note that $outside(top \; ; \; nd_{\beta_j}(x_{\beta_j}))$ looks like a function of x_{β_j}, but in reality it is a function of x_{β_j}'s subset, $x_{\alpha \cap \beta_j}$ by (9). Using parent-to-child messages, we rewrite the equation (9) for generalized outside probability as follows.

$$msg_{\alpha \triangleright \beta_j}(x_{\alpha \cap \beta_j})$$
$$= outside(top \; ; \; nd_{\beta_j}(x_{\beta_j}))$$
$$= \sum_{x_{\alpha \setminus \beta_j}} \left(\phi_\alpha(x_\alpha) \cdot msg_{\gamma \triangleright \alpha}(x_{\gamma \cap \alpha}) \prod_{i \neq j}^{K} msg_{\beta_i \triangleright \alpha}(x_{\beta_i \cap \alpha}) \right). \qquad (11)$$

When α is the root node δ in T, since $outside(top \; ; \; top) = 1$, we have

$$msg_{\delta \triangleright \beta_j}(x_{\delta \cap \beta_j}) = \sum_{x_{\delta \setminus \beta_j}} \left(\phi_\delta(x_\delta) \cdot \prod_{i \neq j}^{K} msg_{\beta_i \triangleright \alpha}(x_{\beta_i \cap \alpha}) \right). \qquad (12)$$

The equations (10), (11) and (12) are identical to the equations for BP on T (without normalization) where (10) specifies the collecting evidence step and (11) and (12) specify the distributing evidence step respectively [12,11,33]. So we conclude

Theorem 2. *The generalized IO probability computation for DB_T describing a junction tree T with respect to the top goal top by (2) coincides with BP on T.*

Let us compute some messages in the case of the program in Fig. 8.

$$msg_{\{4,5\} \triangleright \{2,4\}}(X_4) = outside(top \; ; \; nd(X_2, X_4))$$
$$= \sum_{X_5} P_{DB_{T_0}}(nd(X_4, X_5))$$
$$= \sum_{X_5} P_{BN_0}(X_4 \mid X_5) P_{BN_0}(X_5)$$
$$= P_{BN_0}(X_4)$$
$$msg_{\{1,2\} \triangleright \{2,4\}}(X_2) = inside(nd(X_1 = a, X_2))$$
$$= P_{DB_{T_0}}(nd(X_1 = a, X_2))$$
$$= P_{BN_0}(X_1 = a \mid X_2)$$
$$msg_{\{2,3\} \triangleright \{2,4\}}(X_2) = \sum_{X_3} inside(nd(X_2, X_3))$$
$$= \sum_{X_3} P_{DB_{T_0}}(nd(X_2, X_3))$$
$$= \sum_{X_3} P_{BN_0}(X_3 \mid X_2)$$
$$= 1$$

Using these messages we can confirm Theorem 1.

$$inside(nd(X_2, X_4)) \cdot outside(top \,; \, nd(X_2, X_4))$$
$$= P_{BN_0}(X_2 \mid X_4) \cdot msg_{\{4,5\} \triangleright \{2,4\}}(X_4) \cdot msg_{\{1,2\} \triangleright \{2,4\}}(X_2) \cdot msg_{\{2,3\} \triangleright \{2,4\}}(X_2)$$
$$= P_{BN_0}(X_2 \mid X_4) \cdot P_{BN_0}(X_4) \cdot P_{BN_0}(X_1 = a \mid X_2) \cdot 1$$
$$= P_{BN_0}(X_1 = a, X_2, X_4).$$

Theorem 2 implies that computation of generalized IO probabilities in PRISM can be used to implement the junction tree algorithm. If we do so, we will first unfold a junction tree to a boolean combination of ground node atoms and msw atoms which propositionally represents the junction tree and then compute required probabilities from that boolean combination. This approach is a kind of *propositionalized BN computation*, a recent trend in the Bayesian network computation [17,26,34,35] in which BNs are propositionally represented and computed. We implemented the junction tree algorithm based on generalized IO probability computation. We compare the performance of our implementation with ACE, one of the propositionalized BN computation systems in the next section.

4   Performance Data

In this section we compare the computing performance of PRISM with two recent systems, Dyna [15] and ACE [16,17,18]. Unlike PRISM, they are not a general-purpose programming system that came out of research in PLL (probabilistic logic learning) or SRL (statistical relational learning). Quite contrary they are developed with their own purpose in a specific domain, i.e. statistical NLP (natural language processing) in the case of Dyna and fast probabilistic inference for BNs in the case of ACE. So our comparison can be seen as one between a general-purpose system and a specific-purpose system. We first measure the speed of probabilistic inference for a PCFG by PRISM and compare it with Dyna.

4.1   Computing Performance with PCFGs

Dyna System: Dyna[17] is a high-level declarative language for probabilistic modeling in NLP [15]. The primary purpose of Dyna is to facilitate various types of dynamic programming found in statistical NLP such as IO probability computation and Viterbi parsing to name a few. It is similar to PRISM in the sense that both can be considered as a probabilistic extension of a logic programming language but PRISM takes a top-down approach while Dyna takes a bottom-up approach in their evaluation. Also implementations differ. Dyna programs are compiled to C++ code (and then to native code[18]) while PRISM

---

[17] http://www.dyna.org/

[18] In this comparison, we added --optimize option to the dynac command, a batch command for compiling Dyna programs into native code.

programs are compiled to byte code for B-Prolog.[19] We use in our experiment PRISM version 1.10 and Dyna version 0.3.9 on a PC having Intel Core 2 Duo (2.4GHz) and 4GB RAM. The operating system is 64-bit SUSE Linux 10.0.

Computing Sentence Probability and Viterbi Parsing: To compare the speed of various PCFG-related inference tasks, we have to prepare benchmark data, i.e. a set of sentences and a CFG (so-called a tree-bank grammar [36]) for them. We use the WSJ portion of Penn Treebank III[20] which is widely recognized as one of the standard corpora in the NLP community [37].

We converted all 49,208 raw sentences in the WSJ sections 00–24 to POS (part of speech) tag sequences (the average length is about 20.97) and at the same time extract 11,534 CFG rules from the labeled trees. These CFG rules are further converted to Chomsky normal form[21] to yield 195,744 rules with 5,607 nonterminals and 38 terminals (POS tags). Finally by giving uniform probabilities to CFG rules, we obtain a PCFG we use in the comparison.

Two types of PRISM program are examined for the derived PCFG. In the former, which is referred to as PRISM-A here, an input POS tag sequence $t_1, \ldots, t_N$ is converted into a set of ground atoms $\{\text{input}(0, t_1, 1), \ldots, \text{input}(N-1, t_N, N)\}$ and supplied (by the "assert" predicate) to the program. Each $\text{input}(d-1, t, d)$ means that the input sentence has a POS tag t is at position d $(1 \leq d \leq N)$. The latter type, referred to as PRISM-D, is a probabilistic version of definite clause grammars which use difference lists. Dyna is closer to PRISM-A since in Dyna, we first provide the items equivalent to the above ground atoms to the chart (the inference mechanism of Dyna is based on a bottom-up chart parser). Compared to PRISM-A, PRISM-D incurs computational overhead due to the use of difference list.

We first compared time for computing the probability of a sentence (POS tag sequence) w.r.t. the derived PCFG using PRISM and Dyna. The result is plotted in Fig. 9 (left). Here X-axis shows sentence length (up to 24 by memory limitation of PRISM) and Y-axis shows the average time for probability computation[22] of randomly picked up 10 sentences. The graph clearly shows PRISM runs faster than Dyna. Actually at length 21 (closest to the average length), PRISM-A runs more than 10 times faster than Dyna.

We also conducted a similar experiment on Viterbi parsing, i.e. obtaining the most probable parse w.r.t. the derived CFG. Fig. 9 (right) show the result, where X-axis is the sentence length and Y-axis is the average time for Viterbi parsing of randomly picked up 10 sentences. This time PRISM-A is slightly slower than Dyna until length 13 but after that PRISM-A becomes faster than Dyna and the speed gap seems steadily growing.

One may notice that Dyna has a significant speed-up in Viterbi parsing compared to sentence probability computation while in PRISM the computation time

---

[19] http://www.probp.com/

[20] http://www.ldc.upenn.edu/

[21] We used the Dyna version of the CKY algorithm presented in [15].

[22] In the case of PRISM, the total computation time is the sum of the time for constructing the explanation graph and the time for computing the inside probability.

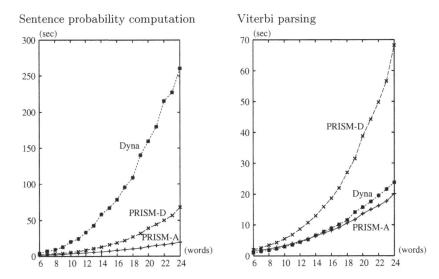

Fig. 9. Comparison between PRISM and Dyna on the speed of PCFG-related inference tasks

remains the same between these two inference tasks. This is because, in Viterbi parsing, Dyna performs a best-first search which utilizes a priority-queue agenda. On the other hand, the difference between PRISM-A and Dyna in sentence probability computation indicates the efficiency of PRISM's basic search engine. Besides, not surprisingly, PRISM-D runs three times slower than PRISM-A at the average sentence length.

Thus in our experiment with a realistic PCFG, PRISM, a general-purpose programming system, runs faster than or competitively with Dyna, a specialized system for statistical NLP. We feel this is somewhat remarkable. At the same time though, we observed PRISM's huge memory consumption which might be a severe problem in the future.[23]

4.2 Computing Performance with BNs

Next we measure the speed of a single marginal probability computation in BNs by PRISM programs DB_T described in Section 3.2 (hereafter called junction-tree PRISM programs) and compare it with ACE [16,17,18].

ACE[24] is a software package to perform probabilistic inference in a BN in three steps. It first encodes the BN by CNF propositionally, then transforms the CNF to yet another form d-DNNF (deterministic, decomposable negation normal

---

[23] In an additional experiment using another computer with 16GB memory, we could compute sentence probability for all of randomly picked up 10 sentences of length 43, but the sentences longer than 43 causes thrashing. It should be noted, however, that the PRISM system did not crash as far as we observed.

[24] http://reasoning.cs.ucla.edu/ace/

Table 1. Comparison between PRISM and ACE on the average computation time [sec] for single marginal probabilities

Network	#Nodes	Junction-tree PRISM			ACE		
		Trans.	Cmpl.	Run	Cmpl.	Read	Run
Asia	8	0.1	0.03	0	0.53	0.14	0.02
Water	32	1.57	1.79	0.38	1.87	0.5	0.8
Mildew	35	6.48	11.2	12.5	4.27	1.95	3.42
Alarm	37	0.34	0.2	0.01	0.33	0.18	0.13
Pigs	441	2.52	9.38	5.38	2.48	0.57	1.95
Munin1	189	1.93	2.54	N/A (1)	2242	0.51	N/A (2)
Munin2	1,003	7.61	85.7	15	9.94	1.0	6.07
Munin3	1,044	8.82	70.3	15.9	8.12	0.95	4.11
Munin4	1,041	8.35	90.7	408	11.2	0.97	6.64

form) and finally extracts from the d-DNNF, a graph called AC (arithmetic circuit). An AC is a directed acyclic graph in which nodes represent arithmetic operations such as addition and multiplication. The extracted AC is used to compute the marginal probability, given evidence. In compilation, we can make the resulting Boolean formulas more compact than a junction tree by exploiting the information in the local CPTs such as determinism (zero probabilities) and parameter equality, and more generally, CSI (context-specific independence) [38].

We picked up benchmark data from Bayesian Network Repository (http://www.cs.huji.ac.il/labs/compbio/Repository/). The network size ranges from 8 nodes to 1,044 nodes. In this experiment, PRISM version 1.11 and ACE 2.0 are used and run by a PC having AMD Opteron254(2.8GHz) with 16GB RAM on SUSE 64bit Linux 10.0. We implemented a Java translator from an XMLBIF network specification to the corresponding junction-tree PRISM program.[25] For ACE, we used the default compilation option (and the most advanced) -cd06 which enables the method defined in [18].

Table 1 shows time for computing a single marginal $P(X_i|e)$.[26] For junction-tree PRISM, the columns "Trans.," "Cmpl.," and "Run" mean respectively the translation time from XMLBIF to junction-tree PRISM, the compilation time from junction-tree PRISM to byte code of B-Prolog and the inference time. For ACE, the column "Cmpl." is compile time from an XBIF file to an AC wheres the column "Read" indicates time to load the compiled AC onto memory. The column "Run" shows the inference time. N/A (1) and N/A (2) in Munin1 means PRISM ran out of memory and ACE stopped by run time error respectively.

---

[25] In constructing a junction-tree, the elimination order basically follows the one specified in a *.num file in the repository. If such a *.num file does not exist, we used MDO (minimally deficient order). In this process, entries with 0 probabilities in CPTs are eliminated to shrink CPT size.

[26] For ACE, the computation time for $P(X_i \mid e)$ is averaged on all variables in the network. For junction-tree PRISM, since it sometimes took too long a time, the computation time is averaged on more than 50 variables for large networks.

When the network size is small, there are cases where PRISM runs faster than ACE, but in general PRISM does not catch up to ACE. One of the possible reasons might be that while ACE thoroughly exploits CSI in a BN to optimize computation, PRISM performs no such optimization when it propositionalizes a junction tree to an explanation graph. As a result, explanation graphs used by PRISM are thought to have much more redundancy than AC used by ACE. Translation to an optimized version of junction-tree PRISM programs using CSI remains as future work.

The results of two benchmark tests in this section show that PRISM is now maturing and *reasonably competitive* with state-of-the-art implementations of existing models in terms of computing performance, considering PRISM is a general-purpose programming language using general data structure – first order terms. We next look into another aspect of PRISM, PRISM as a vehicle for exploring new models.

5 Generative Modeling with Constraints

5.1 Loss of Probability Mass

Probabilistic modeling in PRISM is generative. By generative we mean that a probabilistic model describes a sequential process of generating an outcome in which nondeterminacy is resolved by a probabilistic choice (using msw/2 in the case of PRISM). All of BNs, HMMs and PCFGs are generative in this sense.

A generative model is described by a *generation tree* such that each node represents some state in a generation process. If there are k possible choices at node N and if the i-th choice with probability $p_i > 0$ ($1 \leq i \leq k$, $\sum_{i=1}^{k} p_i = 1$) causes a state transition from N to a next state N_i', there is an edge from N to N_i'. Note that we neither assume the transition is always successful nor the tree is finite. A leaf node is one where an outcome o is obtained. $P(o)$, the probability of the outcome o is that of an occurrence of a path from the root node to a leaf generating o. The generation tree defines a distribution over possible outcomes if $\sum_{o \in obs(DB)} P(o) = 1$ (*tightness condition* [39]).[27]

Generative models are intuitive and popular but care needs to be taken to ensure the tightness condition.[28] There are two cases in which the danger of violating the tightness condition exists. The first case is *probability-loss-to-infinity*. It means infinite computations occur *and* the probability mass assigned to them is non-zero.[29] In fact this happens in PCFGs depending on parameters associated with CFG rules in a grammar [39,40].

---

[27] The tightness condition is part of the *uniqueness condition* introduced in Section 2 [6].

[28] We call distributions *improper* if they do not satisfy the tightness condition.

[29] Recall that mere existence of infinite computation does not necessarily violate the tightness condition. Take a PCFG $\{p : S \rightarrow a, q : S \rightarrow SS\}$ where S is a start symbol, a a terminal symbol, $p + q = 1$ and $p, q \geq 0$. If $q > 0$, infinite computations occur, but the probability mass assigned to them is 0 as long as $q \leq p$.

The second case is *probability-loss-to-failure* which occurs when a transition to the next state fails. Since the probability mass put on a choice causing the transition is lost without producing any outcome, the total sum of probability mass on all outcomes becomes less than one. Probability-loss-to-failure does not happen in BNs, HMMs or PCFGs but can happen in more complex modeling such as probabilistic unification grammars [41,42]. It is a serious problem because it prevents us from using constraints in complex probabilistic modeling, the inherent target of PRISM. In the sequel, we detail our approach to generative modeling with constraints.

5.2   Constraints and Improper Distributions

We first briefly explain constraints. When a computational problem is getting more and more complex, it is less and less possible to completely specify every bit of information flow to solve it. Instead we specify conditions an answer must meet. They are called *constraints*. Usually constraints are binary relations over variables such as equality, ordering, set inclusion and so on. Each of them is just a partial declarative characterization of an answer but their interaction creates information flow to keep global consistency.

Constraints are mixed and manipulated in a constraint handling process, giving the simplest form as an answer just like a polynomial equation is solved by reduction to the simplest (but equivalent) form. Sometimes we find that constraints are inconsistent and there is no answer. When this happens, we stop the constraint handling process with failure.

Allowing the use of constraints significantly enhances modeling flexibility as exemplified by unification grammars such as HPSGs [43]. Moreover since they are declarative, they are easy to understand and easy to maintain. The other side of the coin however is that they can be a source of probability-loss-to-failure when introduced to generative modeling as they are not necessarily satisfiable. The loss of probability mass implies $\sum_{o \in obs(DB)} P(o) < 1$, an improper distribution, and ignoring this fact would destroy the mathematical meaning of computed results.

5.3   Conditional Distributions and Their EM Learning

How can we deal with such improper distributions caused by probability-loss-to-failure? It is apparent that what we can observe is an outcome o of some successful generation process specified by our model and thus should be interpreted as an realization of a conditional distribution, $P(o \mid \mathsf{success})$ where $\mathsf{success}$ denotes the event of generating some outcome.

Let us put this argument in the context of distribution semantics and let $q(\cdot)$ be a target predicate in a program $DB = F \cup R$ defining the distribution $P_{DB}(\cdot)$. Clauses in R may contain constraints such as $X < Y$ in their body. The event $\mathsf{success}$ is representable as $\exists X \, q(X)$ because the latter says there exists some outcome. Our conditional distribution is therefore written as $P_{DB}(q(t) \mid \exists X \, q(X))$. If R in DB satisfies the modeling principles stated in Subsection 5.1, which we assume hereafter, it holds that $P_{DB}(\exists X\, q(X)) = \sum_{q(t)\in obs(DB)} P_{DB}(q(t))$ where $obs(DB)$ is the set of ground target atoms provable from DB, i.e. $obs(DB) = \{q(t) \mid DB \vdash q(t)\}$. Consequently we have

$$P_{DB}(q(t) \mid \texttt{success}) = \frac{P_{DB}(q(t))}{P_{DB}(\exists X\, q(X))} = \frac{P_{DB}(q(t))}{\sum_{q(t')\in obs(DB)} P_{DB}(q(t'))}.$$

So if there occurs probability-loss-to-failure, what we should do is to normalize $P_{DB}(\cdot)$ by computing $P_{DB}(\texttt{success}) = \sum_{q(t')\in obs(DB)} P_{DB}(q(t'))$. The problem is that this normalization is almost always impossible. First of all there is no way to compute $P_{DB}(\texttt{success})$, the normalizing constant, if $obs(DB)$ is infinite. Second even if $obs(DB)$ is finite, the computation is often infeasible since there are usually exponentially many observable outcomes and hence so many times of summation is required.

We therefore abandon unconditional use of constraints and use them only when $obs(DB)$ is finite and an efficient computation of $P_{DB}(\texttt{success})$ is possible by dynamic programming. Still, there remains a problem. The *gEM (graphical EM)* algorithm, the central algorithm in PRISM for EM learning by dynamic programming, is not applicable if failure occurs because it assumes the tightness condition. We get around this difficulty by merging it with the FAM algorithm (failure-adjusted maximization) proposed by Cussens [7]. The latter is an EM algorithm taking failure into account.[30]

Fortunately the difference between the gEM algorithm and the FAM algorithm is merely that the latter additionally computes expected counts of msw atoms in a failed computation of $q(X)$ ($\exists X\, q(X)$). It is therefore straightforward to augment the gEM algorithm with a routine to compute the required expected counts in a failed computation, *assuming a failure program is available* which defines failure predicate that represents all failed computations of $q(X)$ w.r.t. DB. The augmented algorithm, *the fgEM algorithm* [8], works as the FAM algorithm with dynamic programming and implemented in the current PRISM.

So the last barrier against the use of constraints is the construction of a failure program. Failure must be somehow "reified" as a failure program for dynamic programming to be applicable. However how to construct it is not self-evident because there is no mechanism of recording failure in the original program DB. We here apply FOC (*first order compiler*), a program synthesis algorithm based on deductive program transformation [44]. It can derive, though not always, automatically a failure program from the source program DB for the target predicate $q(X)$ [8]. Since the synthesized failure program is a usual PRISM program, $P_{DB}(\texttt{failure})$ is computed efficiently by dynamic programming.

---

[30] The FAM algorithm [7] assumes there occur failed computations before an outcome is successfully generated. It requires to count the number of occurrences of each msw atom in the failed computation paths but [7] does not give how to count them. Usually there are exponentially many failure paths and naive computation would take exponential time. We solved this problem by merging FAM with gEM's dynamic programming.

In summary, in our approach, generative modeling with constraints is possible with the help of the fgEM algorithm and FOC, provided that a failure program is successfully synthesized by FOC and the computation tree (SLD tree) for `failure` is finite (and not too large). We next look at some examples.

5.4   Agreement in Number

A Small Example: We here take a small example of generative modeling with constraints and see its EM learning.

```
values(subj,[sg,pl]).    % introduce msw/2 named sbj and obj
values(obj,[sg,pl]).     % with outcomes = {sg,pl}
target(agree/1).

agree(A):-
  msw(subj,A),           % flip the coin subj
  msw(obj,B),            % flip the coin obj
  A=B.                   % equality constraint
```

Fig. 10. `agree` program describing agreement in number

A program in Fig. 10 models agreement in number in some natural language using two biased coins `subj` and `obj`. `values` clauses declare we use multi-ary random switches `msw(a,v)` and `msw(b,v)` where v is `sg` or `pl`. `target(agree/1)` declares a target predicate and what we observe are atoms of the form `agree(·)`.

For a top-goal `:-sample(agree(X))` which starts a sampling of the defined distribution for `agree/1`, `msw(subj,A)` is executed simulating coin tossing of `subj` which probabilistically instantiates `A` either to `sg` or `pl`. Similarly for `msw(obj,B)`. Hence an outcome is one of {`agree(sg)`, `agree(pl)`} and it is observable only when both coins agree (see the equality constraint `A=B`). When the two coins disagree, `A=B` fails and we have no observable outcome from this model.

Parameter Learning by the fgEM Algorithm: Given the program in Fig. 10 and a list of observations such as [`agree(sg)`,`agree(pl)`,...], we estimate parameters, i.e. probabilities of each coin showing `sg` or `pl`. In what follows, to simplify notation and discussion, we treat logical variables `A` and `B` as random variables and put parameters by $\theta_a = P(A = sg) = P(\texttt{msw(subj, sg)})$, $\theta_b = P(B = sg) = P(\texttt{msw(obj, sg)})$ and $\theta = (\theta_a, \theta_b)$.

Parameters are estimated from the observable distribution $P(A \mid \texttt{success}, \theta)$ $= \sum_{B \in \{sg,pl\}} P(\texttt{agree(A)}, A = B \mid \theta)/P(\texttt{success} \mid \theta)$ (hereafter we omit θ when obvious). Because the normalizing constant $P(\texttt{success}) = P(\texttt{agree(sg)}) + P(\texttt{agree(pl)}) = \theta_a \theta_b + (1 - \theta_a)(1 - \theta_b)$ is not necessarily equal to unity, the defined model becomes log-linear.

Obtaining a Failure Program: Thanks to the deep relationship between failure and negation in logic programming, a failure program can be derived

```
failure :- not(success).
success :- agree(_).        % agree(_) = ∃ X agree(X)
```

Fig. 11. Clauses defining `failure`

```
failure:- closure_success0(f0).    % f0 is initial continuation
closure_success0(C):- closure_agree0(C).
closure_agree0(C):-
    msw(subj,A),
    msw(obj,B),
    \+A=B.                          % \+ is Prolog's negation
```

Fig. 12. Compiled `failure` program

automatically by 'negating' the target atom in the original program. Let q(X) be a target atom. We add `failure:- ∀X not(q(X))` to the original program. `failure` says that there is no observable outcome of q(X). As it is not executable, we 'compile' it using FOC to obtain a negation-free executable program. For the `agree` program in Fig. 10 we add two clauses shown in Fig. 11.[31]

FOC compiles the `failure` predicate into executable code shown in Fig. 12. As can be seen, it eliminates negation in Fig. 11 while introducing two new predicates and one new function symbol.[32] We would like to point out that negation elimination is just one functionality of FOC. It can compile a much wider class of formulas into executable logic programs.

Using the failure program in Fig. 12 we conducted a learning experiment with artificial data sampled from the `agree` program. The sample size is 100 and the original and learned parameters (by fgEM and by gEM) are shown below.

parameters	original	fgEM	gEM
θ_a	0.4	0.4096	0.48
θ_b	0.6	0.6096	0.48

As seen clearly, parameters estimated by the gEM algorithm that does not take failure into account are widely off the mark. Worse yet it cannot even distinguish between two parameters. We suspect that such behavior always occurs when failure is ignored though data is generated from a failure model.

5.5 Constrained HMMs

As an instance of generative modeling with constraints that may fail, we introduce *constrained HMMs*, a new class of HMMs that have constraints over

---

[31] We here decompose `failure :- ∀X not(agree(X))` into two clauses for readability.

[32] They convey 'continuation' (data representing the remaining computation). `f0` is an initial continuation and bound to C in `closure_success0(C)` and `closure_agree0(C)`.

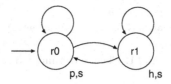

Fig. 13. HMM for the dieting person

```
failure:- not(success).
success:- chmm(L,r0,0,7).

chmm(L,R,C,N):- N>0,
   msw(tr(R),R2),              % choose a restaurant
   msw(lunch(R),D),            % choose lunch
   ( R = r0,
     ( D = p, C2 is C+900      % pizza:900, sandwich:400
     ; D = s, C2 is C+400 )    % hanburger:400, sandwich:500
   ; R = r1,
     ( D = h, C2 is C+400
     ; D = s, C2 is C+500 ) ),
   L = [D|L2],
   N2 is N-1,
   chmm(L2,R2,C2,N2).          % recursion for next day
chmm([],_,C,0):- C < 4000.     % calories must be < 4,000
```

Fig. 14. Constrained HMM program

the states and emitted symbols [8]. Constraints are arbitrary and can be global such as the total number of emitted symbols being equal to a multiple of three. Constrained HMMs define conditional distributions just like CRFs (conditional random fields) [45] but generatively. Our hope is that they contribute to the modeling of complex sequence data and will be applied to, say bioinformatics, computer music etc.

We illustrate constrained HMMs by an example borrowed from [8]. It models the probabilistic behavior of a person on a diet. The situation is like this. He, the person, takes lunch at one of two restaurants 'r0' and 'r1' which he probabilistically chooses at lunch time. He also probabilistically chooses pizza (900) or sandwich (400) at 'r0', and hamburger (500) or sandwich (500) at 'r1' (numbers are calories). He is ordered by his doctor to keep calories for lunch in a week less than 4000 in total. Furthermore he is asked to record what he has eaten in a week like [p,s,s,p,h,s,h] and show the record to the doctor. He however is a smart person and preserves it only when he has succeeded in satisfying the constraint. Our task is to estimate his behavioral probability from the list of preserved records.

An HMM in Fig. 13 represents the probabilistic behavior of the dieting person except the constraint on total calories for lunch in a week. A program in Fig. 14 is

Table 2. Estimated probabilities for the dieting person

sw name	original value	estimation(average)
tr(r0)	r0 (0.7) r1 (0.3)	r0 (0.697) r1 (0.303)
tr(r1)	r1 (0.7) r0 (0.3)	r1 (0.701) r0 (0.299)
lunch(r0)	p (0.4) s (0.6)	p (0.399) s (0.601)
lunch(r1)	h (0.5) s (0.5)	h (0.499) s (0.501)

a usual HMM program corresponding to Fig. 13 but augmented with a constraint atom C < 4000 where C stands for accumulated calories. If the accumulated calories, C, is not less than 4,000 on the seventh day, C < 4000 in the last clause fails and so does the execution of the program (sampling).

The program looks self-explanatory but we add some comments for readability. R is the current restaurant and msw(tr(R),R2) is used for a random choice of the restaurant next day, R2. Likewise msw(lunch(R),D) stands for a random choice of dish 'D' for lunch at 'R'. We accumulate calories in C and record the chosen dish in L. FOC eliminates not in the program and produces a failure program that runs linearly in the number of days ('N').

With this synthesized failure program, we conducted a learning experiment using artificial data sampled from the original program. After setting parameters as shown on the left column of *original value* in Table 2, we generated 500 examples and let the fgEM algorithm estimate parameters from these 500 examples. The right column in Table 2 shows shows averages of 50 experiments. The result says for example that the probability of msw(tr(r0),r0) be true, i.e. the probability of visiting the restaurant r0 from r0, is originally 0.7 and the average of estimation is 0.697. It seems we have reasonably succeeded in recovering original parameters.

In this section, we explained how to overcome probability-loss-to-failure in generative modeling and introduced constrained HMMs, a class of HHMs with constraints, as an application. We also introduced *finite PCFGs* in [46] as a preliminary step toward approximate computation of probabilistic HPSGs. They are PCFGs with a constraint on the size of parse trees and only a finite number of parse trees are allowed for a sentence. Their unique property is that even if the original PCFG suffers probability-loss-to-infinity, parameters can be estimated from the constrained one. We conducted a learning experiment with a real corpus and observed that the deterioration of parsing accuracy by truncating parse trees to a certain depth is small [46].

6   Related Work and Discussion

There are three distinctive features which jointly characterize PRISM as a logic-based probabilistic modeling language. They are the *distribution semantics* [2], *two-staged probability computation*, i.e. probability computation combined with tabled search [6,47] and EM learning by the *fgEM algorithm* [8]. Since each topic has a body of related work of its own, we only selectively state some of them.

6.1 Semantic Aspects

First we deal with semantic aspects. When looking back on probabilistic reasoning in AI, one can observe at least two research streams. One of them focuses on the inference of probabilistic intervals, in a deductive style [48,49,50] or in a logic programming style [51,52,53,54]. The other aims to define distributions. The latter is further divided, for the sake of explanation, into three groups in view of how distributions are defined. The first group uses undirected graphical models [55,56] and define discriminative models. The second group is based on directed graphs, i.e. BNs and their combination with KBs (knowledge bases) called KBMC (knowledge-based model construction) [57,58,59,60,61,62,63,64,65,66,67]. The third group does not rely on graphs but relies on the framework of logic programming [2,6,7,23,24,68,69,70,71,72] to which PRISM belongs, or on the framework of functional programming [73].

Semantically the most unique feature of PRISM's declarative semantics, i.e. the distribution semantics, is that it defines *unconditionally* a global probability measure over the set of uncountably many Herbrand interpretations for any program in a first order language with countably many function and predicate symbols. In short, it always uniquely defines a *global and infinite* distribution for any program, even if it is a looping program such as p(X):-p(X),q(Y) or even if it contains negation in some clause body [8]. Note that we are not claiming that PRISM, an embodiment of the distribution semantics, can always compute whatever probability defined by its semantics. Apparently it is impossible. All it can do is to compute computable part of the distribution semantics.

Unconditional Global Distribution: The unconditional existence of global distributions by the distribution semantics sharply differs from the KBMC approach. Basically in the KBMC approach, a BN is reconstructed from a KB and an input every time the input is given. For example in the case of PRMs (probabilistic relational models) [61], when a new record is added to a RDB, a new BN is constructed. Since different inputs yield different BNs, there is no global distribution defined by the KB, which makes it difficult to consider observed data as iid data from the KB.

Our semantics also sharply differs from SLP (stochastic logic programming) [68,7] which defines a distribution over SLD proofs, not over Herbrand interpretations. Furthermore for a distribution to be definable in SLP, programs must satisfy a syntactic condition called "range-restrictedness" which excludes many ordinary programs such as the member program. The same range-restrictedness condition is imposed on BLPs (Bayesian logic programs) [62] and LBNs (logical Bayesian networks) [67].

Infinite Distribution: Concerning infinite distributions (infinite BNs), there are attempts to construct them from infinitely many BNs. Kersting and De Raedt showed the existence of infinite BNs defined by BLPs. They assume certain conditions including the acyclicity of the ground level caller-callee relation defined by a BLP program. Under the conditions, local BNs are constructed for each ground atom in the least model and distributions defined by them are pasted together to construct an infinite distribution using the Kolmogorov extension theorem [65].

Similar conditions were placed when Laskey proposed a probabilistic logic called MEBN (multi-entity BN) [74]. In MEBN local BNs are defined by schemes containing logical variables and constraints. Assuming local BNs, when combined, create no infinitely many parents (in terms of the CPT) or ancestors for each node, the existence of a global infinite distribution satisfying each scheme is proved by Kolmogorov's extension theorem.

BLOG (Bayesian logic) proposed by Milch et al. [75] defines a distribution over possible worlds consisting of objects through an infinite BN. In a BLOG model (program), objects are generated according to certain statements containing CPTs. These statements generate structured random variables in conjunction with their dependency as a local BN. CBNs (contingent BNs) are introduced to precisely specify such dependency [76]. They are BNs with constraints attached to edges such that edges are removed when constraints are not satisfied. The existence of a global distribution satisfying these local BNs is proved when a CBN satisfies conditions ensuring that every node has finitely many ancestors and parents.

These approaches all unify infinitely many distributions defined by local BNs, and hence need conditions to ensure the possibility of their unification. The distribution semantics of PRISM on the other hand does not attempt to unify local BNs to construct an infinite distribution. Instead it starts from a simple infinite distribution (over msw atoms) that surely exists and extends it by way of fixpoint operation which is always possible, thereby achieving the unconditional definability of infinite distributions.

Independent vs. Dependent Choice: In the distribution semantics, P_{DB}, the distribution defined by a program DB, is parameterized with a base distribution P_F. PRISM implements the simplest P_F, a product measure of independent choices represented by msw atoms like PHA [23], aiming at efficiency of probability computation. At first one may feel that msw atoms are not enough for complex modeling because they cannot represent dependent choices. In reality, however, they can, because the name of an msw switch is allowed to be any term and hence, one choice $msw(s,X)$ can affect another choice $msw(t[X],Y)$ through their common variable X. The realization of dependent choice by this "name trick" is used to write a naive BN program and also used to implement an extension of PCFGs called pseudo context sensitive models [6]. We note that mutually dependent choices can be implementable by substituting a Boltzmann machine for P_F.

6.2   Probability Computation

Two-Staged Probability Computation: The second unique feature of PRISM is a way of probability computation. To compute a probability $P_{DB}(G)$ of a top-goal atom (query) G from a distribution P_{DB} defined by a program DB at the predicate level, we first reduce G by tabled search to an explanation graph

$Expl(G)$ for G such that $Expl(G) \Leftrightarrow G$. $Expl(G)$ is a compact propositional formula logically equivalent to G and consists of ground user atoms and msw atoms. Then we compute $P_{DB}(Expl(G))(= P_{DB}(G))$ as the generalized inside probability by dynamic programming assuming the exclusiveness of disjuncts and the independence of conjuncts in $Expl(G)$.

The reduction from a predicate level expression (goal) to a propositional expression (explanation graph) can have three favorable effects though it sometimes blows up the size of the final expression. The first one is that of pruning, detecting impossible conditions corresponding to zero probability. The second one is to be able to explicitly express exclusiveness as a disjunction and independence as a conjunction. In an attempt by Pynadath and Wellman that formulates a PCFG by a BN [77], they end up in a very loopy BN partly because zero probability and exclusiveness are not easy to graphically express in BNs. To state the last one, we need some terminology.

Value-Wise vs. Variable-Wise: We say that random variables are used *variable-wise* in an expression if they are treated uniformly in the expression regardless of their values. Otherwise we say they are used *value-wise*. For example BNs are variable-wise expressions and their standard computation algorithm, BP, is a variable-wise algorithm because uniform operations are performed on variables regardless of their values whereas, for instance, d-DNNFs after optimization used in ACE are value-wise. The point is that value-wise expressions have a bigger chance of sharing subexpressions (and hence sharing subcomputations in dynamic programming) than variable-wise expressions because for sharing to occur in a value-wise expression, it is enough that only some values, not all values of a random variable, are shared by some subexpressions. Also we should note that value-wise dependency is always sparser than variable-wise dependency because 0 probability cuts off the chain of value-wise dependency.

Returning to PRISM, in a program, a random variable X is expressed as a logical variable X in an msw atom like $\mathtt{msw}(id, X)$, but a search process instantiates it, differently depending on the context of its use, to ground terms like $\mathtt{msw}(id, t_1), \ldots, \mathtt{msw}(id, t_n)$ resulting in a value-wise explanation graph. The reduction to a value-wise expression is a key step to make possible $O(L^3)$ computation of a sentence probability by PRISM [6] where L is the sentence length. Moreover, it was empirically shown that for moderately ambiguous PCFGs, the probability computation by PRISM is much faster than the one by the standard Inside-Outside algorithm [6]. Another example of value-wise computation is the left-to-right HMMs shown in Fig. 1. Their direct implementation by a BN (Fig. 6) would require $O(N^2)$ time for probability computation where N is the number of states though their transition matrix is sparse. Contrastingly PRISM, exploiting this sparseness by value-wise computation, realizes $O(N)$ computation as can be seen from Fig. 4.

Propositionalized BNs: It is interesting to note that probability computation of BNs by way of explanation graphs such as the one in Section 3 agrees with the emerging trend of "propositionalized BNs" which computes probabilities of a BN by converting it to a value-wise propositional level formula [17,34,35]. For example in [17] Chavira and Darwiche considered a BN as a multi-variate polynomial composed of binary-random variables representing individual values of a random variable and compile the BN into an arithmetic circuit. They empirically showed that their approach can efficiently detect and take advantage of CSI (context-specific independence) [38] in the original BN. McAllester et al. proposed CFDs (case factor diagrams) which are formulas representing a "feasible" set of assignments for infinitely many propositional variables. They can compute probabilities of linear Boolean models, a subclass of log-linear models [34]. In a CFD, subexpressions are shared and probabilities are computed by dynamic programming, thus realizing cubic order probability computation for PCFGs. Mateescu and Dechter introduced AND/OR search trees representing variable elimination of BNs propositionally [26]. Value dependency of nodes in a BN is expressed as an AND/OR search tree but identical subtrees are merged to produce a "minimal AND/OR graph" which realizes shared probability computation.

The main difference between PRISM and these approaches is that they manipulate propositional level expressions and predicate level expressions are out of concern. In the case of CFDs for example, programming for a PCFG starts from encoding parse forests of sentences. Contrastingly in PRISM, we do not encode parse forests but encode the PCFG itself using high level expressions in predicate logic. A subsequent search process automatically reduces a sentence to propositional formulas representing parse forests. Our approach thus makes compatible the ease and flexibility of high level programming and the computational efficiency in low level probability computation. However hitting the right balance between generality and efficiency in designing a programming language is always a difficult problem. PRISM is one extreme aiming at generality. A recent proposal of LOHMMs (logical hidden Markov models) by Kersting et al. [71] takes an intermediate approach by specializing in a logical extension of HMMs.

Eliminating Conditions: Finally we discuss the possibility of eliminating some conditions in Subsection 2.2. The first candidate is the exclusiveness condition on disjunctions. It can be eliminated by appealing to general computation schemes such as the inclusion-exclusion principle generalizing $P(A \vee B)$ $= P(A) + P(B) - P(A \wedge B)$ and the sum-of-disjoint products generalizing $P(A \vee B) = P(A) + P(\neg A \wedge B)$, or BDDs (binary decision diagrams) [78]. ProbLog, seeking efficiency, uses BDDs to compute probabilities of nonexclusive disjunctions [72]. Although it seems possible in principle to introduce BDDs to PRISM's explanation graphs at the cost of increasing time and memory, details are left as future work.

The second candidate is the acyclicity condition. When eliminated, we might have a "loopy" explanation graph. Such a graph makes mathematical sense if, like loopy BP, loopy probability computation guided by the graph converges. There is a general class of loopy probability computation that looks relatively simple and useful; prefix computation of PCFGs. Given a string $s = w_1, \ldots, w_k$ and a PCFG, we would like to compute the probability that s is an initial string of some complete sentence $s = w_1, \ldots, w_k, \ldots, w_n$ derived from the PCFG. There already exists an algorithm for that purpose [79] and we can imagine a generalized prefix computation in the context of PRISM. We however need to consider computation cost as the resulting algorithm will heavily use matrix operations to compute "loopy inside probability."

6.3   EM Learning

The advantage of EM learning by PRISM is made clear when we are given the task of EM learning for N new probabilistic model classes like BN, HMMs, PCFGs etc. We write N different programs and apply the same algorithm, the (f)gEM algorithm, to all of them, instead of deriving a new EM algorithm N times. The differences in model classes are subsumed by those in their explanation graphs and do not affect the gEM algorithm itself. The cost we have to pay for this uniformity however is time and space inefficiency due to the use of predetermined data structure, explanation graphs, for all purposes. For example, HMMs in PRISM require memory proportional to the input length to compute forward-backward probabilities while a specialized implementation only needs a constant space.

Another problem is that when we attempt EM learning of a generative model with failure, we have to synthesize a failure program that can represent all failed computation paths of the original program for the model. When models are variants of HMMs like constrained HMMs in Section 5, this synthesis is always possible. However for other cases including PCFGs with constraints, the synthesis is future work.

7   Conclusion

PRISM is a full programming language system equipped with rich functionalities and built-in predicates of Prolog enhanced by three components for probabilistic modeling. The first one is the *distribution semantics* [2], a measure theoretical semantics for probabilistic logic programs. The second one is *two-staged probability computation* [6,47], i.e. generalized IO computation after tabled-search for explanation graphs. The third one is an EM algorithm, *the fgEM algorithm* [8], for generative models allowing failure. PRISM not only uniformly subsumes three representative model classes, i.e. BNs, HMMs, and PCFGs as instances of the distribution semantics at the semantic level but uniformly subsumes their probability computation with the same time complexity, i.e. BP on junction trees for BNs [10], the forward-backward algorithm for HMMs, and IO probability computation for PCFGs respectively as instances of generalized IO probability computation for logic programs [6].

Despite the generality of computational architecture, PRISM runs reasonably fast compared to the state-of-art systems as demonstrated in Section 4 as long as we accept memory consumption for tabling. We also emphasize that PRISM facilitates the creation and exploration of new models such as constrained HMMs as exemplified in Section 5. Hence we believe PRISM is now a viable tool for prototyping of various probabilistic models.

There remains much to be done. The biggest problem is memory consumption. Currently terms are created dynamically by pointers and every pointer occupies 64 bits. This is a very costly approach from a computational viewpoint though it gives us great flexibility. Restricting the class of admissible programs to make it possible to introduce array is one way to avoid the memory problem. The second one is to make PRISM more Bayesian. Currently only MAP estimation is possible though we are introducing built-in predicates for BIC [80] and the Cheeseman-Stutz criterion [81]. Probably we need a more powerful Bayesian computation such as variational Bayes to cope with data sparseness. Also parallelism is inevitable to break computational barrier. Although an initial step was taken toward that direction [82], further investigation is needed.

Acknowledgments

We are grateful to Yusuke Izumi and Tatsuya Iwasaki for generously providing all graphs in this chapter.

References

1. Sato, T., Kameya, Y.: PRISM: A language for symbolic-statistical modeling. In: Proceedings of the 15th International Joint Conference on Artificial Intelligence (IJCAI 1997), pp. 1330–1335 (1997)
2. Sato, T.: A statistical learning method for logic programs with distribution semantics. In: Proceedings of the 12th International Conference on Logic Programming (ICLP 1995), pp. 715–729 (1995)
3. Dempster, A.P., Laird, N.M., Rubin, D.B.: Maximum likelihood from incomplete data via the EM algorithm. Royal Statistical Society B39(1), 1–38 (1977)
4. Tamaki, H., Sato, T.: OLD resolution with tabulation. In: Shapiro, E. (ed.) ICLP 1986. LNCS, vol. 225, pp. 84–98. Springer, Heidelberg (1986)
5. Zhou, N.F., Sato, T.: Efficient fixpoint computation in linear tabling. In: Proceedings of the 5th ACM-SIGPLAN International Conference on Principles and Practice of Declarative Programming (PPDP 2003), pp. 275–283 (2003)
6. Sato, T., Kameya, Y., Abe, S., Shirai, K.: Fast EM learning of a family of PCFGs. Technical Report (Dept. of CS) TR01-0006, Tokyo Institute of Technology (2001)
7. Cussens, J.: Parameter estimation in stochastic logic programs. Machine Learning 44(3), 245–271 (2001)
8. Sato, T., Kameya, Y., Zhou, N.F.: Generative modeling with failure in PRISM. In: Proceedings of the 19th International Joint Conference on Artificial Intelligence (IJCAI 2005), pp. 847–852 (2005)
9. Pearl, J.: Probabilistic Reasoning in Intelligent Systems. Morgan Kaufmann, San Francisco (1988)
10. Sato, T.: Inside-Outside probability computation for belief propagation. In: Proceedings of the 20th International Joint Conference on Artificial Intelligence (IJCAI 2007), pp. 2605–2610 (2007)

11. Lauritzen, S., Spiegelhalter, D.: Local computations with probabilities on graphical structures and their applications to expert systems. Journal of the Royal Statistical Society, B 50, 157–224 (1988)
12. Jensen, F.V.: An Introduction to Bayesian Networks. UCL Press (1996)
13. Rabiner, L.R.: A tutorial on hidden Markov models and selected applications in speech recognition. Proceedings of the IEEE 77(2), 257–286 (1989)
14. Baker, J.K.: Trainable grammars for speech recognition. In: Proceedings of Spring Conference of the Acoustical Society of America, pp. 547–550 (1979)
15. Eisner, J., Goldlust, E., Smith, N.: Compiling Comp Ling: Weighted dynamic programming and the Dyna language. In: Proceedings of Human Language Technology Conference and Conference on Empirical Methods in Natural Language Processing (HLT-EMNLP)., pp. 281–290 (2005)
16. Darwiche, A.: A compiler for deterministic, decomposable negation normal form. In: Proceedings of the 18th national conference on Artificial intelligence (AAAI 2002), pp. 627–634 (2002)
17. Chavira, M., Darwiche, A.: Compiling Bayesian networks with local structure. In: Proceedings of the 19th International Joint Conference on Artificial Intelligence (IJCAI 2005), pp. 1306–1312 (2005)
18. Chavira, M., Darwiche, A., Jaeger, M.: Compiling relational bayesian networks for exact inference. International Journal of Approximate Reasoning 42, 4–20 (2006)
19. Doets, K.: From Logic to Logic Programming. The MIT Press, Cambridge (1994)
20. Manning, C.D., Schütze, H.: Foundations of Statistical Natural Language Processing. The MIT Press, Cambridge (1999)
21. Castillo, E., Gutierrez, J.M., Hadi, A.S.: Expert Systems and Probabilistic Network Models. Springer, Heidelberg (1997)
22. Chow, Y., Teicher, H.: Probability Theory, 3rd edn. Springer, Heidelberg (1997)
23. Poole, D.: Probabilistic Horn abduction and Bayesian networks. Artificial Intelligence 64(1), 81–129 (1993)
24. Poole, D.: The independent choice logic for modeling multiple agents under uncertainty. Artificial Intelligence 94(1-2), 7–56 (1997)
25. Clark, K.: Negation as failure. In: Gallaire, H., Minker, J. (eds.) Logic and Databases, pp. 293–322. Plenum Press, New York (1978)
26. Mateescu, R., Dechter, R.: The relationship between AND/OR search spaces and variable elimination. In: Proceedings of the 21st Conference on Uncertainty in Artificial Intelligence (UAI 2005), pp. 380–387 (2005)
27. Sato, T.: Modeling scientific theories as PRISM programs. In: Proceedings of ECAI 1998 Workshop on Machine Discovery, pp. 37–45 (1998)
28. Mitomi, H., Fujiwara, F., Yamamoto, M., Sato, T.: Bayesian classification of human custom based on stochastic context-free grammar (in Japanese). IEICE Transaction on Information and Systems J88-D-II(4), 716–726 (2005)
29. Wang, S., Wang, S., Greiner, R., Schuurmans, D., Cheng, L.: Exploiting syntactic, semantic and lexical regularities in language modeling via directed Markov random fields. In: Proceedings of the 22th International Conference on Machine Learning (ICML 2005), pp. 948–955 (2005)
30. Sato, T., Kameya, Y.: Parameter learning of logic programs for symbolic-statistical modeling. Journal of Artificial Intelligence Research 15, 391–454 (2001)
31. Smyth, P., Heckerman, D., Jordan, M.: Probabilistic independence networks for hidden Markov probability models. Neural Computation 9(2), 227–269 (1997)
32. Kask, K., Dechter, R., Larrosa, J., Cozman, F.: Bucket-tree elimination for automated reasoning. ICS Technical Report Technical Report No.R92, UC Irvine (2001)

33. Shafer, G., Shenoy, P.: Probability propagation. Annals of Mathematics and Artificial Intelligence 2, 327–352 (1990)
34. McAllester, D., Collins, M., Pereira, F.: Case-factor diagrams for structured probabilistic modeling. In: Proceedings of the 20th Annual Conference on Uncertainty in Artificial Intelligence (UAI2004), Arlington, Virginia, pp. 382–391. AUAI Press (2004)
35. Minato, S., Satoh, K., Sato, T.: Compiling bayesian networks by symbolic probability calculation based on zero-suppressed bdds. In: Proceedings of the 20th International Joint Conference on Artificial Intelligence (IJCAI 2007), pp. 2550–2555 (2007)
36. Charniak, E.: Tree-bank grammars. In: Proceedings of the 13th National Conference on Artificial Intelligence(AAAI 1996), pp. 1031–1036 (1996)
37. Marcus, M., Santorini, B., Marcinkiewicz, M.: Building a large annotated corpus of English: The Penn Treebank. Computational Linguistics 19, 313–330 (1993)
38. Boutilier, C., Friedman, N., Goldszmidt, M., Koller, D.: Context-specific independence in Bayesian networks. In: Procceding of the 12th Conference on Uncertainty in Artificial Intelligence (UAI 1996), pp. 115–123 (1996)
39. Chi, Z., Geman, S.: Estimation of probabilistic context-free grammars. Computational Linguistics 24(2), 299–305 (1998)
40. Wetherell, C.S.: Probabilistic languages: A review and some open questions. Computing Surveys 12(4), 361–379 (1980)
41. Abney, S.: Stochastic attribute-value grammars. Computational Linguistics 23(4), 597–618 (1997)
42. Schmid, H.: A generative probability model for unification-based grammars. In: Proceedings of the 21st International Conference on Computational Linguistics (COLING 2002, pp. 884–896 (2002)
43. Sag, I., Wasow, T.: Syntactic Theory: A Formal Introduction. CSLI Publications, Stanford (1999)
44. Sato, T.: First Order Compiler: A deterministic logic program synthesis algorithm. Journal of Symbolic Computation 8, 605–627 (1989)
45. Lafferty, J., McCallum, A., Pereira, F.: Conditional random fields: Probabilistic models for segmenting and labeling sequence data. In: Proceedings of the 18th International Conference on Machine Learning (ICML 2001, pp. 282–289 (2001)
46. Sato, T., Kameya, Y.: Negation elimination for finite PCFGs. In: Etalle, S. (ed.) LOPSTR 2004. LNCS, vol. 3573, pp. 119–134. Springer, Heidelberg (2005)
47. Kameya, Y., Sato, T.: Efficient EM learning for parameterized logic programs. In: Palamidessi, C., Moniz Pereira, L., Lloyd, J.W., Dahl, V., Furbach, U., Kerber, M., Lau, K.-K., Sagiv, Y., Stuckey, P.J. (eds.) CL 2000. LNCS (LNAI), vol. 1861, pp. 269–294. Springer, Heidelberg (2000)
48. Nilsson, N.J.: Probabilistic logic. Artificial Intelligence 28, 71–87 (1986)
49. Frish, A., Haddawy, P.: Anytime deduction for probabilistic logic. Journal of Artificial Intelligence 69, 93–122 (1994)
50. Lukasiewicz, T.: Probabilistic deduction with conditional constraints over basic events. Journal of Artificial Intelligence Research 10, 199–241 (1999)
51. Ng, R., Subrahmanian, V.S.: Probabilistic logic programming. Information and Computation 101, 150–201 (1992)
52. Lakshmanan, L.V.S., Sadri, F.: Probabilistic deductive databases. In: Proceedings of the 1994 International Symposium on Logic Programming (ILPS 1994), pp. 254–268 (1994)

53. Dekhtyar, A., Subrahmanian, V.S.: Hybrid probabilistic programs. In: Proceedings of the 14th International Conference on Logic Programming (ICLP 1997), pp. 391–405 (1997)
54. Saad, E., Pontelli, E.: Toward a more practical hybrid probabilistic logic programming framework. In: Hermenegildo, M.V., Cabeza, D. (eds.) PADL 2004. LNCS, vol. 3350, pp. 67–82. Springer, Heidelberg (2005)
55. Taskar, B., Abbeel, P., Koller, D.: Discriminative probabilistic models for relational data. In: Proceedings of the 18th Conference on Uncertainty in Artificial Intelligence (UAI 2002), pp. 485–492 (2002)
56. Richardson, M., Domingos, P.: Markov logic networks. Machine Learning 62, 107–136 (2006)
57. Breese, J.S.: Construction of belief and decision networks. Computational Intelligence 8(4), 624–647 (1992)
58. Wellman, M., Breese, J., Goldman, R.: From knowledge bases to decision models. Knowledge Engineering Review 7(1), 35–53 (1992)
59. Koller, D., Pfeffer, A.: Learning probabilities for noisy first-order rules. In: Proceedings of the 15th International Joint Conference on Artificial Intelligence (IJCAI 1997), pp. 1316–1321 (1997)
60. Ngo, L., Haddawy, P.: Answering queries from context-sensitive probabilistic knowledge bases. Theoretical Computer Science 171, 147–177 (1997)
61. Friedman, N., Getoor, L., Koller, D., Pfeffer, A.: Learning probabilistic relational models. In: Proceedings of the 16th International Joint Conference on Artificial Intelligence (IJCAI 1999), pp. 1300–1309 (1999)
62. Kristian Kersting, K., De Raedt, L.: Bayesian logic programs. In: Cussens, J., Frisch, A.M. (eds.) ILP 2000. LNCS (LNAI), vol. 1866, pp. 138–155. Springer, Heidelberg (2000)
63. Jaeger, J.: Complex probabilistic modeling with recursive relational Bayesian networks. Annals of Mathematics and Artificial Intelligence 32(1-4), 179–220 (2001)
64. Getoor, L., Friedman, N., Koller, D.: Learning probabilistic models of relational structure. In: Proceedings of the 18th International Conference on Machine Learning (ICML 2001), pp. 170–177 (2001)
65. Kersting, K., De Raedt, L.: Basic principles of learning bayesian logic programs. Technical Report Technical Report No. 174, Institute for Computer Science, University of Freiburg (2002)
66. Chavira, M., Darwiche, A., Jaeger, M.: Compiling relational bayesian networks for exact inference. In: Proceedings of the Second European Workshop on Probabilistic Graphical Models (PGM 2004), pp. 49–56 (2004)
67. Fierens, D., Blockeel, H., Bruynooghe, M., Ramon, J.: Logical Bayesian networks and their relation to other probabilistic logical models. In: Kramer, S., Pfahringer, B. (eds.) ILP 2005. LNCS (LNAI), vol. 3625, pp. 121–135. Springer, Heidelberg (2005)
68. Muggleton, S.: Stochastic logic programs. In: de Raedt, L. (ed.) Advances in Inductive Logic Programming, pp. 254–264. IOS Press, Amsterdam (1996)
69. Vennekens, J., Verbaeten, S., Bruynooghe, M.: Logic programs with annotated disjunctions. In: Demoen, B., Lifschitz, V. (eds.) ICLP 2004. LNCS, vol. 3132, pp. 431–445. Springer, Heidelberg (2004)
70. Baral, C., Gelfond, M., Rushton, N.: Probabilistic reasoning with answer sets. In: Lifschitz, V., Niemelä, I. (eds.) LPNMR 2004. LNCS (LNAI), vol. 2923, pp. 21–33. Springer, Heidelberg (2003)
71. Kersting, K., De Raedt, L., Raiko, T.: Logical hidden Markov models. Journal of Artificial Intelligence Research 25, 425–456 (2006)

72. De Raedt, L., Angelika, K., Toivonen, H.: ProbLog: A probabilistic Prolog and its application in link discoverry. In: Proceedings of the 20th International Joint Conference on Artificial Intelligence (IJCAI 2007) (2007)
73. Pfeffer, A.: IBAL: A probabilistic rational programming language. In: Proceedings of the 17th International Conference on Artificial Intelligence (IJCAI 2001), pp. 733–740 (2001)
74. Laskey, K.: MEBN: A logic for open-world probabilistic reasoning. C4I Center Technical Report C4I06-01, George Mason University Department of Systems Engineering and Operations Research (2006)
75. Milch, B., Marthi, B., Russell, S., Sontag, D., Ong, D., Kolobov, A.: BLOG: Probabilistic models with unknown objects. In: Proceedings of the 19th International Joint Conference on Artificial Intelligence (IJCAI 2005), pp. 1352–1359 (2005)
76. Milch, B., Marthi, B., Sontag, D., Russell, S., Ong, D., Kolobov, A.: Approximate Inference for Infinite Contingent Bayesian Networks. In: Proceedings of the 10th International Workshop on Artificial Intelligence and Statistics (AISTATS 2005), pp. 1352–1359 (2005)
77. Pynadath, D.V., Wellman, M.P.: Generalized queries on probabilistic context-free grammars. IEEE Transaction on Pattern Analysis and Machine Intelligence 20(1), 65–77 (1998)
78. Rauzy, A., Chatelet, E., Dutuit, Y., Berenguer, C.: A practical comparison of methods to assess sum-of-products. Reliability Engineering and System Safety 79, 33–42 (2003)
79. Stolcke, A.: An efficient probabilistic context-free parsing algorithm that computes prefix probabilities. Computational Linguistics 21(2), 165–201 (1995)
80. Schwarz, G.: Estimating the dimension of a model. Annals of Statistics 6(2), 461–464 (1978)
81. Cheeseman, P., Stutz, J.: Bayesian classification (AutoClass): Theory and results. In: Fayyad, U., Piatesky, G., Smyth, P., Uthurusamy, R. (eds.) Advances in Knowledge Discovery and Data Mining, pp. 153–180. The MIT Press, Cambridge (1995)
82. Izumi, Y., Kameya, Y., Sato, T.: Parallel EM learning for symbolic-statistical models. In: Proceedings of the International Workshop on Data-Mining and Statistical Science (DMSS 2006), pp. 133–140 (2006)

CLP($\mathcal{BN}$): Constraint Logic Programming for Probabilistic Knowledge

Vítor Santos Costa[1], David Page[2], and James Cussens[3]

[1] DCC-FCUP and LIACC
Universidade do Porto
Portugal
vsc@dcc.fc.up.pt
[2] Dept. of Biostatistics and Medical Informatics
University of Wisconsin-Madison
USA
page@biostat.wisc.edu
[3] Department of Computer Science
University of York UK
jc@cs.york.ac.uk

Abstract. In Datalog, missing values are represented by Skolem constants. More generally, in logic programming missing values, or existentially quantified variables, are represented by terms built from Skolem functors. The CLP($\mathcal{BN}$) language represents the joint probability distribution over missing values in a database or logic program by using constraints to represent Skolem functions. Algorithms from inductive logic programming (ILP) can be used with only minor modification to learn CLP($\mathcal{BN}$) programs. An implementation of CLP($\mathcal{BN}$) is publicly available as part of YAP Prolog at http://www.ncc.up.pt/~vsc/Yap.

1 Introduction

One of the major issues in knowledge representation is how to deal with incomplete information. One approach to this problem is to use probability theory in order to represent the likelihood of an event. More specifically, advances in representation and inference with Bayesian networks have generated much interest and resulted in practical systems, with significant industrial applications [1]. A Bayesian network represents a joint distribution over a set of random variables where the network structure encapsulates conditional independence relations between the variables.

A Bayesian network may be seen as establishing a set of relations between events. This presents a clear analogy with propositional calculus, as widely discussed in the literature [2], and raises the question of whether one could move one step forward towards a Bayesian network system based on the more powerful predicate calculus. Arguably, a more concise representation of Bayesian Networks would avoid wasted work and possible mistakes. Moreover, it would make it easier to learn interesting patterns in data. Work such as Koller's Probabilistic Relational Models (PRMs) [3], Sato's PRISM [4], Ngo and Haddawy's

L. De Raedt et al. (Eds.): Probabilistic ILP 2007, LNAI 4911, pp. 156–188, 2008.
© Springer-Verlag Berlin Heidelberg 2008

Probabilistic Logic Programs [5], Muggleton and Cussens' Stochastic Logic Programs [6], and Kersting and De Raedt's Bayesian Logic Programs [7] have shown that such a goal is indeed attainable.

The purpose of probabilistic first order languages is to propose a concise encoding of probability distributions for unobserved variables. Note that manipulating and reasoning on unknown values is a well-known problem in first-order representations. As an example, First-Order Logic is often used to express existential properties, such as:

$$\forall x \exists y, Make(x) \rightarrow OwnsCar(y, x)$$

A natural interpretation of this formula is that every make of car has at least one owner. In other words, for every make x there is an individual y that owns a car of this make. Notice that the formula does not state who the owner(s) may be, just that one exists. In some cases, e.g., for inference purposes, it would be useful to refer to the individual, even if we do not know its actual name. A process called *skolemization* can replace the original formula by a formula without existential quantifiers:

$$\forall x, Make(x) \rightarrow OwnsCar(y, s(x))$$

where $y = s(x)$ and $s(x)$ is called a Skolem function: we know the function describes an individual for each x, but we do not know which individual.

Skolem functions have an interesting analogy in probabilistic relational models (PRMs) [3]. PRMs express probability distributions of a field in the database by considering related fields, thus encoding a Bayesian network that represents the joint probability distribution over all the fields in a relational database. The Bayes network constructed by PRMs can then be used to infer probabilities about *missing values* in the database. We know that the field must take one value, we know that the value will depend on related fields, and we know the values for at least some of these related fields. As for Skolem functions, PRMs refer to fields that are unknown function of other fields. But, in contrast with First Order Logic, PRMs do allow us to estimate probabilities for the different outcomes of the function: they allow us to represent partial information on Skolem functions.

Can we take this process a step further and use a Bayesian network to represent the joint probability distribution over terms constructed from the Skolem functors in a logic program? We extend the language of logic programs to make this possible. Our extension is based on the idea of defining a language of Skolem functions where we can express properties of these functions. Because Skolem functions benefit from a special interpretation, we use Constraint Logic Programming (CLP), so we call the extended language CLP($\mathcal{BN}$). We show that any PRM can be represented as a CLP($\mathcal{BN}$) program.

Our work in CLP($\mathcal{BN}$) has been motivated by our interest in multi-relational data mining, and more specifically in inductive logic programming (ILP). Because CLP($\mathcal{BN}$) programs are a kind of logic program, we can use existing ILP systems to learn them, with only simple modifications to the ILP systems. Induction of clauses can be seen as model generation, and parameter fitting can be seen as generating the CPTs for the constraint of a clause. We show that the ILP system ALEPH [8] is able to learn CLP($\mathcal{BN}$) programs.

Next, we present the design of CLP($\mathcal{BN}$) through examples. We then discuss the foundations of CLP($\mathcal{BN}$), including detailed syntax, proof theory (or operational semantics), and model-theoretic semantics. We next discuss important features of CLP($\mathcal{BN}$), namely its ability to support aggregation and recursion. Finally, we present the results of experiments in learning CLP($\mathcal{BN}$) programs using ILP. Lastly, we relate CLP($\mathcal{BN}$) with PRMs and with other related work.

2   CLP($\mathcal{BN}$) Goes to School

We introduce CLP($\mathcal{BN}$) through a simplified version of the school database originally used to explain Probabilistic Relational Models [3] (PRMs). We chose this example because it stems from a familiar background and because it clearly illustrates how CLP($\mathcal{BN}$) relates to prior work on PRMs. Figure 2 shows a simplified fragment of the school database. The schema consists of three relations: *students*, *courses*, and *grades*. For each student, we have a primary key, Student, and its Skill. To simplify, the value for skill is the expected final grade of the student: an A student would thus be a top student. For each course, Course is the primary key and Difficulty gives the course's difficulty: an A difficulty course would be a course where we would expect even the average student to do very well. Lastly, the *Registration* records actual student participation in a course. This table's key is a registration key. Both Student and Course are foreign keys giving student and course data. The last field in the table gives the grade for that registration.

Figure 1 shows an example database with these 3 tables. Notice that some non-key data is missing in the database. For example, we do not know what was mary's grade on c0, maybe because the grade has not been input yet. Also, we

Course	Difficulty
c0	A
c2	?
c3	C

Reg	Student	Course	Grade
r0	John	c0	B
r1	Mary	c0	?
r2	Mary	c2	A
r3	John	c2	?
r4	Mary	c3	A

Student	Skill
John	A
Mary	?

Fig. 1. A Simplified School Database with Tables on Students, Courses and Grades

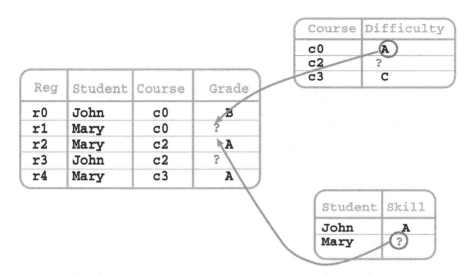

Fig. 2. Direct Dependency Between Random Variables in Simplified School Database

do have john's skill, but we could not obtain data on mary. This is a common problem in databases: often, the database only contains partial data for some items. A fundamental observation in the PRM design is that such missing data can be represented as *random variables*. The idea is that columns for which there is missing data should be seen as sets of random variables. If a specific value is known, we say that the database includes *evidence* on this item. For example, we may say that the Skill attribute in the Student table is a set of random variables, and that we have evidence for the skill variable corresponding to key john. Other sets of random variables are for Difficulty and Grade.

An immediate step in the PRMs is that we can estimate probabilities on the missing values through considering other items in the database. Such items may be on the same relation, or may also be found at a different relations. In our database example, for instance, it would make sense to assume that better skilled students would have better grades, e.g., an A level student would have an higher probability of achieving an A. Moreover, it makes sense to assume that a grade will also depend on the course's difficulty: the easier the course, the better the grades. We can go one step further and argue that given course and student information, a grade should be *conditionally independent* on the other elements on the database. This reasoning suggests that all the random variables in our PRM-extended database should form a Bayesian network.

Figure 2 shows a fragment of the Bayesian network induced by this rule. At first sight, our expectations for mary's grade on course c0 depend on data we have on mary and course c0.

CLP($\mathcal{BN}$) is grounded on the idea that such beliefs can easily be expressed in Logic. Namely, the previous example can be expressed in a variety of ways, but one approach (using Prolog Notation) would be as follows:

```
grade(r1,Grade) :-
    skill(mary,Mskill),
    difficulty(c0,CODifficulty),
    Grade = 'G'(MSkill,CODifficulty).
```

Grade is as an attribute of registration r1. We know that its actual value will depend on mary's skill and course c0's difficulty. The clause says exactly that: Grade is a random variable that can also be described as an unknown function, G(), of r1, mary's skill, Mskill, and c0's difficulty, CODifficulty. Such unknown functions are often used in Logic, where they are usually called *Skolem functions*: thus, in our formalism we shall say that a random variable is given by a Skolem function of its arguments.

Note that we do have some expectations on G(). Such data can be expressed in a variety of ways, say through the fact:

```
random_variable('G'(_,_),['A','B','C','D'],[0.4,0.3,0.2,0.1]).
```

that states that the random variable has domain A,B,C,D and a discrete conditional probability table that we represent as a Prolog list with a number of floating point values between 0 and 1.

Of course, one of the main advantages of the PRMs (and of using first-order representations), is that we can generalize. In this case, we could write a single rule for Grade by lifting the constants in our grade clause and making the individual registration an argument to the Skolem function. We will need to access the foreign keys, giving the following clause:

```
grade(Registration,Grade) :-
    registration_student(Registration, Student),
    registration_course(Registration, Course),
    difficulty(Course,Difficulty),
    skill(Student,Skill),
    Grade = 'S'(Registration,Skill,Difficulty).
```

```
random_variable('S'(_,_,_),['A','B','C','D'],[0.4,0.3,0.2,0.1]).
```

Next, we need rules for difficulty and skill. In our model, we do not have helpful dependencies for Skill and Difficulty, so the two columns should be given from priors. We thus just write:

```
skill(Student,'S1'(Student)).
difficulty(Course,'S2'(Course)).
```

```
random_variable('S1'(_),['A','B','C','D'],[0.25,0.25,0.25,0.25]).
random_variable('S2'(_),['A','B','C','D'],[0.25,0.25,0.25,0.25]).
```

At this point we have a small nice little logic program that fully explains the database. We believe this representation is very attractive (indeed, a similar approach was proposed independently by Blockeel [9]), but it does have one major limitation: it hides the difference between doing inference in first order logic and in Bayesian network, as we discuss next.

Evidence. We have observed that `mary`'s grade on `c0` depends on two factors: course `c0`'s difficulty, and `mary`'s skill. In practice, the actual database does have some extra information that can help in refining the probabilities for this grade.

First, we have actual data on an item. Consider `c0`'s difficulty: we actually know that `c0` is an easy course. We thus have two sources of information about `c0`'s difficulty: we know that it is a random function, `'S1'(c0)`; but we also know that it takes the value `A`. Logically, this evidence can be seen as setting up the equation `S1(c0) = 'A'`. Unfortunately, unification cannot be easily redefined in Prolog.

One simple solution would be to add evidence through an extra fact:

```
evidence('S1'(c0),'A').
```

We assume that somehow this evidence is going to be used when we actually run the program. This solution is not entirely satisfactory, as we now have two separate sources of data on skill: the `skill/2` relation and some facts for `evidence/2`.

Evidence plays indeed a very important role in Bayesian networks. Imagine we want to know the probability distribution for `mary`'s grade on course `c0`. We have more accurate probabilities knowing `c0` is an easy course. And, even though we do not have actual evidence on `mary`'s skill, `Mskill`, we can achieve a better estimate for its probability distribution if we consider evidence relevant to `Mskill`. Namely, `mary` has attended two other courses, `c2` and `c3`, and that she had good grades on both. In fact, course `c3` should be quite a good indicator, as we know grades tend to be bad (a `C`). We do not know the difficulty of course `c2`, but we can again make an estimate by investigating evidence on the students that attended this course. Following all sources of evidence in a network can be quite complex [10], even for such simple examples. In this case, the result is the network shown in Figure 3. Bayesian networks have developed a number of both exact and approximate methods to estimate the probabilities for `Grade` given all the evidence we have on this graph [1].

Evaluating all the relevant evidence is a complex process: first, we need to track down all relevant sources of evidence, through algorithms such as knowledge based model construction [11]. Next, we need to perform probabilistic inference on this graph and marginalize the probabilities on the query variables. To do so would require an extra program, which would have to process both the original query and every source of evidence.

Constraints. The previous approach suggests that a Prolog only approach can be used to represent all the properties of Bayesian networks, but that it does expose the user to the mechanisms used by the Bayesian network to accept and propagate evidence. The user would have the burden of knowing which random variables have evidence, and she would be responsible to call a procedure for probabilistic inference.

Such difficulties suggest that we may want to work at an higher abstraction level. Constraint Logic Programming is an important framework that was designed in order to allow specific *interpretations* on some predicates of interest. These interpretations can then be used to implement specialized algorithms over

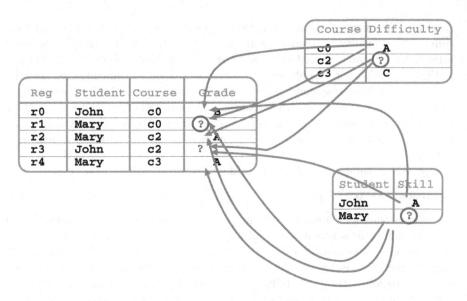

Fig. 3. Full Bayesian Network Induced by Random Variables in the Network

the standard Prolog inference. For example, CLP($\mathcal{R}$) defines constraints over reals: it redefines equality and arithmetic operations to create constraints, or *equations*, that are manipulated by a separate *constraint solver*. The constraint solver maintains a *Store* of all active arithmetic constraints, and calls specialized algorithms to process the equations into some canonical form.

We believe the constraint framework is a natural answer to the problems mentioned above. First, through constraints we can extend *equality* to support evidence. More precisely we can redefine the equality:

```
{ S1(c0) = 'A' }
```

to state that random variable S1(c0) has received evidence A. In constraint programming, we are not forced to bind S1(c0) to A. Instead, we can add S1(c0) = 'A' to a store for later evaluation.

The second advantage of using constraint programming is that it is natural to see a Bayesian Network as a store: both constraints stores and Bayesian networks are graphs; in fact, it is well known that there is a strong connection between both [12]. It is natural to see the last step of probabilistic inference as constraint solving. And it is natural to see marginalization as projection.

Moreover, because constraint stores are opaque to the actual inference process, it is possible to have a *global constraint store* that accommodates all evidence so far. In other words, any time we add evidence to the database we can add this evidence to a global constraint store. Probabilistic inference will then be called having this global Bayesian network as a backdrop.

We have presented the key ideas of CLP($\mathcal{BN}$). Next, we discuss the CLP($\mathcal{BN}$) language in some more detail.

3   The CLP(*BN*) Language

A CLP(*BN*) program is a constraint logic program that can encode Bayesian constraints. Thus, CLP(*BN*) programs are sets of Prolog clauses, where some clauses may contain *BN* constraints. *BN* constraints are of the form {X = F with P}, where X must be a Prolog variable, F a term, and P a probability distribution.

As an example, consider the following clause:

```
skill(S,Skill) :-
 {Skill = skill(S) with p([ 'A', 'B', 'C', 'D'],
                           [0.25,0.25,0.25,0.25],[])}.
```

This clause sets a constraint on Skill. The constraint declares that Skill should be constrained to the term skill(S), an unknown function, or *Skolem function*, of S. Throughout the paper we refer to this term that uniquely identifies a random variable as the *Skolem term*. The constraint declares some further information on skill(S) through the *with* construct. In this case, the right hand side of with declares that *skill(S)* is a discrete random variable with 4 possible values and a prior distribution:

1. skill(S) has domain A, B, C and D;
2. it has an uniform probability distribution over those values;
3. and that skill(S) has no parent nodes.

The right-hand-side of the with is a Prolog term. Thus, the same constraint could be written as:

```
skill(S,Skill) :-
 cpt(skill(S), CPT),
 {Skill = skill(S) with CPT }.

cpt(skill(_), p([ 'A', 'B', 'C', 'D'],
                [0.25,0.25,0.25,0.25],[])).
```

One advantage of this approach is that it makes it straightforward to represent different CPTs for different students with a single constraint. Imagine we have extra information on student's academic story: in this case, we could expect senior students to have better skills than first-year students.

```
skill(S,Skill) :-
 cpt(skill(S), CPT),
 {Skill = skill(S) with CPT }.

cpt(skill(S), p(['A','B','C','D'],[PA, PB, PC, PD],[])) :-
  skill_table(S, PA, PB, PC, PD).
```

```
skill_table(S, 0.25, 0.25, 0.25, 0.25) :-
  freshman(S).
skill_table(S, 0.35, 0.30, 0.20, 0.15) :-
  sophomore(S).
skill_table(S, 0.38, 0.35, 0.17, 0.10) :-
  junior(S).
skill_table(S, 0.40, 0.45, 0.15,0.00) :-
  senior(S).
```

In general, all CLP($\mathcal{BN}$) objects are first class objects. They can be specified as compile-time constants, but they can also be computed through arbitrary logic programs. And they can be fully specified before or *after* setting up the constraint, so

```
skill(S,Skill) :-
 {Skill = skill(S) with CPT },
 cpt(skill(S), CPT).
```

is a legal program, and so is:

```
skill(S,CPT,Skill) :-
 {Skill = skill(S) with CPT }.
```

Conditional Probabilities. Let us next consider an example of a conditional probability distribution (CPT). We may remember from Figure 2 that a registration's grade depends on the course's difficulty and on the student's intelligence. This is encoded in the following clause:

```
grade(Registration, Grade) :-
  registration_student(Registration, Student),
  registration_course(Registration, Course),
  difficulty(Course,Dif),
  intelligence(Student,Skill),
  grade_table(TABLE),
  {
     Grade = grade(Course, Dif, Skill) with
        p(['A','B','C','D'],TABLE,[Dif,Skill])
  }.
```

The constraint says that *Grade* is a Skolem function of *Reg*, *Dif*, and *Skill*. We know that *Grade* must be unique for each *Reg*, and we know that the probability distribution for the possible values of *Grade* depend on the random variables *Dif* and *Skill*. These variables are thus the parents in *Grades*'s CPT, i.e., the third argument in the `with` term. The actual table must include 4^3 cases: we save some room in the text by assuming it was given by an auxiliary predicate `grade_table/1`.

Figure 4 shows an alternative, pictorial, representation for a CLP($\mathcal{BN}$) clause in this example. The representation clearly shows the clause as having two components: the logical component sets all variables of interest, and the Bayesian constraint connects them in a sub-graph.

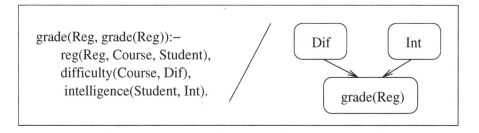

grade(Reg, grade(Reg)):–
    reg(Reg, Course, Student),
    difficulty(Course, Dif),
    intelligence(Student, Int).

Fig. 4. Pictorial representation of a grade clause

3.1   Execution

The evaluation of a CLP(BN) program results in a network of constraints. In the previous example, the evaluation of

```
?- grade(r1,Grade).
```

will set up a constraint network with grade(r2) depending on dif(course) and int(student). CLP(BN) will output the marginal probability distribution on grade(r2).

Table 1 shows in some detail the actual execution steps for this query in the absence of prior evidence. The binding store and the query store grow along as we execute a query on grade r1. The leftmost column shows the step number, the middle column shows new bindings, and the rightmost column shows new constraints. We represent each binding as an equality, and we represent a constraint as the corresponding Skolem term. For space considerations, we abbreviate names of constants and variables, and we do not write the full CPTs, only the main functor and arguments for each Skolem term.

Table 1. A Query on Grade

Step	Bindings	Skolem Terms
0	$\{R = \mathtt{r1}\}$	{}
1	$\cup\{S = \mathtt{mary}\}$	
2	$\cup\{C = \mathtt{c0}\}$	
3		$\cup\{D(\mathtt{c0})\}$
4		$\cup\{S(\mathtt{mary})\}$
5		$\cup\{G(r1, D(\mathtt{c0}), I(\mathtt{mary}))\}$

Each step in the computation introduces new bindings or BN constraints. In step 1 the call to `registration_student/2` obtains a student, mary. In step 2 the call to `registration_course/2` obtains a course, c0. The course's difficulty is obtained from c0 in step 3. Step 4 gives mary's skill. We then link the two variables together to obtain the CPT for *Grade*.

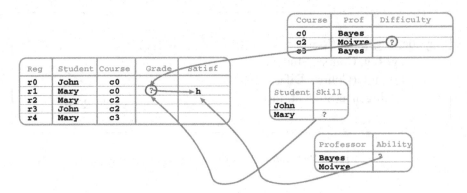

Fig. 5. School Database Extended to Include Satisfaction and Professor Data

Execution: Conditioning on Evidence. One major application of Bayesian network systems is conditioning on evidence. To give this next example, we will add some extra information to the database, as shown in Fig 5. First, we assume that now we have some information on the professors that actually taught the course. We shall need an extra table for professors, and an extra column on courses saying which professor teaches each course. Second, we are interested in how happy students were in our courses. Thus, we add an extra field for courses saying how happy, or *Satisfied*, the student was.

Satisfaction is a random variable. We do not always know it. We do know that it depends on grade and that it even with bad grades, students will be happy to attend courses taught by very good professors. Notice that in order to obtain a professor's ability, we need to navigate in the database: we find the course associated with the registration, the professor who taught the course, and finally the professor's ability. The corresponding program is shown next:

```
satisfaction(Reg, Sat) :-
   registration_course(Reg, Course),
   professor(Course, Prof),
   ability(Prof, Abi),
   grade(Reg, Grade),
   sat_table(Abi, Grade, Table),
   { Sat = satisfaction(Reg) with Table }.
```

Next, imagine that mary was the only student who actually gave her satisfaction, and that she was highly satisfied with registration r1. Given this extra evidence on satisfaction for r1, can we compute the new marginal for grade?

We need to be able to query for *Grade*, given that r2's satisfaction is bound to *h*. In CLP($\mathcal{BN}$) the random variable for satisfaction can be obtained by asking a query,and evidence can be introduced by unifying the answer to the query. The full query would thus be:

```
?- grade(r1,X), satisfaction(r1,h).
```

Table 2. A Query on Grade and Satisfaction

Step	Bindings	Skolem Terms
0	$\{R = \mathtt{r1}\}$	$\{\}$
1	$\cup\{S = \mathtt{mary}\}$	
2	$\cup\{C = \mathtt{c0}\}$	
3		$\cup\{D(\mathtt{c0})\}$
4		$\cup\{S(\mathtt{mary})\}$
5		$\cup\{G(r1, D(\mathtt{c0}), S(\mathtt{mary}))\}$
6	$\{C' = \mathtt{c0}\}$	
7	$\cup\{P' = \mathtt{Bayes}\}$	
8		$\cup\{A(\mathtt{Bayes})\}$
9 – 13		
14		$\cup\{S(\mathtt{r1}, A(\mathtt{Bayes}), G(\mathtt{r1}, \ldots))\}$
15		$\cup\{S(\mathtt{r1}, \ldots) = \mathtt{h}\}$

Table 2 shows how the execution steps update the stores in this case.

The first five steps repeat the computation for grade. Step 5 and 6 find the professor for the course. Step 8 finds its ability. Next, we recompute *Grade*. The computation will in fact be redundant, as the Skolem term for *Grade* was already in the store. The final step introduces evidence. Unification in CLP($\mathcal{BN}$) implements evidence through updating the original constraint in the store. The full store is then marginalized against the query variables by the constraint solver.

Evidence in Store. Imagine again we ask *grade(r1, X)* but now given the database shown in Figure 1. The actual query should now be:

```
?- grade(r1,X),
   grade(r0,'B'), grade(r2,'A'), grade(r4, 'A'),
   difficulty(c0, 'A'), difficulty(c3, 'C'),
   skill(john, 'A').
```

Writing such long queries is cumbersome, to say the least. It may be unclear which evidence is relevant, whereas giving all the evidence in a database may be extremely expensive computationally.

Table 3. A Query on Grade

Step	Bindings	Skolem Terms
0	$\{R = \mathtt{r1}\}$	$\{G(\mathtt{r0}, D(\mathtt{c0}), S(\mathtt{john})), D(\mathtt{c0}), S(\mathtt{john}), \ldots\}$
1	$\cup\{S = \mathtt{mary}\}$	
2	$\cup\{C = \mathtt{c0}\}$	
3		$\cup\{D(\mathtt{c0})\}$
4		$\cup\{S(\mathtt{mary})\}$
5		$\cup\{G(\mathtt{r1}, D(\mathtt{c0}), S(\mathtt{mary}))\}$

CLP($\mathcal{BN}$) allows the user to declare evidence in the program. This is simply performed by stating evidence as a fact for the predicate. Currently, we use the construct {} to inform CLP($\mathcal{BN}$) that a fact introduces evidence:

```
grade(r0, 'B') :- {}.
grade(r2, 'A') :- {}.
grade(r4, 'A') :- {}.
```

This *global evidence* is processed at compile-time, by running the evidence data as goals and adding the resulting constraints to the *Global Store*. Execution of grade(1,X) would thus be as shown in Table 3.

4   Foundations

We next present the basic ideas of CLP($\mathcal{BN}$) more formally. For brevity, this section necessarily assumes prior knowledge of first-order logic, model theory, and resolution.

First, we remark that CLP($\mathcal{BN}$) programs are constraint logic programs, and thus inherit the well-known properties of logic programs. We further interpret a CLP($\mathcal{BN}$) program as defining a set of probability distributions over the models of the underlying logic program. Any Skolem function sk of variables $X_1, ..., X_n$, has an associated CPT specifying a probability distribution over the possible denotations of $sk(X_1, ..., X_n)$ given the values, or bindings, of X_1, ..., X_n. The CPTs associated with a clause may be thought of as a Bayes net fragment, where each node is labeled by either a variable or a term built from a Skolem function. Figure 4 illustrates this view using a clause that relates a registration's grade to the course's difficulty and to the student's intelligence.

4.1   Detailed Syntax

The alphabet of CLP($\mathcal{BN}$) is the alphabet of logic programs. We shall take a set of functors and call these functors *Skolem functors*; *Skolem constants* are simply Skolem functors of arity 0. A Skolem term is a term whose primary functor is a Skolem functor. We assume that Skolem terms have been introduced into the program during a Skolemization process to replace the existentially-quantified variables in the program. It follows from the Skolemization process that any Skolem functor sk appears in only one Skolem term, which appears in only one clause, though that Skolem term may have multiple occurrences in that one clause. Where the Skolem functor sk has arity n, its Skolem term has the form $sk(W_1, ..., W_n)$, where $W_1, ..., W_n$ are variables that also appear outside of any Skolem term in the same clause.

A CLP($\mathcal{BN}$) program in canonical form is a set of *clauses* of the form $H \leftarrow A/B$. We call H the head of the clause. H is a literal and A is a (possibly empty) conjunction of literals. Together they form the logical portion of the clause, C. The probabilistic portion, B, is a (possibly empty) conjunction of atoms of the form: $\{V = Sk$ **with** $CPT\}$. We shall name these atoms *constraints*. Within a constraint, we refer to Sk as the Skolem term and CPT as the conditional probability table. We focus on discrete variables in this paper. In this case, CPT may be an unbound variable or a term or the form $\mathbf{p}(D, T, P)$. We refer to D as the domain, T as the table, and P as the parent nodes.

A CLP($\mathcal{BN}$) constraint B_i is well-formed if and only if:

1. All variables in B_i appear in C;
2. $Sk's$ functor is unique in the program; and,
3. There is at least one substitution σ such that $CPT\sigma = \mathbf{p}(D\sigma, T\sigma, P\sigma)$, and **(a)** $D\sigma$ is a ground list, all members of the list are different, and no sub-term of a term in the list is a Skolem term; **(b)** $P\sigma$ is a ground list, all members of the list are different, and all members of the list are Skolem terms; and **(c)** $T\sigma$ is a ground list, all members of $T\sigma$ are numbers p such that $0 \leq p \leq 1$, and the size of $T\sigma$ is a multiple of the size of $D\sigma$.

If the probabilistic portion of a clause is empty, we also call the clause a *Prolog clause*. According to this definition, every Prolog program is a CLP($\mathcal{BN}$) program.

4.2 Operational Semantics

A query for CLP($\mathcal{BN}$) is an ordinary Prolog query, which is a conjunction of positive literals. In logic programming, a query is answered by one or more proofs constructed through resolution. At each resolution step, terms from two different clauses may be unified. If both of the terms being unified also participate in CPTs, or Bayes net constraints, then the corresponding nodes in the Bayes net

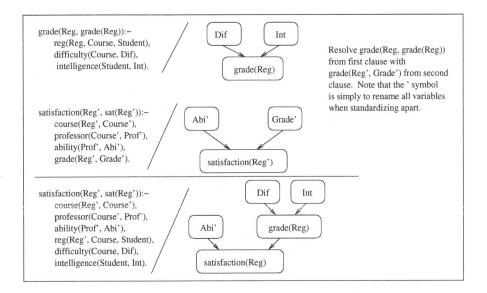

Fig. 6. Resolution

constraints must be *unified* as illustrated in Figure 6. In this way we construct a large Bayes net consisting of all the smaller Bayes nets that have been unified during resolution.

A cycle may arise in the Bayes Net if we introduce a constraint such that Y is a parent of X, and X is an ancestor of Y. In this case, when unifying Y to an argument of the CPT constraint for X, X would be a sub-term of the CPT constraint for Y which causes unification failure. To detect this failure it is necessary to do a proper unification test using the 'occur-check', something standard Prolog does not do (for efficiency reasons).

To be rigorous in our definition of the distribution defined by a Bayes net constraint, let C_i/B_i, $1 \leq i \leq n$, be the clauses participating in the proof, where C_i is the ordinary logical portion of the clause and B_i is the attached Bayes net, in which each node is labeled by a term. Let θ be the answer substitution, that is, the composition of the most general unifiers used in the proof. Note that during resolution a clause may be used more than once but its variables always are renamed, or standardized apart from variables used earlier. We take each such renamed clause used in the proof to be a distinct member of $\{C_i/B_i|1 \leq i \leq n\}$. We define the application of a substitution θ to a Bayes net as follows. For each node in the Bayes net, we apply θ to the label of that node to get a new label. If some possible values for that node (according to its CPT) are not instances of that new label, then we marginalize away those values from the CPT.

4.3   Model-Theoretic Semantics

A CLP($\mathcal{BN}$) program denotes a set of probability distributions over models. We begin by defining the probability distribution over ground Skolem terms that is specified by the probabilistic portion of a CLP($\mathcal{BN}$) program. We then specify the probability distribution over models, consistent with this probability distribution over ground Skolem terms, that the full CLP($\mathcal{BN}$) program denotes.

A CLP($\mathcal{BN}$) program P defines a unique joint probability distribution over ground Skolem terms as follows. Consider each ground Skolem term to be a random variable whose domain is a finite set of non-Skolem constants.[1] We now specify a Bayes net $\mathcal{BN}$ whose variables are these ground Skolem terms. Each ground Skolem term s is an instance of exactly one Skolem term t in the program P. To see this recall that, from the definition of Skolemization, any Skolem functor appears in only one term in the program P, and this one term appears in only one clause of P, though it may appear multiple times in that clause. Also from the definition of Skolemization, t has the form $sk(W_1, ..., W_m)$, where sk is a Skolem functor and $W_1, ..., W_m$ are distinct variables. Because s is a ground instance of t, $s = t\sigma$ for some substitution σ that grounds t. Because $t = sk(W_1, ..., W_n)$ appears in only one clause, t has exactly one associated (generalized) CPT, T, conditional on the Skolem terms in $W_1, ..., W_n$. Let the

---

[1] This can be extended to a finite subset of the set of ground terms not containing Skolem symbols (functors or constants). We restrict ourselves to constants here merely to simplify the presentation.

parents of s in $\mathcal{BN}$ be the Skolem terms in $W_1\sigma, ..., W_m\sigma$, and let the CPT be $T\sigma$. Note that for any node in $\mathcal{BN}$ its parents are sub-terms of that node. It follows that the graph structure is acyclic and hence that $\mathcal{BN}$ is a properly defined Bayes net, though possibly infinite. Therefore $\mathcal{BN}$ uniquely defines a joint distribution over ground Skolem terms; we take this to be the distribution over ground Skolem terms defined by the program P.

The meaning of an ordinary logic program typically is taken to be its least Herbrand model. Recall that the individuals in a Herbrand model are themselves ground terms, and every ground term denotes itself. Because we wish to consider cases where ground Skolem terms denote (non-Skolem) constants, we instead consider Herbrand quotient models [13]. In a Herbrand quotient model, the individuals are equivalence classes of ground terms, and any ground term denotes the equivalence class to which it belongs. Then two ground terms are equal according to the model if and only if they are in the same equivalence class. We take the set of minimal Herbrand quotient models for P to be those derived as follows.[2] Take the least Herbrand model of the logical portion of P, and for each non-Skolem constant, merge zero or more ground Skolem terms into an equivalence class with that constant. This equivalence class is a new individual, replacing the merged ground terms, and it participates in exactly the relations that at least one of its members participated in, in the same manner. It follows that each resulting model also is a model of P. The set of models that can be constructed in this way is the set S of minimal Herbrand quotient models of P. Let D be any probability distribution over S that is consistent with the distribution over ground Skolem terms defined by P. By consistent, we mean that for any ground Skolem term t and any constant c, the probability that $t = c$ according to the distribution defined by P is exactly the sum of the probabilities according to D of the models in which $t = c$. At least one such distribution D exists, since S contains one model for each possible combination of equivalences. We take such $\langle D, S \rangle$ pairs to be the models of P.

4.4   Agreement Between Operational and Model-Theoretic Semantics

Following ordinary logic programming terminology, the negation of a query is called the "goal," and is a clause in which every literal is negated. Given a program and a goal, the CLP($\mathcal{BN}$) operational semantics will yield a derivation of the empty clause if and only if every model $\langle D, S \rangle$ of the CLP($\mathcal{BN}$) program falsifies the goal and hence satisfies the query for some substitution to the variables in the query. This follows from the soundness and refutation-completeness of SLD-resolution. But in contrast to ordinary Prolog, the proof will be accompanied by a Bayes net whose nodes are labeled by Skolem terms appearing in the query or proof. The following theorem states that the answer to any query of

---

[2] For brevity, we simply define these minimal Herbrand quotient models directly. Alternatively, we can define an ordering based on homomorphisms between models and prove that what we are calling the minimal models are indeed minimal with respect to this ordering.

this attached Bayes net will agree with the answer that would be obtained from the distribution D, or in other words, from the distribution over ground Skolem terms defined by the program P. Therefore the operational and model-theoretic semantics of CLP($\mathcal{BN}$) agree in a precise manner.

Theorem 1. *For any CLP($\mathcal{BN}$) program P, any derivation from that program, any grounding of the attached Bayes net, and any query to this ground Bayes net,[3] the answer to the query is the same as if it were asked of the joint distribution over ground Skolem terms defined by P.*

Proof. Assume there exists some program P, some derivation from P and associated ground Bayes net B, and some query $Pr(q|E)$ such that the answer from B is not the same as the answer from the full Bayes net $\mathcal{BN}$ defined by P. For every node in B the parents and CPTs are the same as for that same node in $\mathcal{BN}$. Therefore there must be some path through which evidence flows to q in $\mathcal{BN}$, such that evidence cannot flow through that path to q in B. But by Lemma 1, below, this is not possible.

Lemma 1. *Let B be any grounding of any Bayes net returned with any derivation from a CLP($\mathcal{BN}$) program P. For every query to B, the paths through which evidence can flow are the same in B and in the full Bayes net $\mathcal{BN}$ defined by P.*

Proof. Suppose there exists a path through which evidence can flow in $\mathcal{BN}$ but not in B. Consider the shortest such path; call the query node q and call the evidence node e. The path must reach q through either a parent of q or a child of q in $\mathcal{BN}$. Consider both cases. *Case 1*: the path goes through a parent p of q in $\mathcal{BN}$. Note that p is a parent of q in B as well. Whether evidence flows through p in a linear or diverging connection in $\mathcal{BN}$, p cannot itself have evidence—otherwise, evidence could not flow through p in $\mathcal{BN}$. Then the path from e to p is a shorter path through which evidence flows in $\mathcal{BN}$ but not B, contradicting our assumption of the shortest path. *Case 2*: the path from e to q flows through some child c of q in $\mathcal{BN}$. Evidence must flow through c in either a linear or converging connection. If a linear connection, then c must not have evidence; otherwise, evidence could not flow through c to q in a linear connection. Then the path from e to c is a shorter path through which evidence flows in $\mathcal{BN}$ but not B, again contradicting our assumption of the shortest path. Therefore, evidence must flow through c in a converging connection in $\mathcal{BN}$. Hence either c or one of its descendants in $\mathcal{BN}$ must have evidence; call this additional evidence node n. Since n has evidence in the query, it must appear in B. Therefore its parents appear in B, and their parents, up to q. Because evidence can reach c from e in B (otherwise, we contradict our shortest path assumption again), and a descendant of c in B (possibly c itself) has evidence, evidence can flow through c to q in B.

---

[3] For simplicity of presentation, we assume queries of the form $Pr(q|E)$ where q is one variable in the Bayes net and the evidence E specifies the values of zero or more other variables in the Bayes net.

5   Non-determinism and Aggregates

One important property of relational databases is that they allow users to query for properties of sets of elements, or *aggregates*. Aggregates are also particularly important in the application of Probabilistic Relational Models, as they allow one to state that the value of a random variable depends on a set of elements that share some properties [14].

To clarify this concept, imagine that we run a private school, and we want to find out which courses are most attractive. To do so, we would want one extra attribute on the *Courses* table giving how popular the course is, as shown in Figure 7. Ideally, one would ask students who have attended the course and average the results. On the other hand, if we cannot obtain a representative sample, we can try to estimate popularity from the average of student satisfaction for that course.

Towards this goal, an extension to the Bayesian network is shown in Figure 8. We need an operator to aggregate on the set of satisfactions for a course, and then we can estimate the field's value from the aggregate.

Course	Prof	Difficulty	Popularity
c0	Bayes		
c2	Moivre		
c3	Bayes		

Reg	Student	Course	Grade	Satisf
r0	John	c0		
r1	Mary	c0		
r2	Mary	c2		
r3	John	c2		
r4	Mary	c3		

Student	Skill
John	
Mary	

Professor	Ability
Bayes	
Moivre	

Fig. 7. School Database Extended to Include a Field on Course Popularity

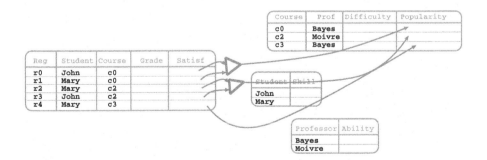

Fig. 8. School Database: Popularity is a random variable, and depends on the average of student satisfaction

CLP($\mathcal{BN}$) can deal with such aggregates in a straightforward way, by taking advantage of the second order features in Prolog, as shown in the next clause:

```
rating(C, Rat) :-
  setof(S,R^(registration(R,C),
            satisfaction(R,S)), Sats),
  average(Sats, Average),
  rating_table(Table),
  { Rat = rating(C) with prob([a,b,c,d],Table,[Average])
```

The call to `setof` obtains the satisfactions of all students registered in the course. The procedure `average/3` generates a the conditional probability of their average as a new random variable, *Average*. The course's rating, *Rat*, is assumed to be highly dependent on *Average*.

5.1   Building Aggregates

Aggregates are deterministic functions. Given n discrete random variables that range over k possible values, the aggregate value will take one well defined value. Hence, the probability of that value will be 1, and 0 for the remaining $k-1$ values. Writing the CPTs for aggregates should therefore be quite straightforward.

Unfortunately, aggregates create two problems. First, CPTs are most often represented as tables, where the size of the table grows exponentially with the number of dimensions. As the number of nodes n in the aggregate grows, table size grows exponentially. The current implementation of CLP($\mathcal{BN}$) uses *divorcing* [1]. The idea is to introduce hidden nodes, also called *mediating* variables, between a node and its parents, so that the total number of parents for every node can be guaranteed to never exceed a small number. The current system implements an aggregate node through a binary tree of mediating variables.

Figure 9 shows a fragment of an example network for an artificially generated School database with 4096 students. The query node is the gray node below. The gray node above is a evidence node for `course_rating`. The node is an aggregate of 68 `student_satisfaction` nodes, hence building the full table would require $3 * 3^{68}$ entries. Figure 9 shows the hierarchy of mediating nodes constructed by CLP($\mathcal{BN}$): note that the value of each node is deterministic on the ancestor nodes.

Figure 9 clearly shows the second problem we need to address in making CLP($\mathcal{BN}$) effective for real data-bases. The Bayes network shown here was created to answer a query on `course_difficulty`, shown as the gray node below. Given the original query, the algorithm searches for related evidence (shown as the other gray nodes). The knowledge-based model-construction algorithm searches parents recursively, eventually finding a number of nodes with evidence. Next, it needs to consider the Markov Blanket for these nodes, thus leading to searching other nodes. In this case, eventually almost every random variable in the database was included in the actual Bayes net (even though most of the nodes will have little relevancy to the original query).

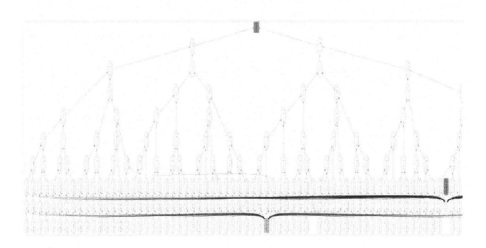

Fig. 9. A Bayesian Network Generated for an Ability Query. The Artificial network includes 4096 students, 128 professors, and 256 courses.

We observed that it is often the case that some nodes in a database are highly connected and take a central position in the graph. If evidence reaches these central nodes probabilistic inference will end up involving most of the network.

1. Exact inference is very expensive, and may not be possible at all.
2. Standard approximate inference such as Gibbs sampling may not always converge as often these networks include deterministic operations, such as **average** in current the example.

Processing such large networks effectively [15,16,17,18] and choosing the best strategy for different networks is one of the major challenges in the development of CLP($\mathcal{BN}$).

6   Recursion and Sequences

Recursion in Logic provides an elegant framework for modeling sequences of events, such as Markov Models. Next we discuss how the main ideas of CLP($\mathcal{BN}$) can be used to represent Hidden Markov Models (HMMs) [19], which are used for a number of applications ranging from Signal Processing, to Natural Language Processing, to Bioinformatics, and Dynamic Bayes Networks (DBNs). This was inspired by prior work on combining the advantages of multi-relational approaches with HMMs and DBNs: evaluation and learning of HMMs is part of PRISM [20,21], Dynamic Probabilistic Relational Models combine PRMs and DBNs [22], Logical HMMs have been used to model protein structure

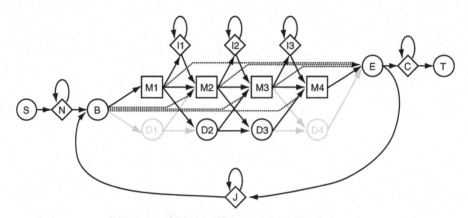

Fig. 10. Plan7 (From the HMMer Manual)

data [23,24]. More recently, non-directed models such as LogCRFs have also been proposed toward this goal [25].

Next, we discuss how to model HMMs and DBNs in CLP($\mathcal{BN}$). We present our experience in modeling profile-HMMs (pHMMs), an HMM structure widely used in Bioinformatics for homology detection between a sequence and a family of sequences. We chose pHMMs because they are extremely important in practice, and because they are not a trivial application. We focus on HMMer, an open-source tool that implements the Plan7 model, and which is one of the most widely used tools [26]. HMMer was used to build the well-known Pfam protein database [27].

HMMer is based on the Plan7 model, shown in Figure 10. The model describes a number of related sequences that share the same *profile*: a number of columns, each one corresponding to a well-preserved amino-acid. The example shows a relatively small profile, we can have profiles with hundreds of columns. A *match* state corresponds to an amino-acid in the sequence being a good match to the amino-acids found at the same position in the profile. *Insert* and *delete* states correspond to gaps: inserts are new material inserted in the sequence, and deletes removed material. There also three other character emitting-states: N, E, and J. The N states corresponds to material preceding the match, the E states to material after the match, and the J states allow several matches on the same sequence.

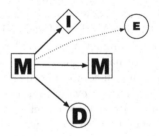

Fig. 11. M-State in Plan7

We model a pHMM by writing rules for each type of state. First, each state has two indexes: I refers to the current position in the sequence, and J to the column we are at. Second, whereas in a standard Bayesian Network we write how a variable depends on its parents, in an HMM we think in terms of *transitions* between *states*. As an example, consider the case of a typical M state, shown in Figure 11.

If we are at M state we can next move to an I state, (meaning a match is followed by a gap), to a D state, meaning the sequence will skip the next match state, or to the next M state. The model also makes it possible to jump directly to the end of the match. The CLP($\mathcal{BN}$) clause is as follows:

```
m_state(I,J,M) :-
    I1 is I+1,
    J1 is J+1,
    i_state(I1,J,NI),
    m_state(I1,J1,M1),
    d_state(I1,J1,ND),
    e_state(I1,NE),
    m_i_prob(J,MIP),
    m_m_prob(J,MMP),
    m_d_prob(J,MDP),
    m_e_prob(J,MEP),
    { M = m(I,J) with p([0,1],trans([MIP,MMP,MDP,MEP]),
                        [ NI, M1, ND, NE])) },
    emitting_state(m, I, J, M).
```

The M variable refers to the random variable for the current state. The rule is not very complex:

1. We can move from $M(I, J)$ to $I(I + 1, J)$, $M(I + 1, J + 1)$, $D(I + 1, J + 1)$, or $E(I + 1)$;
2. The transition probabilities at column I are $P_{M \to I} = MIP$, $P_{M \to M} = MMP$ $P_{M \to D} = MDP$, $P_{M \to E} = MEP$, such that

$$MIP + MMP + MDP + MEP = 1$$

3. M is a binary random variable with the given transition probabilities;
4. *trans* indicates we are setting up a constraint with transition probabilities; such constraints need specialized solvers, such as Viterbi or forward propagation.
5. `emitting_state/3`: if the state emits a symbol, access evidence for sequence element I.

Implementation. One can observe that HMMs are highly-recursive programs, and executing in the standard Prolog way would result in calling the same goal repeatedly over and over again. This problem can be addressed by *tabling* calls so that only the first one is actually executed, and repeated calls just need to lookup a data-base [28]. Tabled execution of these programs has the same complexity as standard dynamic programming algorithms. To the best of our knowledge, PRISM was the first language to use tabling for this task [20]. The CLP($\mathcal{BN}$) implementation originally relied on YAP's tabling mechanism [29]. Unfortunately, the YAP implementation of tabling is optimized for efficient evaluation of non-deterministic goals; we have achieved better performance through a simple program transformation.

Given this tabling mechanism, implementing algorithms such as Viterbi is just a simple walk over the constraint store.

Experiments. We tried this model with a number of different examples. The most interesting example was the *Globin* example from the standard HMMer distribution. The example matches a Plan7 model of the Globin family of proteins against an actual globin from Artemia. The Globin model has 159 columns, and the protein has 1452 amino-acids. The run generates 692 k states (random variables) and is about two orders of magnitude slower than the highly optimized C-code in HMMer. HMMer uses a much more compact and specialized representation than CLP($\mathcal{BN}$). Also, CLP($\mathcal{BN}$) actually creates the complete graph; in contrast, HMMer only needs to work with a column at a time. On the other hand, CLP($\mathcal{BN}$) has some important advantages: it provides a very clear model of the HMM, and it relatively straightforward to experiment and learn different structures.

7   Learning with CLP(BN)

We have performed some experiments on learning with CLP($\mathcal{BN}$). In both cases the goal is learn a model of a database as a CLP($\mathcal{BN}$) program.

The learning builds upon work performed for learning in Bayesian networks and in Inductive Logic Programming. We leverage on the Aleph ILP system. Essentially, we use Aleph to generate clauses which are then rewritten as CLP($\mathcal{BN}$) clauses. The rewriting process is straightforward for deterministic goals. If non-deterministic goal are allowed, we aggregate over the non-deterministic goals. We assume full data in these experiments, hence the parameters can be learned by maximum likelihood estimation. Next, we score the network with this new clause. Note that the score is used to control search in Aleph.

7.1   The School Database

We have so far used the school database as a way to explain some basic concepts in CLP($\mathcal{BN}$), relating them to PRMs. The school database also provides a good example of how to learn CLP($\mathcal{BN}$) programs.

First, we use an interpreter to generate a sample from the CLP($\mathcal{BN}$) program. The smallest database has 16 professors, 32 courses, 256 students and 882 registrations; the numbers roughly double in each successively larger database. We have no missing data. Can we, given this sample, relearn the original CLP($\mathcal{BN}$) program?

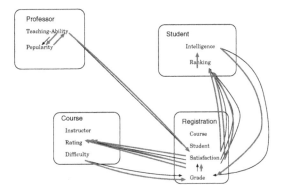

Fig. 12. Pictorial representation of the CLP(ℬ𝒩) clauses learned from the largest schools database, before removal of cycles

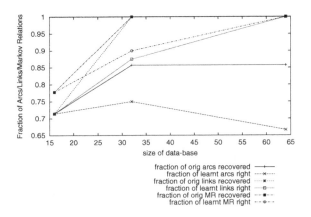

Fig. 13. Graph of results of CLP(ℬ𝒩)-learning on the three sizes of schools databases. Links are arcs with direction ignored. A Markov relation (MR) holds between two nodes if one is in the Markov blanket of the other.

From the ILP point of view, this is an instance of multi-predicate learning. To simplify the problem we assume each predicate would be defined by a single clause. We use the Bayesian Information Criterion (BIC) score to compare alternative clauses for the same predicate. Because ALEPH learns clauses independently, cycles may appear in the resulting CLP(ℬ𝒩) program. We therefore augment ALEPH with a post-processing algorithm that simplifies clauses until no cycles remain; the algorithm is greedy, choosing at each step the simplification that will least affect the BIC score of the entire program.

The following is one of the learned CLP(ℬ𝒩) clauses; to conserve space, we do not show the full conditional probability table.

```
registration_grade(A,B) :-
    registration(A,C,D), course(C,E),
    course_difficulty(C,F), student_intelligence(D,G),
    {F = registration_grade(A,F,G) with
        p(['A','B','C','D'],...,[F,G] }.
```

Figure 12 illustrates, as a PRM-style graph, the full set of clauses learned for the largest of the databases before simplification; this would be the best network according to BIC, if not for the cycles. Figure 13 plots various natural measures of the match between the learned program *after cycles have been removed* and the original program, as the size of the database increases. By the time we get to the largest of the databases, the only measures of match that do not have a perfect score are those that deal with the directions of arcs.

7.2   EachMovie

Next, we experiment our learning algorithm on the EachMovie data-set. This data-set includes three tables: there is data on 1628 movies, including movie type, store-info, and a link to the IMDB database. There is data on 72000 people who voted on the movies. Input was voluntary, and may include age, gender and ZIP code. From ZIP code it is possible to estimate geographical location and to get a good approximation of average income. Lastly, there are 2.8 million votes. Votes can be organized by class and range from 0 to 5. Our task is to predict how every non-key column in the database depends on the other non-key fields. That is we try to predict individual voting patterns, movie popularity, and people information. Given that there is a large amount of data, we use log-likelihood to score the network.

The data-set introduces a number of challenges. Firstly, there is missing data, especially in the people table. Following Domingos, we cannot assume that the individuals who refused to give their ZIP address or their age follow the same distribution as the ones who do [30]. Instead, we introduce an *unknown* evidence value, which says the individual refused to provide the information.

Aggregates are fundamental in these models because we often want to predict characteristics of groups of entities. In the School work we build aggregates *dynamically* during clause-construction by aggregating over non-deterministic goals. but doing so is just too expensive for this larger database. In this work, we use pre-computed aggregates:

– For each person, we compute how many votes and average score.
– For each movie, we compute how many people voted on this movie and average score.

A first result on the full data-set is shown in Figure 14. As for the school data-base, predicates are defined by a single clause. Learning proceeded greedily in this experiment: we first learn the predicate that best improves global log-likelihood. Next, we use this predicate plus the database to learn the other predicates. The process repeats until every predicate has been learned.

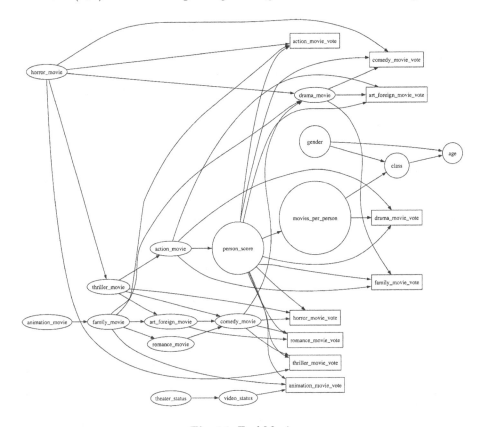

Fig. 14. EachMovie

Figure 14 was generated using the dot program. To the left it shows connections between movies of the different types (e.g., being an animation movie affects whether you are a family movie). The center/left of the network is about people. The system inferred that the average person score affects the number of movies seen per individual, and this in turn affects class. Last, the network includes voting patterns for movies. As an example, votes on family movies seem to depend on whether it is an action movie also, on whether it is also a drama, and on the person's average vote.

8   Relationship to PRMs

Clearly from the preceding discussion the CLP(BN) representation owes an intellectual debt to PRMs. As the reader might suspect at this point, any PRM can be represented as a CLP(BN) program. In this section we present an algorithm to convert any PRM into a CLP(BN) program. But before that, we address the natural question, "given that we already have PRMs, of what possible utility is the CLP(BN) representation?"

First, there has been much work on incorporating probabilities into first-order logic (see Section 9). Hence while there is great interest in the relationship between PRMs and these various probabilistic logics [31,32,33,34], this relationship is difficult to characterize. Approaches such as CLP($\mathcal{BN}$) and BLPs are closely related to PRMs, and can help us to better understand the relationship between PRMs and various probabilistic logics. Second, because CLP($\mathcal{BN}$)s are an extension of logic programming, they permit recursion and the use of function symbols, e.g., to construct data structures such as lists or trees. This expressivity may be useful for a variety of probabilistic applications and is not available in PRMs. Of course we must note that the uses of recursion and recursive data structures are not unlimited. CLP($\mathcal{BN}$)s disallow resolution steps that introduce a cycle into a Bayes net constraint. Third, and most importantly from the authors' viewpoint, the CLP($\mathcal{BN}$) representation is amenable to learning using techniques from inductive logic programming (ILP). Hence CLP($\mathcal{BN}$)s provide a way of studying the incorporation of probabilistic methods into ILP, and they may well give insight into novel learning algorithms for PRMs. The methods of learning in PRMs [3] are based upon Bayes net structure learning algorithms and hence are very different from ILP algorithms. The CLP($\mathcal{BN}$) representation provides a bridge through which useful ideas from ILP might be transferred to PRMs.

The remainder of this section presents an algorithm to convert any PRM into a CLP($\mathcal{BN}$) program. Because of space limits, we necessarily assume the reader already is familiar with the terminology of PRMs.

We begin by representing the skeleton of the PRM, i.e., the database itself with (possibly) missing values. For each relational table R of n fields, one field of which is the key, we define $n-1$ binary predicates $r_2, ..., r_n$. Without loss of generality, we assume the first field is the key. For each tuple or record $\langle t_1, ..., t_n \rangle$ our CLP($\mathcal{BN}$) program will contain the fact $r_i(t_1, t_i)$ for all $2 \leq i \leq n$. If t_i is a missing value in the database, then the corresponding fact in the CLP($\mathcal{BN}$) program is $r_i(t_1, skr_i(t_1))$, where skr_i is a Skolem function symbol. It remains to represent the Bayes net structure over this skeleton and the CPTs for this structure.

For each field in the database, we construct a clause that represents the parents and the CPT for that field within the PRM. The head (consequent) of the clause has the form $r_i(Key, Field)$, where the field is the ith field of relational table R, and Key and $Field$ are variables. The body of the clause is constructed in three stages, discussed in the following three paragraphs: the relational stage, the aggregation stage, and the CPT stage.

The relational stage involves generating a translation into logic of each slot-chain leading to a parent of the given field within the PRM. Recall that each step in a slot chain takes us from the key field of a relational table R to another field, i, in that table, or vice-versa. Each such step is translated simply to the literal $r_i(X, Y)$, where X is a variable that represents the key of R and Y is a variable that represents field i of R, regardless of directionality. If the next step in the slot chain uses field i of table R, then we re-use the variable Y; if the next step instead uses the key of table R then we instead re-use variable X. Suppose field i is the foreign key of another table S, and the slot chain next takes us to field j of S. Then the slot chain is translated as $r_i(X, Y), s_j(Y, Z)$. We can use the same translation to move from field j of S to the key of R, although we would re-order the literals for efficiency. For example, suppose we are given a student key *StudentKey* and want to follow the slot chain through registration and course to find the teaching abilities of the student's professor(s). Assuming that the course key is the second field in the registration table and the student key is the third field, while the professor key is the second field of the course table, and ability is the second field of the professor table, the translation is as below. Note that we use the first literal to take us from *StudentKey* to *RegKey*, while we use the second literal to take us from *RegKey* to *CourseKey*.

$registration_3(RegKey, StudentKey)$,
$registration_2(RegKey, CourseKey)$,
$course_2(CourseKey, ProfKey)$,
$professor_2(ProfKey, Ability)$

In the preceding example, the variable *Ability* may take several different bindings. If this variable is a parent of a field, then the PRM will specify an aggregation function over this variable, such as *mean*. Any such aggregation function can be encoded in a CLP($\mathcal{BN}$) program by a predicate definition, as in ordinary logic programming, i.e. in Prolog. We can collect all bindings for *Ability* into a list using the Prolog built-in function *findall* or *bagof*, and then aggregate this list using the appropriate aggregation function such as *mean*. For the preceding example, we would use the following pair of literals to bind the variable X to the mean of the abilities of the student's professors.

$findall(Ability,$ $(registration_2(RegKey, CourseKey)$,
                  $course_2(CourseKey, ProfKey)$,
                  $professor_2(ProfKey, Ability)$, $L)$,
$mean(L, X)$

At this point, we have constructed a clause body that will compute binding for all the variables that correspond to parents of the field in question. It remains only to add a literal that encodes the CPT for this field given these parents.

9   Other Related Work

The key idea in CLP($\mathcal{BN}$)s is that they provide joint probability distributions over the variables in the answer to a query, i.e., in a single proof. Hence it is not necessary to reconcile various probabilities obtained from different clauses or through different proofs. We combine information using aggregation (see Section 5), and the predicates for aggregation are part of the CLP($\mathcal{BN}$) program. This contrasts with the approach taken in both [35] and [7] where a combining rule is added on top of the logical representation.

CLP($\mathcal{BN}$) implements *Knowledge-based model construction (KBMC)* in that it uses logic "as a basis for generating Bayesian networks tailored to particular problem instances" [11]. However, in contrast to many KBMC approaches

[11,36], a probability in a CLP($\mathcal{BN}$) program does not give the probability that some first-order rule is *true*. Instead it is a (soft) constraint on possible instantiations of a variable in a rule. This also distinguishes it from the work in [4,37]. In these approaches instead of ground atomic formulas (*atoms*) being true or false as in normal logic programming semantics, they are true with a certain probability. In PRISM programs [4] a *basic distribution* gives the probabilities for the msw ground atoms mentioned in the PRISM program; this is then extended to define probabilities for all atoms which can be derived using rules in the program. In contrast, in a Bayesian logic program (BLP) [7] the distribution associated with a ground atom is unrestricted; it need not be always be over the two values {true, false}. In this respect BLPs are closer to CLP($\mathcal{BN}$) than, say, PRISM programs. The central difference is that BLPs represent random variables with ground atoms—in CLP($\mathcal{BN}$) they are represented by (Bayesian) variables.

In Angelopoulos's probabilistic finite domain *Pfd* model [38] hard constraints between variables and probability distributions over the same variables are kept deliberately separate, thereby allowing a normal CLP constraint solver to find variable instantiations permitted by the hard constraints. However, in addition to normal CLP, each such instantiation is returned with its probability. The main difference to our work is that we do not put hard constraints on Bayesian variables. Also CLP($\mathcal{BN}$) exploits conditional independence to permit efficient inference, whereas currently computation within *Pfd* is exponential in the number of variables involved.

10   Conclusions and Future Work

We have presented CLP($\mathcal{BN}$), a novel approach to integrating probabilistic information in logic programs. Our approach is based on the key idea that constraints can be used to represent information on undefined variables. Logical inference is used to define a Bayesian network that can be processed by a Bayesian solver. CLP($\mathcal{BN}$)s are closely related to PRMs, but they permit recursion, the use of functor symbols, and the representation is amenable to learning using techniques from inductive logic programming. Our first implementation of CLP($\mathcal{BN}$) system used Yap as the underlying Prolog system and the Kevin Murphy's Bayesian Network Toolbox as the Bayesian solver [39]. This allowed flexibility in choosing different engines. The newer versions include specialized solvers written in Prolog. The solvers implement variable elimination, Gibbs sampling, and Viterbi. We have successfully experimented the system with both database style and recursive programs.

The main focus of our future work will be in learning with CLP($\mathcal{BN}$) programs. Namely, we are now working with CLP($\mathcal{BN}$) on inducing regulatory networks [40,41]. We are also looking forward at integrating CLP($\mathcal{BN}$) with some of recent work in generating statistical classifiers [42,43,44,45,46]. Last, it would be interesting to study whether the ideas of CLP($\mathcal{BN}$) also apply to undirected models [47]. We are also considering directions to improve CLP(($\mathcal{BN}$). Regarding implementation, most effort will focus on tabling [28,29] that avoids repeated invocation of the same literal and can be quite useful in improving performance of logic programs, namely for database applications. As we have seen, CLP($\mathcal{BN}$) will directly benefit from this work. At present a CLP($\mathcal{BN}$) program generates a query-specific BN, and then standard BN algorithms (e.g. junction tree propagation) are used to compute the desired probabilities. Given the well-known connections between constraint processing and probabilistic computations as given by Dechter [12] it would be interesting to bring the probabilistic computations inside CLP($\mathcal{BN}$).

In common with many logical-probabilistic models [7,36,4,37], CLP($\mathcal{BN}$) exploits its logical framework to quantify over random variables, thereby facilitating the definition of large and complex BNs. An alternative approach, not explicitly based on first-order logic, is the BUGS language [48]. BUGS programs permit Bayesian statistical inference by defining large BNs with one (instantiated) node for each data point. It would be interesting to see if CLP($\mathcal{BN}$) could also be used for such statistical inference, particularly since CLP($\mathcal{BN}$), unlike BUGS, allows recursion.

Acknowledgments

This work was supported by the NSF Colleague project and by the DARPA EELD program. We gratefully acknowledge discussions with Jesse Davis, Inês Dutra, Peter Haddawy, Kristian Kersting, and Irene Ong. Maleeha Qazi participated in the development of learning algorithms for CLP($\mathcal{BN}$) and obtained the learning results for the Artificial School Database. V. Santos Costa was partially supported by CNPq and Fundação para a Ciência e Tecnologia.

References

1. Jensen, F.V.: Bayesian Networks and Decision Graphs. Springer, Heidelberg (2001)
2. Russel, S., Norvig, P.: Artificial intelligence (1996)
3. Getoor, L., Friedman, N., Koller, D., Pfeffer, A.: Learning probabilistic relational models. In: Dzeroski, S., Lavrac, N. (eds.) Relational Data Mining, pp. 307–335. Springer, Berlin (2001)
4. Sato, T., Kameya, Y.: Parameter learning of logic programs for symbolic-statistical modeling. Journal of Artificial Intelligence Research 15, 391–454 (2001)
5. Ngo, L., Haddawy, P.: Probabilistic logic programming and bayesian networks. In: Algorithms, Concurrency and Knowledge, pp. 286–300. Springer, Heidelberg (1995)
6. Muggleton, S.: Stochastic logic programs. In: De Raedt, L. (ed.) Advances in Inductive Logic Programming. Frontiers in Artificial Intelligence and Applications, vol. 32, pp. 254–264. IOS Press, Amsterdam (1996)
7. Kersting, K., De Raedt, L.: Bayesian logic programs. Technical Report 151, Institute for Computer Science, University of Freiburg, Germany (2001)
8. Srinivasan, A.: The Aleph Manual (2001)
9. Blockeel, H.: Prolog for Bayesian networks: A meta-interpreter approach. In: Proceedings of the 2nd International Workshop on Multi-Relational Data Mining (MRDM-2003), pp. 1–13 (2003),
http://www.cs.kuleuven.ac.be/cgi-bin-dtai/publ_info.pl?id=40881

10. Pearl, J.: Probabilistic Reasoning in Intelligent Systems: Networks of Plausible Inference. Morgan Kaufmann Publishers Inc., San Francisco (1988)
11. Koller, D., Pfeffer, A.: Learning probabilities for noisy first-order rules. In: IJCAI 1997, Nagoya, Japan (1997)
12. Dechter, R.: Bucket elimination: A unifying framework for reasoning. Artificial Intelligence 113, 41–85 (1999)
13. Page, C.D.: Anti-unification in constraint logics. PhD thesis, University of Illinois at Urbana-Champaign, UIUCDCS-R-93-1820 (1993)
14. Getoor, L., Friedman, N., Koller, D., Pfeffer, A.: Learning probabilistic relational models. In: Relational Data Mining, pp. 307–335. Springer, Heidelberg (2001)
15. Jaakkola, T., Jordan, M.I.: Variational probabilistic inference and the QMR-DT network. Journal of Artificial Intelligence Research 10, 291–322 (1999)
16. Choi, A., Darwiche, A.: A variational approach for approximating Bayesian networks by edge deletion. In: Proceedings of the 22nd Conference on Uncertainty in Artificial Intelligence (UAI), pp. 80–89 (2006)
17. Poon, H., Domingos, P.: Sound and efficient inference with probabilistic and deterministic dependencies. In: Proceedings, The Twenty-First National Conference on Artificial Intelligence and the Eighteenth Innovative Applications of Artificial Intelligence Conference, Boston, Massachusetts, USA, July 16-20, 2006, AAAI Press, Menlo Park (2006)
18. Chavira, M., Darwiche, A.: Compiling Bayesian networks using variable elimination. In: Proceedings of the 20th International Joint Conference on Artificial Intelligence (IJCAI), pp. 2443–2449 (2007)
19. Rabiner, L.R.: A tutorial on hidden Markov models and selected apllications in speech recognition. In: Waibel, A., Lee, K.-F. (eds.) Readings in Speech Recognition, pp. 267–296. Kaufmann, San Mateo (1990)
20. Sato, T., Kameya, Y., Zhou, N.-F.: Generative modeling with failure in PRISM. In: IJCAI 2005, Proceedings of the Nineteenth International Joint Conference on Artificial Intelligence, Edinburgh, Scotland, UK, July 30-August 5, 2005, pp. 847–852 (2005)
21. Sato, T., Kameya, Y.: New Advances in Logic-Based Probabilistic Modeling. In: Probabilistic Inductive Logic Programming, Springer, Heidelberg (2007)
22. Sanghai, S., Domingos, P., Weld, D.S.: Dynamic probabilistic relational models. In: IJCAI 2003, Proceedings of the Eighteenth International Joint Conference on Artificial Intelligence, Acapulco, Mexico, August 9-15, 2003, pp. 992–1002 (2003)
23. Kersting, K., Raiko, T., Kramer, S., De Raedt, L.: Towards Discovering Structural Signatures of Protein Folds based on Logical Hidden Markov Models. In: Proceedings of the Pacific Symposium on Biocomputing (PSB 2003), Kauai, Hawaii, pp. 3–7 (2003)
24. Kersting, K., de Raedt, L., Raiko, T.: Logical Hidden Markov Models. Journal of Artificial Intelligence Research 25, 425–456 (2006)
25. Kersting, K., de Raedt, L., Gutmann, B., Karwath, A., Landwehr, N.: Relational Sequence Learning. In: Probabilistic Inductive Logic Programming, Springer, Heidelberg (2007)
26. Eddy, S.: Profile hidden Markov models. Bioinformatics 14, 755–763 (1998)
27. Bateman, A., Coin, L., Durbin, R., Finn, R., Hollich, V., Griffiths, S., Khanna, A., Marshall, M., Moxon, S., Sonnhammer, E., Studholme, D., Yeats, C., Eddy, S.: The pfam protein families database. Nucleic Acids Research 32, 138–141 (2004)
28. Ramakrishnan, I.V., et al.: Efficient Tabling Mechanisms for Logic Programs. In: 12th ICLP, Tokyo, Japan, pp. 687–711 (1995)

29. Rocha, R., Silva, F., Santos Costa, V.: On Applying Or-Parallelism and Tabling to Logic Programs. Theory and Practice of Logic Programming Systems 5(1–2), 161–205 (2005)

30. Domingos, P.: Personal communication (December 2002)

31. De Raedt, L., Kersting, K.: Probabilistic inductive logic programming. In: Ben-David, S., Case, J., Maruoka, A. (eds.) ALT 2004. LNCS (LNAI), vol. 3244, pp. 19–36. Springer, Heidelberg (2004)

32. Muggleton, S.: Learning structure and parameters of stochastic logic programs. In: Matwin, S., Sammut, C. (eds.) ILP 2002. LNCS (LNAI), vol. 2583, pp. 198–206. Springer, Heidelberg (2003)

33. Cussens, J.: Logic-based Formalisms for Statistical Relational Learning. In: Introduction to Statistical Relational Learning, MIT Press, Cambridge (2007)

34. de Raedt, L., Kersting, K.: Introduction. In: Probabilistic Inductive Logic Programming, Springer, Heidelberg (2007)

35. Clark, K.L., McCabe, F.G.: PROLOG: A language for implementing expert systems. Machine Intelligence 10, 455–470 (1982)

36. Haddawy, P.: An overview of some recent developments in Bayesian problem solving techniques. AI Magazine (1999)

37. Poole, D.: Probabilistic Horn abduction and Bayesian networks. Artificial Intelligence 64, 81–129 (1993)

38. Angelopoulos, N.: Probabilistic Finite Domains. PhD thesis, Dept of CS, City University, London (2001)

39. Murphy, K.P.: The Bayes Net Toolbox for Matlab. Computing Science and Statistics (2001)

40. Ong, I.M., Glasner, J.D., Page, D.: Modelling regulatory pathways in e. coli from time series expression profiles. In: Proceedings of the Tenth International Conference on Intelligent Systems for Molecular Biology, Edmondon, Alberta, Canada, August 3-7, 2002, pp. 241–248 (2002)

41. Ong, I.M., Topper, S.E., Page, D., Santos Costa, V.: Inferring regulatory networks from time series expression data and relational data via inductive logic programming. In: Proceedings of the Sixteenth International Conference on Inductive Logic Programming, Santiago de Compostela, Spain (2007)

42. Davis, J., Burnside, E.S., Dutra, I., Page, D., Ramakrishnan, R., Santos Costa, V., Shavlik, J.W.: View learning for statistical relational learning: With an application to mammography. In: Kaelbling, L.P., Saffiotti, A. (eds.) IJCAI 2005, Proceedings of the Nineteenth International Joint Conference on Artificial Intelligence, Edinburgh, Scotland, UK, July 30-August 5, 2005, pp. 677–683. Professional Book Center (2005)

43. Landwehr, N., Kersting, K., De Raedt, L.: nFOIL: Integrating naïve Bayes and FOIL. In: Veloso, M.M., Kambhampati, S. (eds.) Proceedings, The Twentieth National Conference on Artificial Intelligence and the Seventeenth Innovative Applications of Artificial Intelligence Conference, Pittsburgh, Pennsylvania, USA, July 9-13, 2005, pp. 795–800 (2005)

44. Davis, J., Burnside, E.S., de Castro Dutra, I., Page, D., Santos Costa, V.: An integrated approach to learning bayesian networks of rules. In: Gama, J., Camacho, R., Brazdil, P.B., Jorge, A.M., Torgo, L. (eds.) ECML 2005. LNCS (LNAI), vol. 3720, pp. 84–95. Springer, Heidelberg (2005)

45. De Raedt, L., Kimmig, A., Toivonen, H.: ProbLog: A Probabilistic Prolog and Its Application in Link Discovery. In: Proceedings of the 20th International Joint Conference on Artificial Intelligence (IJCAI) (2007)

46. Davis, J., Ong, I., Struyf, J., Burnside, E., Page, D., Santos Costa, V.: Change of Representation for Statistical Relational Learning. In: Proceedings of the 20th International Joint Conference on Artificial Intelligence (IJCAI) (2007)
47. Richardson, M., Domingos, P.: Markov logic networks. Machine Learning 62, 107–136 (2006)
48. Spiegelhalter, D., Thomas, A., Best, N., Gilks, W.: BUGS 0.5 Bayesian inference using Gibbs Sampling Manual. MRC Biostatistics Unit, Cambridge (1996)

Basic Principles of Learning Bayesian Logic Programs*

Kristian Kersting[1] and Luc De Raedt[2]

[1] CSAIL, Massachusetts Institute of Technologie,
32 Vassar Street, Cambridge, MA 02139-4307, USA
kersting@csail.mit.edu
[2] Departement Computerwetenschappen, K.U. Leuven,
Celestijnenlaan 200A - bus 2402, B-3001 Heverlee, Belgium
Luc.DeRaedt@cs.kuleuven.be

Abstract. Bayesian logic programs tightly integrate definite logic programs with Bayesian networks in order to incorporate the notions of objects and relations into Bayesian networks. They establish a one-to-one mapping between ground atoms and random variables, and between the *immediate consequence* operator and the *directly influenced by* relation. In doing so, they nicely separate the qualitative (i.e. logical) component from the quantitative (i.e. the probabilistic) one providing a natural framework to describe general, probabilistic dependencies among sets of random variables. In this chapter, we present results on combining Inductive Logic Programming with Bayesian networks to learn both the qualitative and the quantitative components of Bayesian logic programs from data. More precisely, we show how the qualitative components can be learned by combining the inductive logic programming setting learning from interpretations with score-based techniques for learning Bayesian networks. The estimation of the quantitative components is reduced to the corresponding problem of (dynamic) Bayesian networks.

1 Introduction

In recent years, there has been an increasing interest in integrating probability theory with first order logic. One of the research streams [42,40,24,19,29] concentrates on first order extensions of Bayesian networks [41]. The reason why this has attracted attention is, that even though Bayesian networks are one of the most important, efficient and elegant frameworks for representing and reasoning with probabilistic models, they suffer from an inherently propositional character. A single Bayesian network specifies a joint probability density over a finite set of random variables and consists of two components:

---

* The is a slightly modified version of *Basic Principles of Learning Bayesian Logic Programs*, Technical Report No. 174, Institute for Computer Science, University of Freiburg, Germany, June 2002. The major change is an improved section on parameter estimation. For historical reasons, all other parts are left unchanged (next to minor editorial changes).

L. De Raedt et al. (Eds.): Probabilistic ILP 2007, LNAI 4911, pp. 189–221, 2008.

- a *qualitative* one that encodes the local influences among the random variables using a directed acyclic graph, and
- a *quantitative* one that encodes the probability densities over these local influences.

Imagine a Bayesian network modelling the localization of genes/proteins. Every gene would be a single random variable. There is no way of formulating general probabilistic regularities among the localizations of the genes such as *the protein P encoded by gene G has localization L if P interacts with another protein P' that has localization L.*

Bayesian logic programs are a language that overcomes this propositional character by tightly integrating definite logic programs with Bayesian networks to incorporate the notions of objects and relations. In doing so, they can naturally be used to do first order classification, clustering, and regression. Their underlying idea is to establish a one-to-one mapping between ground atoms and random variables, and between the *immediate consequence operator* and the *directly influences by* relation. In doing so, they nicely separate the qualitative (i.e. logical) component from the quantitative (i.e. the probabilistic) one providing a natural framework to describe general, probabilistic dependencies among sets of random variables such as the rule stated above.

It is, however, well-known that determining the structure of a Bayesian network, and therefore also of a Bayesian logic program, can be difficult and expensive. In 1997, Koller and Pfeffer [33] addressed the question *"where do the numbers come from?"* for similar frameworks. So far, this issue has not yet attracted much attention in the context of first order extensions of Bayesian networks (with the exception of [33,19]). In this context, we present for the first time how to calculate the *gradient* for a maximum likelihood estimation of the parameters of Bayesian logic programs. Together with the EM algorithm which we will present, this gives one a rich class of optimization techniques such as conjugate gradient and the possibility to speed up the EM algorithm, see e.g. [38].

Moreover, Koller and Pfeffer [33] rose the question whether techniques from inductive logic programming (ILP) could help to learn the logical component of first order probabilistic models. In [30], we suggested that the ILP setting *learning from interpretations* [13,14,6] is a good candidate for investigating this question. In this chapter we would like to make our suggestions more concrete. We present a novel scheme to learn intensional clauses within Bayesian logic programs [28,29]. It combines techniques from inductive logic programming with techniques for learning Bayesian networks. More precisely, we will show that *learning from interpretations* can indeed be integrated with score-based Bayesian network learning techniques in order to learn Bayesian logic programs. Thus, we answer Koller and Pfeffer's question affirmatively.

We proceed as follows. After briefly reviewing the framework of Bayesian logic programs in Section 2, we define the learning problem in Section 3. Based on this, we then present a score-based greedy algorithm called SCOOBY solving the learning problem. More precisely, Section 4 presents SCOOBY first in the context of a special class of propositional Bayesian logic programs, i.e. Bayesian networks, and then on general Bayesian logic programs. In Section 5, we formulate the likelihood of the parameters of a Bayesian logic program given some data and, based on this, we present a gradient-based and an EM method to find that parameters which maximize the likelihood. Section 6 reports on first experiments. Before concluding the paper, we touch upon related work.

We assume some familiarity with logic programming or Prolog (see e.g. [45,37]) as well as with Bayesian networks (see e.g. [41,10,27]).

2   Bayesian Logic Programs

Throughout the paper we will use an example from genetics which is inspired by Friedman et al. [19]: "it is a genetic model of the inheritance of a single gene that determines a person's X blood type bt(X). Each person X has two copies of the chromosome containing this gene, one, mc(Y), inherited from her mother m(Y,X), and one, pc(Z), inherited from her father f(Z,X)." We will use **P** to denote a probability distribution, e.g. $\mathbf{P}(x)$, and the normal letter P to denote a probability value, e.g. $P(x = v)$, where v is a state of x.

2.1   Representation Language

The basic idea underlying our framework is that each Bayesian logic program specifies a (possibly infinite) Bayesian network, with one node for each (Bayesian) ground atom (see below). A Bayesian logic program B consist of two components:

- a *qualitative* or *logical* one, a set of Bayesian clauses (cf. below), and
- a *quantitative* one, a set of conditional probability distributions and combining rules (cf. below) corresponding to that logical structure.

Definition 1 (*Bayesian Clause*). *A* Bayesian (definite) clause c is an expression of the form

$$A \mid A_1, \ldots, A_n$$

where $n \geq 0$, the $A, A_1, \ldots, A_n$ are Bayesian atoms and all Bayesian atoms are (implicitly) universally quantified. We define $head(c) = A$ and $body(c) = \{A_1, \ldots, A_n\}$.

So, the differences between a *Bayesian clause* and a *logical* one are:

1. The atoms $p(t_1, ..., t_n)$ and predicates p arising are Bayesian, which means that they have an associated (finite) domain[1] $\mathbf{S}(p)$, and
2. We use " $\mid$ " instead of ":-".

For instance, consider the Bayesian clause c

    bt(X) | mc(X), pc(X).

where $\mathbf{S}(bt) = \{a, b, ab, 0\}$ and $\mathbf{S}(mc) = \mathbf{S}(pc) = \{a, b, 0\}$. It says that the blood type of a person X depends on the inherited genetical information of X. Note

---

[1] For the sake of simplicity we consider finite random variables, i.e. random variables having a finite set $\mathbf{S}$ of states. However, the ideas generalize to discrete and continuous random variables.

that the domain $\mathbf{S}(p)$ has nothing to do with the notion of a domain in the logical sense. The domain $\mathbf{S}(p)$ defines the states of random variables. Intuitively, a Bayesian predicate p generically represents a set of (finite) random variables. More precisely, each Bayesian ground atom g over p represents a (finite) random variable over the states $\mathbf{S}(g) := \mathbf{S}(p)$. E.g. $bt(ann)$ represents the blood type of a person named Ann as a random variable over the states $\{a, b, ab, 0\}$. Apart from that, most other *logical* notions carry over to Bayesian logic programs. So, we will speak of Bayesian predicates, terms, constants, substitutions, ground Bayesian clauses, Bayesian Herbrand interpretations etc. We will assume that all Bayesian clauses are range-restricted. A clause is *range-restricted* iff all variables occurring in the head also occur in the body. Range restriction is often imposed in the database literature; it allows one to avoid derivation of non-ground true facts.

In order to represent a probabilistic model we associate with each Bayesian clause c a conditional probability distribution $cpd(c)$ encoding $\mathbf{P}(head(c) \mid body(c))$. To keep the expositions simple, we will assume that $cpd(c)$ is represented as table, see Figure 1. More elaborate representations like decision trees or rules are also possible. The distribution $cpd(c)$ generically represents the conditional probability distributions of all ground instances $c\theta$ of the clause c. In general, one may have many clauses, e.g. clauses c_1 and the c_2

```
bt(X) | mc(X).
bt(X) | pc(X).
```

and corresponding substitutions θ_i that ground the clauses c_i such that $head(c_1\theta_1) = head(c_2\theta_2)$. They specify $cpd(c_1\theta_1)$ and $cpd(c_2\theta_2)$, but not the distribution required: $\mathbf{P}(head(c_1\theta_1) \mid body(c_1) \cup body(c_2))$. The standard solution to obtain the distribution required are so called *combining rules*.

Definition 2 (*Combining Rule*). *A combining rule is a functions which maps finite sets of conditional probability distributions* $\{\mathbf{P}(A \mid A_{i1}, \ldots, A_{in_i}) \mid i = 1, \ldots, m\}$ *onto one (combined) conditional probability distribution* $\mathbf{P}(A \mid B_1, \ldots, B_k)$ *with* $\{B_1, \ldots, B_k\} \subseteq \bigcup_{i=1}^{m}\{A_{i1}, \ldots, A_{in_i}\}$.

We assume that for each Bayesian predicate p there is a corresponding combining rule cr, such as *noisy or* (see e.g. [27]) or *average*. The latter assumes $n_1 = \ldots = n_m$ and $\mathbf{S}(A_{ij}) = \mathbf{S}(A_{kj})$, and computes the average of the distributions over $\mathbf{S}(A)$ for each joint state over $\bigotimes_j \mathbf{S}(A_{ij})$.

To summarize, we could define Bayesian logic program in the following way:

Definition 3 (*Bayesian Logic Program*). *A Bayesian logic program B consists of a (finite) set of Bayesian clauses. To each Bayesian clause c there is exactly one conditional probability distribution $cpd(c)$ associated, and for each Bayesian predicate p there is exactly one associated combining rule $cr(p)$.*

2.2   Declarative Semantics

The declarative semantics of Bayesian logic programs is given by the annotated *dependency graph*. The *dependency graph* $DG(B)$ is that directed graph whose nodes

```
m(ann,dorothy).
f(brian,dorothy).
pc(ann).
pc(brian).
mc(ann).
mc(brian).

mc(X) | m(Y,X),mc(Y),pc(Y).
pc(X) | f(Y,X),mc(Y),pc(Y).
bt(X) | mc(X),pc(X).
      (1)
```

$mc(X)$	$pc(X)$	$\mathbf{P}(bt(X))$
a	a	$(0.97, 0.01, 0.01, 0.01)$
b	a	$(0.01, 0.01, 0.97, 0.01)$
...	...	...
0	0	$(0.01, 0.01, 0.01, 0.97)$

(2)

Fig. 1. (1) The Bayesian logic program *bloodtype* encoding our genetic domain. To each Bayesian predicate, the identity is associated as combining rule. (2) A conditional probability distribution associated to the Bayesian clause `bt(X) | mc(X), pc(X)` represented as a table.

correspond to the ground atoms in the least Herbrand model LH(B) (cf. below). It encodes the *directly influenced by* relation over the random variables in LH(B):

> there is an edge from a node x to a node y if and only if there exists a clause $c \in B$ and a substitution θ, s.t. $y = head(c\theta)$, $x \in body(c\theta)$ and for all atoms z in $c\theta : z \in$ LH(B).

The direct predecessors of a graph node x are called its parents, $\mathbf{Pa}(x)$. The Herbrand base HB(B) is the set of all random variables we can talk about. It is defined as if B were a logic program (cf. [37]). The least Herbrand model LH(B) $\subseteq$ HB(B) consists of all *relevant* random variables, the random variables over which a probability distribution is well-defined by B, as we will see. It is the least fix point of the *immediate consequence operator* applied on the empty interpretation. Therefore, a ground atom which is true in the logical sense corresponds to a relevant random variables. Now, to each node x in $DG(B)$ we associate the combined conditional probability distribution which is the result of the combining rule $cr(p)$ of the corresponding Bayesian predicate p applied to the set of $cpd(c\theta)$'s where $head(c\theta) = x$ and $\{x\} \cup body(c\theta) \subseteq$ LH(B). Thus, if $DG(B)$ is acyclic and not empty then it encodes a (possibly infinite) Bayesian network, because the least Herbrand model always exists and is unique. Therefore, the following independence assumption holds:

Independence Assumption 1. *Each node x is independent of its non-descendants given a joint state of its parents $\mathbf{Pa}(x)$ in the dependency graph.*

E.g. in the program in Figure 1, the random variable $bt(dorothy)$ is independent from $pc(brian)$ given a joint state of $pc(dorothy), mc(dorothy)$. Using this assumption the following proposition holds:

Proposition 1. *Let B be a Bayesian logic program. If*

1. LH(B) $\neq \emptyset$,
2. $DG(B)$ *is acyclic, and*
3. *each node in $DG(B)$ is influenced by a finite set of random variables*

then B specifies a unique probability distribution $\mathbf{P}_B$ over LH(B).

```
m(ann,dorothy).
f(brian,dorothy).
pc(ann).
pc(brian).
mc(ann).
mc(brian).
mc(dorothy) | m(ann, dorothy),mc(ann),pc(ann).
pc(dorothy) | f(brian, dorothy),mc(brian),pc(brian).
bt(ann)     | mc(ann), pc(ann).
bt(brian)   | mc(brian), pc(brian).
bt(dorothy) | mc(dorothy),pc(dorothy).
```

Fig. 2. The grounded version of the Bayesian logic program of Figure 1. It (directly) encodes a Bayesian network.

The proof of the proposition can be sketched as follows (for a detailed proof see [29]). The least Herbrand $LH(B)$ always exists, is unique and countable. Thus, $DG(B)$ exists and is unique, and due to condition (3) the combined probability distribution for each node of $DG(B)$ is computable. Furthermore, because of condition (1) a total order π on $DG(B)$ exists, so that one can see B together with π as a stochastic process over $LH(B)$. An induction "along" π together with condition 2 shows that the family of finite-dimensional distribution of the process is projective (cf. [2]), i.e the joint probability density over each finite subset $s \subseteq LH(B)$ is uniquely defined and $\int_y p(s, x = y)\, dy = p(s)$. Thus, the preconditions of *Kolmogorov's theorem* [2, page 307] hold, and it follows that B given π specifies a probability density function p over $LH(B)$. This proves the proposition because the total order π used for the induction is arbitrary.

A program B satisfying the conditions (1), (2) and (3) of proposition 1 is called *well-defined*. The program *bloodtype* in Figure 1 is an example of a well-defined Bayesian logic program. It encodes the regularities in our genetic example. Its grounded version, which is a Bayesian network, is shown in Figure 2. This illustrates that Bayesian networks [41] are well-defined propositional Bayesian logic programs. Each node-parents pair uniquely specifies a propositional Bayesian clause; we associate the identity as combining rule to each predicate; the conditional probability distributions are those of the Bayesian network.

Some interesting insights follow from the proof sketch. We interpreted a Bayesian logic program as a stochastic process. This places them in a wider context of what Cowell et al. call *highly structured stochastic systems* (HSSS), cf. [10], because Bayesian logic programs represent discrete-time stochastic processes in a more flexible manner. Well-known probabilistic frameworks such as dynamic Bayesian networks, first order hidden Markov models or Kalman filters are special cases of Bayesian logic programs. Moreover, the proof in [29] indicates the important *support network* concept. Support networks are a graphical representation of the finite-dimensional distribution, cf. [2], and are needed for the formulation of the likelihood function (see below) as well as for answering probabilistic queries in Bayesian logic programs.

Definition 4 (*Support Network*). *The* support network N *of a variable* $x \in$ LH(B) *is defined as the induced subnetwork of* $S = \{x\} \cup \{y \mid y \in$ LH(B) *and* y *is influencing* $x\}$. *The support network of a finite set* $\{x_1, \ldots, x_k\} \subseteq$ LH(B) *is the union of the networks of each single* x_i.

Because we consider well-defined Bayesian logic programs, each $x \in$ LH(B) is influenced by a finite subset of LH(B). So, the support network N of a finite set $\{x_1, \ldots, x_k\} \subseteq$ LH(B) of random variables is always a finite Bayesian network and computable in finite time. The distribution factorizes in the usual way, i.e. $\mathbf{P}_N(x_1 \ldots, x_n) = \prod_{i=1}^{n} \mathbf{P}_N(x_i \mid \mathbf{Pa}\, x_i)$, where $\{x_1 \ldots, x_n\} = S$, and $\mathbf{P}(x_i \mid \mathbf{Pa}\, x_i)$ is the combined conditional probability distribution associated to x_i. Because N models the finite-dimensional distribution specified by S, any interesting probability value over subsets of S is specified by N. For the proofs and an effective inference procedure (together with a Prolog implementation) we refer to [29].

3   The Learning Problem

So far, we have assumed that there is an expert who provides both the structure and the conditional probability distributions of the Bayesian logic program. This is not always easy. Often, there is no-one possessing necessary the expertise or knowledge. However, instead of an expert we may have access to data. In this section, we investigate and formally define the problem of learning Bayesian logic programs. While doing so, we exploit analogies with Bayesian network learning as well as with inductive logic programming.

3.1   Data Cases

In the last section, we have introduced Bayesian logic programs and argued that they contain two components, the quantitative (the combining rules and the conditional probability distributions) and the qualitative ones (the Bayesian clauses). Now, if we want to learn Bayesian logic programs, we need to employ data. Hence, we need to formally define the notions of a data case.

Let B be a Bayesian logic program consisting of the Bayesian clauses $c_1, \ldots, c_n$, and let $\mathbf{D} = \{D_1, \ldots, D_m\}$ be a set of data cases.

Definition 5 (*Data Case*). *A data case* $D_i \in \mathbf{D}$ *for a Bayesian logic program* B *consists of a*

Logical part: *Which is a Herbrand interpretation* $Var(D_i)$ *such that* $Var(D_i)$ = LH($B \cup Var(D_i)$), *and a*
Probabilistic part: *Which is a partially observed joint state of some variables, i.e. an assignment of values to some of the facts in* $Var(D_i)$.

Examples of data cases are

$$D_1 = \{m(\textit{cecily}, \textit{fred}) = \textit{true}, f(\textit{henry}, \textit{fred}) =?, pc(\textit{cecily}) = a,$$
$$pc(\textit{henry}) = b, pc(\textit{fred}) =?, mc(\textit{cecily}) = b, mc(\textit{henry}) = b,$$
$$mc(\textit{fred}) =?, bt(\textit{cecily}) = ab, bt(\textit{henry}) = b, bt(\textit{fred}) =?\},$$

$$D_2 = \{m(ann, dorothy) = true, f(brian, dorothy) = true, pc(ann) = b,$$
$$mc(ann) = ?, mc(brian) = a, mc(dorothy) = a,$$
$$pc(dorothy) = a, pc(brian) = ?, bt(ann) = ab, bt(brian) = ?,$$
$$bt(dorothy) = a\},$$

where '?' stands for an unobserved state. Notice that – for ease of writing – we merged the two components of a data case into one. Indeed, the *logical part* of a data case $D_i \in \mathbf{D}$, denoted as $Var(D_i)$, is a Herbrand interpretation, such as

$$Var(D_1) = \{m(cecily, fred), f(henry, fred), pc(cecily), pc(henry),$$
$$pc(fred), mc(cecily), mc(henry), mc(fred), bt(cecily),$$
$$bt(henry), bt(fred)\},$$
$$Var(D_2) = \{m(ann, dorothy), f(brian, dorothy), pc(ann),$$
$$mc(ann), mc(brian), mc(dorothy), pc(dorothy),$$
$$pc(brian), bt(ann), bt(brian), bt(dorothy)\},$$

satisfy this logical property w.r.t. the target Bayesian logic program B

```
mc(X)  | m(Y,X),mc(Y),pc(Y).
pc(X)  | f(Y,X),mc(Y),pc(Y).
bt(X)  | mc(X),pc(X).
```

Indeed, $Var(B \cup Var(D_i)) = Var(D_i)$ for all $D_i \in \mathbf{D}$.

So, the logical components of the data cases should be seen as the least Herbrand models of the target Bayesian logic program. They specify different sets of *relevant* random variables, depending on the given "extensional context". If we accept that the genetic laws are the same for both families then a learning algorithm should find regularities among the Herbrand interpretations that can be to compress the interpretations. The key assumption underlying any inductive technique is that the rules that are valid in one interpretation are likely to hold for any interpretation. This is exactly what the *learning from interpretations* in inductive logic programming [14,6] is doing. Thus, we will adapt this setting for learning the structure of the Bayesian logic program, cf. Section 4.

There is one further logical constraints to take into account while learning Bayesian logic programs. It is concerned with the acyclicity requirement (cf. property 2 in proposition 1) imposed on Bayesian logic programs. Thus, we require that for each $D_i \in \mathbf{D}$ the induced Bayesian network over $\mathrm{LH}(B \cup Var(D_i))$ has to be acyclic.

At this point, the reader should also observe that we require that the logical part of a data case is a *complete* model of the target Bayesian logic program and not a *partial* one[2]. This is motivated by 1) Bayesian network learning and 2) the problems with learning from partial models in inductive logic programming. First, data cases as they have been used in Bayesian network learning are the

---

[2] Partial models specify the truth-value (false or true) of *some* of the elements in the Herbrand Base.

propositional equivalent of the data cases that we introduced above. Indeed, if we have a Bayesian network B over the propositional Bayesian predicates $\{p_1, ..., p_k\}$ then $\mathrm{LH}(B) = \{p_1, ..., p_k\}$ and a data case would assign values to some of the predicates in B. This also shows that the second component of a data case is pretty standard in the Bayesian network literature. Second, it is well-known that learning from partial models is harder than learning from complete models (cf. [12]). More specifically, learning from partial models is akin to multiple predicate learning, which is a very hard problem in general. These two points also clarify why the semantics of the set of relevant random variables coincided with the least Herbrand domain and at the same time why we do not restrict the domain of Bayesian predicates to $\{true, false\}$.

Before we are able to fully specify the problem of learning Bayesian logic programs, let us introduce the hypothesis space and scoring functions.

3.2 The Hypothesis Space

The *hypothesis space* $\mathcal{H}$ explored consists of Bayesian logic programs, i.e. finite set of Bayesian clauses to which conditional probability distributions are associated. More formally, let $\mathcal{L}$ be the language, which determines the set $\mathcal{C}$ of clauses that can be part of a hypothesis. It is common to impose syntactic restrictions on the space $\mathcal{H}$ of hypotheses.

> *Language Assumption*: In this paper, we assume that the alphabet of $\mathcal{L}$ only contains constant and predicate symbols that occur in one of the data cases, and we restrict $\mathcal{C}$ to range-restricted, constant-free clauses containing maximum $k = 3$ atoms in the body. Furthermore, we assume that the combining rules associated to the Bayesian predicates are given.

E.g. given the data cases D_1 and D_2, $\mathcal{C}$ looks like

```
mc(X)   |  m(Y,X).
mc(X)   |  mc(X).
mc(X)   |  pc(X).
mc(X)   |  m(Y,X),mc(Y).
...
pc(X)   |  f(Y,X),mc(Y),pc(Y).
...
bt(X)   |  mc(X),pc(X).
```

Not every element $H \in \mathcal{H}$ has to be a candidate. The logical parts of the data cases constraint the set of possible candidates. To be a candidate, H has to be

- (logically) valid on the data, and
- acyclic on the data i.e. the induced Bayesian network over $\mathrm{LH}(H \cup Var(D_i))$ has to be acyclic.

E.g. given the data cases D_1 and D_2, the Bayesian clause

```
mc(X)   |  mc(X)
```

is not included in any candidate, because the Bayesian network induced over the data cases would be cyclic.

3.3 Scoring Function

So far, we mainly exploit the logical part of the data cases. The probabilistic part of the data cases are partially observed joint states. They induce a joint distribution over the random variables of the logical parts of the data cases. A candidate $H \in \mathcal{H}$ should reflect this distribution. We assume that there is a scoring function $score_{\mathbf{D}} : \mathcal{H} \mapsto \mathbb{R}$ which expresses how well a given candidate H fits the data $\mathbf{D}$. Examples of scoring functions are the likelihood (see Section 5) or the *minimum description length* score (which bases on the likelihood).

Putting all together, we can define the basic learning problem as follows:

Definition 6 (*Learning Problem*). **Given** *a set* $\mathbf{D} = \{D_1, \ldots, D_m\}$ *of data cases, a set* $\mathcal{H}$ *of sets of Bayesian clauses according to some language bias, and a scoring function* $score_{\mathbf{D}} : \mathcal{H} \mapsto \mathbb{R}$, **find** *a hypothesis* $H^* \in \mathcal{H}$ *such that for all* $D_i \in \mathbf{D} : \mathrm{LH}(H^* \cup Var(D_i)) = Var(D_i)$, H^* *is acyclic on the data, and* H^* *maximizes* $score_{\mathbf{D}}$.

As usual, we assume the all data cases are independently sampled from identical distributions. In the following section we will present an algorithm solving the learning problem.

4   Scooby: An Algorithm for Learning Intensional Bayesian Logic Programs

In this section, we introduce SCOOBY (structural learning of intensional Bayesian logic programs), cf. Algorithm 1. Roughly speaking, SCOOBY performs a heuristic search using traditional inductive logic programming refinement operators on clauses. The hypothesis currently under consideration is evaluated using some score as heuristic. The hypothesis that scores best is selected as the final hypothesis.

First, we will illustrate how SCOOBY works for the special case of Bayesian networks. As it will turn out, SCOOBY coincides in this case with well-known and effective score-based techniques for learning Bayesian networks [22]. Then, we will show that SCOOBY works for first-order Bayesian logic programs, too. For the sake of readability, we assume the existence of a method to compute the parameters maximizing the score given a candidate and data cases. Methods to do this will be discussed in Section 5. They assume that the combining rules are decomposable, a concept which we will introduce below. Furthermore we will discuss the basic framework only; extensions are possible.

4.1   The Propositional Case: Bayesian Networks

Let us first explain how SCOOBY works on Bayesian networks. and show that well-known score-based methods for structural learning of Bayesian networks are special cases of SCOOBY.

Let $\mathbf{x} = \{x_1, \ldots, x_n\}$ be a fixed set of random variables. The set $\mathbf{x}$ corresponds to a least Herbrand model of an unknown propositional Bayesian logic program

Algorithm 1. A simplified skeleton of a greedy algorithm for structural learning of intensional Bayesian logic programs (Scooby). Note that we have omitted the initialization of the conditional probability distributions associated with Bayesian clauses with random values. The operators ρ_g and ρ_s are generalization and specialization operators.

Let H be an initial (valid) hypothesis;
$S(H) := score_D(H)$;
repeat
    $H' := H$;
    $S(H') := S(H)$;
    foreach $H'' \in \rho_g(H') \cup \rho_s(H')$ **do**
        if H'' *is (logically) valid on* D **then**
            if *the Bayesian networks induced by* H'' *on the data are acyclic* **then**
                if $score_D(H'') > S(H)$ **then**
                    $H := H''$;
                    $S(H) := S(H'')$;

until $S(H') = S(H)$;
Return H;

representing a Bayesian network. The probabilistic dependencies among the relevant random variables are not known, i.e. the propositional Bayesian clauses are unknown. Therefore, we have to select such a propositional Bayesian logic program as a *candidate* and estimate its parameters. Assume the data cases $D = \{D_1, \ldots, D_m\}$ look like

$$\{m(ann, dorothy) = true, f(brian, dorothy) = true, pc(ann) = a,$$
$$mc(ann) =?, mc(brian) =?, mc(dorothy) = a, mc(dorothy) = a,$$
$$pc(brian) = b, bt(ann) = a, bt(brian) =?, bt(dorothy) = a\}$$

which is a data case for the Bayesian network in Figure 2. Note, that the atoms have to be interpreted as propositions. Each H in the hypothesis space $\mathcal{H}$ is a Bayesian logic program consisting of n propositional clauses: for each $x_i \in \mathbf{x}$ a single clause c with $head(c) = x_i$ and $body(c) \subseteq \mathbf{x} \setminus \{x_i\}$. To traverse $\mathcal{H}$ we specify two *refinement* operators $\rho_g : \mathcal{H} \mapsto 2^{\mathcal{H}}$ and $\rho_s : \mathcal{H} \mapsto 2^{\mathcal{H}}$, that take a hypothesis and modify it to produce a set of possible candidates. In the case of Bayesian networks the operator $\rho_g(H)$ deletes a Bayesian proposition from the body of a Bayesian clause $c_i \in H$, and the operator $\rho_s(H)$ adds a Bayesian proposition to the body of $c_i \in H$ (cf Figure 3). The search algorithm performs an greedy, informed search in $\mathcal{H}$ based on $score_D$.

As a simple illustration we consider a greedy hill-climbing algorithm incorporating $score_D(H) := LL(\mathbf{D}, H)$, the log-likelihood of the data $\mathbf{D}$ given a candidate structure H with the best parameters. We pick an initial candidate $S \in \mathcal{H}$

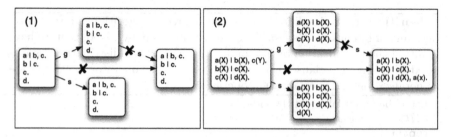

Fig. 3. (1) The use of refinement operators during structural search for Bayesian networks. We can add (ρ_s) a proposition to the body of a clause or delete (ρ_g) it from the body. (2) The use of refinement operators during structural search within the framework of Bayesian logic programs. We can add (ρ_s denoted as 's'') an atom to the body of a clause or delete (ρ_g denoted as 'g') it from the body. Candidates crossed out in (1) and (2) are illegal because they are cyclic.

as starting point (e.g. the set of all propositions) and compute the likelihood $LL(\mathbf{D}, S)$ with the best parameters. Then, we use $\rho(S)$ to compute the legal "neighbors" (candidates being acyclic) of S in $\mathcal{H}$ and score them. All neighbors are valid (see below for a definition of validity). E.g. replacing pc(dorothy) with pc(dorothy) | pc(brian) gives such a "neighbor". We take that $S' \in \rho(S)$ with the best improvements in the score. The process is continued until no improvements in score are obtained.

4.2   The First Order Case: Bayesian Logic Programs

Let us now explain how Scooby works in the first order case. The key differences with the propositional case are The key difference to the propositional case are

1. That some Bayesian logic programs will be logically invalid (see below for an example), and
2. That the traditional first order refinement operators must be used.

Difference 1 is the most important one, because it determines the hypotheses that are candidate Bayesian logic programs. To account for this difference, two modifications of the traditional Bayesian network algorithm are needed.

The first modification concerns the initialization phase where we have to choose a logically valid, acyclic Bayesian logic program. Such a program can be computed using a CLAUDIEN like procedure ([13,14,6]). CLAUDIEN is an ILP-program that computes a logically valid hypothesis H from a set of data cases. Furthermore, all clauses in H will be maximally general (w.r.t. θ-subsumption), and CLAUDIEN will compute all such clauses (within $\mathcal{L}$). This means none of the clauses in H can be generalized without violating the logical validity requirement (or leaving $\mathcal{L}$). Consider again the data cases

$$D_1 = \{m(\textit{cecily, fred}) = \textit{true}, f(\textit{henry, fred}) =?, pc(\textit{cecily}) = a,$$
$$pc(\textit{henry}) = b, pc(\textit{fred}) =?, mc(\textit{cecily}) = b, mc(\textit{henry}) = b,$$
$$mc(\textit{fred}) =?, bt(\textit{cecily}) = ab, bt(\textit{henry}) = b, bt(\textit{fred}) =?\},$$
$$D_2 = \{m(\textit{ann, dorothy}) = \textit{true}, f(\textit{brian, dorothy}) = \textit{true}, pc(\textit{ann}) = b,$$
$$mc(\textit{ann}) =?, mc(\textit{brian}) = a, mc(\textit{dorothy}) = a,$$
$$pc(\textit{dorothy}) = a, pc(\textit{brian}) =?, bt(\textit{ann}) = ab, bt(\textit{brian}) =?,$$
$$bt(\textit{dorothy}) = a\},$$

The clause bt(X) is not a member of $\mathcal{L}$. The clause bt(X) | mc(X),pc(X) is valid but not maximally general because the literal pc(X) can be deleted without violating the logical validity requirement. Any hypothesis including m(X,Y) | mc(X),pc(Y) would be logically invalid because *cecily* is not the mother of *henry*. Examples of maximally general clauses are

```
mc(X)   | m(Y, X).
pc(X)   | f(Y, X).
bt(X)   | mc(X).
bt(X)   | pc(X).
...
```

Roughly speaking, CLAUDIEN works as follows (for a detailed discussion we refer to [14]). It keeps track of a list of candidate clauses Q, which is initialized to the maximally general clause (in $\mathcal{L}$). It repeatedly deletes a clause c from Q, and tests whether c is valid on the data. If it is, c is added to the final hypothesis, otherwise, all maximally general specializations of c (in $\mathcal{L}$) are computed (using a so-called refinement operator ρ, see below) and added back to Q. This process continues until Q is empty and all relevant parts of the search space have been considered. The clauses generated by CLAUDIEN can be used as an initial hypothesis.

In the experiments, for each predicate, we selected one of the clause generated by CLAUDIEN for inclusion in the initial hypothesis such that the valid Bayesian logic program was also acyclic on the data cases (see below). An initial hypothesis is e.g.

```
mc(X)   | m(Y, X).
pc(X)   | f(Y, X).
bt(X)   | mc(X).
```

The second modification concerns filtering out those Bayesian logic programs that are logically invalid during search. This is realized by the first **if**-condition in the loop. The second **if**-condition tests whether cyclic dependencies are induced on the data cases. This can be done in time $O(s \cdot r^3)$ where r is the number of random variables of the largest data case in **D** and s is the number of clauses in H. To do so, we build the Bayesian networks induced by H over each $Var(D_i)$ by computing the ground instances for each clause $c \in H$ where the ground atoms are members of $Var(D_i)$. Thus, ground atoms, which are not appearing

```
m(ann,dorothy).    m(cecily,fred).
f(brian,dorothy). f(henry,fred).
pc(ann). pc(brian). pc(cecily). pc(henry).
mc(ann). mc(brian). mc(cecily). mc(henry).
mc(dorothy) | m(ann,dorothy).    mc(fred) | m(cecily,fred).
pc(dorothy) | f(brian,dorothy). pc(fred) | f(cecily,fred).
bt(ann)     | mc(ann).     bt(brian)  | mc(brian).
bt(dorothy) | mc(dorothy). bt(cecily) | mc(cecily).
bt(henry)   | mc(henry).   bt(fred)   | mc(fred).
```

Fig. 4. The support network induced by the initial hypothesis S (see text) over the the data cases D_1 and D_2

as a head atom of a valid ground instance, are apriori nodes, i.e. nodes with an empty parent set. This takes $O(s \cdot r_i^3)$. Then, we test in $O(r_i)$ for a topological order of the nodes in the induced Bayesian network. If it exists, then the Bayesian network is acyclic. Otherwise, it is cyclic. Figure 4 shows the support network induced by the initial hypothesis over D_1 and D_2.

For Difference 2, i.e. the refinements operators, we employ the traditional ILP refinement operators. In our approach we use the two refinement operators $\rho_s : 2^{\mathcal{H}} \mapsto \mathcal{H}$ and $\rho_g : 2^{\mathcal{H}} \mapsto \mathcal{H}$. The operator $\rho_s(H)$ adds constant-free atoms to the body of a single clause $c \in H$, and $\rho_g(H)$ deletes constant-free atoms from the body of a single clause $c \in H$. Figure 3 shows the different refinement operators for the first order case and the propositional case for learning Bayesian networks. Instead of adding (deleting) propositions to (from) the body of a clause, they add (delete) according to our language assumption constant-free atoms. Furthermore, Figure 3 shows that using the refinement operators each hypothesis can in principle be reached.

Finally, we need to mention that whereas the maximal general clauses are the most interesting ones from the logical point of view, this is not necessarily the case from the probabilistic point of view. E.g. having data cases D_1 and D_2 (see Section 3.1), the initial candidate S

```
mc(X)  | m(Y, X).
pc(X)  | f(Y, X).
bt(X)  | mc(X).
```

is likely not to score maximally on the data cases. E.g. the blood type does not depend on the fatherly genetical information.

As a simple instantiation of Algorithm 1, we consider a greedy hill-climbing algorithm incorporating $score_{\mathbf{D}}(H) := LL(\mathbf{D}, H)$ with $\mathbf{D} = \{D_1, D_2\}$. It takes $S \in \mathcal{H}$ (see above) as starting point and computes $LL(\mathbf{D}, S)$ with the best parameters. Then, we use $\rho_s(S)$ and $\rho_g(S)$ to compute the legal "neighbors" of S in $\mathcal{H}$ and score them. E.g. one such a "neighbor" is given by replacing bt(X) | mc(X) with bt(X) | mc(X), pc(X). Let S' be that valid and acyclic neighbor which scores best. If $LL(\mathbf{D}, S) < LL(\mathbf{D}, S')$, then we take S' as new hypothesis. The process is continued until no improvements in score are obtained.

4.3   Discussion

The algorithm presented serves as a basic, unifying framework. Several extensions and modifications based on ideas developed in both fields, inductive logic programming and Bayesian networks are possible. These include: lookaheads, background knowledge, mode declarations and improved scoring functions. Let us briefly address some of these.

Lookahead: In some cases, an atom might never be chosen by our algorithm because it will not – in itself – result in a better score. However, such an atom, while not useful in itself, might introduce new variables that make a better score possible by adding another atom later on. Within inductive logic programming this is solved by allowing the algorithm to *look ahead* in the search space. Immediately after refining a clause by putting some atom A into the body, the algorithm checks whether any other atom involving some variable of A results in a better score [5]. The same problem is encountered when learning Bayesian networks [47].

Background   Knowledge: Inductive   logic   programming   emphasizes background knowledge, i.e. predefined, fixed regularities which are common to all examples. Background knowledge can be incorporated into our approach in the following way. It is expressed as a fixed Bayesian logic program BK. Now, we search for a candidate H^* which is together with BK acyclic on the data such that for all $D_i \in \mathbf{D} : \mathrm{LH}(BK \cup H^* \cup Var(D_i)) = Var(D_i)$, and $BK \cup H^*$ matches the data $\mathbf{D}$ best according to $score_{\mathbf{D}}$. Therefore, all the Bayesian facts that can be derived from the background knowledge and an example are part of the corresponding "extended" example. This is particularly interesting to specify deterministic knowledge as in inductive logic programming. In [29], we showed how pure Prolog programs can be represented as Bayesian logic programs w.r.t. the conditions 1,2 and 3 of Proposition 1.

Improved Scoring Function: Using the likelihood directly as scoring function, score-based algorithm to learn Bayesian networks prefer fully connected networks. To overcome the problem advanced scoring functions were developed. On of these is the *minimum description length* (MDL) score which trades off the fit to the data with complexity. In the context of learning Bayesian networks, the whole Bayesian network is encoded to measure the compression [34]. In the context of learning clause programs, other compression measure were investigated such as the average length of proofs [44]. For Bayesian logic programs, a combination of both seems to be appropriate.

Finally, an extension for learning predicate definitions consisting of more than one clause is in principle possible. The refinement operators could be modified in such a way that for a clause $c \in H'$ with head predicate p another (valid) clause c' (e.g. computed by CLAUDIEN) with head predicate p is added or deleted.

5   Learning Probabilities in a Bayesian Logic Program

So far, we have assumed that there is a method estimating the parameters of an candidate program given data. In this section, we show how to learn the quantitative component of a Bayesian logic program, i.e. the conditional probability distributions. The learning problem can be stated as follows:

Definition 7 (*Parameter Estimation*). **Given** *a set* $\mathbf{D} = \{D_1, \ldots, D_m\}$ *of data cases*[3], *a set* H *of Bayesian clauses according to some language bias, which is logically valid and acyclic on the data, and a scoring function* $score_{\mathbf{D}} : \mathcal{H} \mapsto \mathbb{R}$, **find** *the parameters of* H *maximizing* $score_{\mathbf{D}}$.

We will concentrate on *maximum likelihood estimation* (MLE).

5.1   Maximum Likelihood Estimation

Maximum likelihood is a classical method for parameter estimation. The likelihood is the probability of the observed data as a function of the unknown parameters with respect to the current model. Let B be a Bayesian logic program consisting of the Bayesian clauses $c_1, \ldots, c_n$, and let $\mathbf{D} = \{D_1, \ldots, D_m\}$ be a set of data cases. The parameters $cpd(c_i)_{jk} = P(u_j \mid \mathbf{u}_k)$, where $u_j \in \mathbf{S}(\text{head}(c_i))$ and $\mathbf{u}_k \in \mathbf{S}(\text{body}(c_i))$, affecting the associated conditional probability distributions $cpd(c_i)$ constitute the set $\boldsymbol{\lambda} = \bigcup_{i=1}^n cpd(c_i)$. The version of B where the parameters are set to $\boldsymbol{\lambda}$ is denoted by $B(\boldsymbol{\lambda})$, and as long as no ambiguities occur we will not distinguish between the parameters $\boldsymbol{\lambda}$ themselves and a particular instance of them.

Now, the likelihood $L(\mathbf{D}, \boldsymbol{\lambda})$ is the probability of the data $\mathbf{D}$ as a function of the unknown parameters $\boldsymbol{\lambda}$:

$$L(\mathbf{D}, \boldsymbol{\lambda}) := P_B(\mathbf{D} \mid \boldsymbol{\lambda}) = P_{B(\boldsymbol{\lambda})}(\mathbf{D}). \tag{1}$$

Thus, the search space $\mathcal{H}$ is spanned by the product space over the possible values of $\lambda(c_i)$ and we seek to find the parameter values $\boldsymbol{\lambda}^*$ that maximize the likelihood, i.e.

$$\boldsymbol{\lambda}^* = \max_{\lambda \in \mathcal{H}} P_{B(\boldsymbol{\lambda})}(\mathbf{D}).$$

Usually, B specifies a distribution over a (countably) infinite set of random variables namely $\text{LH}(B)$ and hence we cannot compute $P_{B(\boldsymbol{\lambda})}(\mathbf{D})$ by considering the whole dependency graph. But as we have argued it is sufficient to consider the support network $N(\boldsymbol{\lambda})$ of the random variables occurring in $\mathbf{D}$ to compute $P_{B(\boldsymbol{\lambda})}(\mathbf{D})$. Thus, using the monotonicity of the logarithm, we seek to find

$$\boldsymbol{\lambda}^* = \max_{\lambda \in \mathcal{H}} \log P_{N(\boldsymbol{\lambda})}(\mathbf{D}) \tag{2}$$

---

[3] Given a well-defined Bayesian network B, the logical part of a data case D_i can also be a partial model only if we only estimate the parameters and do not learn the structure, i.e. $\text{RandVar}(D_i) \subseteq \text{LH}(B)$. The given Bayesian logic program will fill in the missing random variables.

where $P_{N(\lambda)}$ is the probability distribution specified by the support network $N(\lambda)$ of the random variables occurring in $\mathbf{D}$. Equation (2) expresses the original problem in terms of the maximum likelihood parameter estimation problem of Bayesian networks:

A Bayesian logic program together with data cases induces a Bayesian network over the variables of the data cases.

This is not surprising because the learning setting is an instance of the probabilistic learning from interpretations. More important, due to the reduction, all techniques for maximum likelihood parameter estimation within Bayesian networks are in principle applicable. We only need to take the following issues into account:

1. Some of the nodes in $N(\lambda)$ are hidden, i.e., their values are not observed in $\mathbf{D}$.
2. We are not interested in the conditional probability distributions associated to ground instances of Bayesian clauses, but in those associated to the Bayesian clauses themselves.
3. Not only $L(\mathbf{D}, \lambda)$ but also $N(\lambda)$ itself depends on the data, i.e. the data cases determine the subnetwork of $DG(B)$ that is sufficient to calculate the likelihood.

The available data cases may not be complete, i.e., some values may not be observed. For instance in medical domains, a patient rarely gets all of the possible tests. In presence of missing data, the maximum likelihood estimate typically cannot be written in closed form. Unfortunately, it is a numerical optimization problem, and all known algorithms involve nonlinear, iterative optimization and multiple calls to a Bayesian inference procedures as subroutines, which are typically computationally infeasible. For instance the inference within Bayesian network has been proven to be NP-hard [9]. Typical ML parameter estimation techniques (in the presence of missing data) are the Expectation-Maximization (EM) algorithm and gradient-based approaches. We will now discuss both approaches in turn.

5.2   Gradient-Based Approach

We will adapt Binder *et al.*'s solution for dynamic Bayesian networks based on the chain rule of differentiation [3]. For simplicity, we fix the current instantiation of the parameters λ and, hence, we write B and $N(\mathbf{D})$. Applying the chain rule to (2) yields

$$\frac{\partial \log P_N(\mathbf{D})}{\partial \operatorname{cpd}(c_i)_{jk}} = \sum_{\substack{\text{subst. } \theta \text{ s.t.} \\ \text{sn}(c_i\theta)}} \frac{\partial \log P_N(\mathbf{D})}{\partial \operatorname{cpd}(c_i\theta)_{jk}} \qquad (3)$$

where θ refers to grounding substitutions and $\operatorname{sn}(c_i\theta)$ is true iff $\{\operatorname{head}(c_i\theta)\} \cup \operatorname{body}(c_i\theta) \subset N$. Assuming that the data cases $D_l \in \mathbf{D}$ are independently sampled from the same distribution we can separate the contribution of the different data cases to the partial derivative of a single ground instance $c\theta$:

$$\frac{\partial \log P_N(\mathbf{D})}{\partial \operatorname{cpd}(c_i\theta)_{jk}} = \frac{\partial \log \prod_{l=1}^{m} P_N(D_l)}{\partial \operatorname{cpd}(c_i\theta)_{jk}} \qquad \text{by independence}$$

$$= \sum_{l=1}^{m} \frac{\partial \log P_N(D_l)}{\partial \operatorname{cpd}(c_i\theta)_{jk}} \qquad \text{by } \log \prod = \sum \log$$

$$= \sum_{l=1}^{m} \frac{\partial P_N(D_l)/\partial \operatorname{cpd}(c_i\theta)_{jk}}{P_N(D_l)} . \qquad (4)$$

In order to obtain computations local to the parameter $\operatorname{cpd}(c_i\theta)_{jk}$ we introduce the variables $\operatorname{head}(c_i\theta)$ and $\operatorname{body}(c_i\theta)$ into the numerator of the summand of (4) and average over their possible values, i.e.,

$$\frac{\partial P_N(D_l)}{\partial \operatorname{cpd}(c_i\theta)_{jk}} = \frac{\partial}{\partial \operatorname{cpd}(c_i\theta)_{jk}} \left(\sum_{j',k'} P_N(D_l, \operatorname{head}(c_i\theta) = u_{j'}, \operatorname{body}(c_i\theta) = \mathbf{u}_{k'}) \right)$$

Applying the chain rule yields

$$\frac{\partial P_N(D_l)}{\partial \operatorname{cpd}(c_i\theta)_{jk}} = \frac{\partial}{\partial \operatorname{cpd}(c_i\theta)_{jk}} \left(\sum_{j',k'} P_N(D_l \mid \operatorname{head}(c_i\theta) = u_{j'}, \operatorname{body}(c_i\theta) = \mathbf{u}_{k'}) \right.$$

$$\left. \cdot P_N(\operatorname{head}(c_i\theta) = u_{j'}, \operatorname{body}(c_i\theta) = \mathbf{u}_{k'}) \right)$$

$$= \frac{\partial}{\partial \operatorname{cpd}(c_i\theta)_{jk}} \left(\sum_{j',k'} P_N(D_l \mid \operatorname{head}(c_i\theta) = u_{j'}, \operatorname{body}(c_i\theta) = \mathbf{u}_{k'}) \right.$$

$$\cdot P_N(\operatorname{head}(c_i\theta) = u_{j'} \mid \operatorname{body}(c_i\theta) = \mathbf{u}_{k'})$$

$$\left. \cdot P_N(\operatorname{body}(c_i\theta) = \mathbf{u}_{k'}) \right) \qquad (5)$$

where $u_j \in \mathbf{S}(\operatorname{head}(c_i))$, $\mathbf{u}_k \in \mathbf{S}(\operatorname{body}(c_i))$ and j, k refer to the corresponding entries in $\operatorname{cpd}(c_i)$, respectively $\operatorname{cpd}(c_i\theta)$. In (5), $\operatorname{cpd}(c_i\theta)_{jk}$ appears only in linear form. Moreover, it appears only when $j' = j$, and $k' = k$. Therefore, (5) simplifies two

$$\frac{\partial P_N(D_l)}{\partial \operatorname{cpd}(c_i\theta)_{jk}} = P_N(D_l \mid \operatorname{head}(c_i\theta) = u_j, \operatorname{body}(c_i\theta) = \mathbf{u}_k) \cdot P_N(\operatorname{body}(c_i\theta) = \mathbf{u}_k).$$

$$(6)$$

Substituting (6) back into (4) yields

$$\sum_{l=1}^{m} \frac{\partial \log P_N(D_l)/\partial \operatorname{cpd}(c_i\theta)_{jk}}{P_N(D_l)}$$

$$= \sum_{l=1}^{m} \frac{P_N(D_l \mid \operatorname{head}(c_i\theta) = u_j, \operatorname{body}(c_i\theta) = \mathbf{u}_k) \cdot P_N(\operatorname{body}(c_i\theta) = \mathbf{u}_k)}{P_N(D_l)}$$

$$= \sum_{l=1}^{m} \frac{P_N(\text{head}(c_i\theta) = u_j, \text{body}(c_i\theta) = \mathbf{u}_k \mid D_l) \cdot P_N(D_l) \cdot P_N(\text{body}(c_i\theta) = \mathbf{u}_k)}{P_N(\text{head}(c_i\theta) = u_j, \text{body}(c_i\theta) = \mathbf{u}_k) \cdot P_N(D_l)}$$

$$= \sum_{l=1}^{m} \frac{P_N(\text{head}(c_i\theta) = u_j, \text{body}(c_i\theta) = \mathbf{u}_k \mid D_l)}{P_N(\text{head}(c_i\theta) = u_j \mid \text{body}(c_i\theta) = \mathbf{u}_k)}$$

$$= \sum_{l=1}^{m} \frac{P_N(\text{head}(c_i\theta) = u_j, \text{body}(c_i\theta) = \mathbf{u}_k \mid D_l)}{\text{cpd}(c_i\theta)_{jk}} .$$

Combining all these, (3) can be rewritten as

$$\frac{\partial \log P_N(\mathbf{D})}{\partial \, \text{cpd}(c_i)_{jk}} = \sum_{\substack{\text{subst. } \theta \text{ with} \\ \text{sn}(c_i\theta)}} \frac{\text{en}(c_{ijk} \mid \theta, \mathbf{D})}{\text{cpd}(c_i\theta)_{jk}} \tag{7}$$

where

$$\text{en}(c_{ijk} \mid \theta, \mathbf{D}) := \text{en}(\text{head}(c_i\theta) = u_j, \text{body}(c_i\theta) = \mathbf{u}_k \mid \mathbf{D})$$

$$:= \sum_{l=1}^{m} P_N(\text{head}(c_i\theta) = u_j, \text{body}(c_i\theta) = \mathbf{u}_k \mid D_l) \tag{8}$$

are the so-called *expected counts* of the joint state $\text{head}(c_i\theta) = u_j, \text{body}(c_i\theta) = \mathbf{u}_k$ given the data $\mathbf{D}$.

Equation (7) shows that $P_N(\text{head}(c_i\theta) = u_j, \text{body}(c_i\theta) = \mathbf{u}_k \mid D_l)$ is all what is needed. This can essentially be computed using any standard Bayesian network inference engine. This is not surprising because (7) differs from the one for Bayesian networks given in [3] only in that we sum over all ground instances of a Bayesian clause holding in the data. To stress this close relationship, we rewrite (7) in terms of expected counts of clauses instead of ground clauses. They are defined as follows:

Definition 8 (Expected Counts of Bayesian Clauses). *The expected counts of a Bayesian clauses c of a Bayesian logic program B for a data set* **D** *are defined as*

$$\text{en}(c_{ijk} \mid \mathbf{D}) := \text{en}(\text{head}(c_i) = u_j, \text{body}(c_i) = \mathbf{u}_k \mid \mathbf{D})$$

$$:= \sum_{\substack{subst. \ \theta \ with \\ \text{sn}(c_i\theta)}} \text{en}(\text{head}(c_i\theta) = u_j, \text{body}(c_i\theta) = \mathbf{u}_k \mid \mathbf{D}) . \tag{9}$$

Reading (7) in terms of Definition 8 proves the following proposition:

Proposition 1 (Partial Derivative of Log-Likelihood). *Let B be a Bayesian logic program with parameter vector* $\boldsymbol{\lambda}$. *The partial derivative of the log-likelihood of B with respect to* $\text{cpd}(c_i)_{jk}$ *for a given data set* **D** *is*

$$\frac{\partial LL(\mathbf{D}, \boldsymbol{\lambda})}{\partial \, \text{cpd}(c_i)_{jk}} = \frac{\text{en}(c_{ijk} \mid \mathbf{D})}{\text{cpd}(c_i)_{jk}} . \tag{10}$$

Algorithm 2. A simplified skeleton of the algorithm for *adaptive Bayesian logic programs* estimating the parameters of a Bayesian logic program

---

input : B, a Bayesian logic program; associated cpds are parameterized by $\boldsymbol{\lambda}$;
　　　　 $\mathbf{D}$, a finite set of data cases
output: a modified Bayesian logic program

$\boldsymbol{\lambda} \leftarrow$ INITIALPARAMETERS
$N \leftarrow$ SUPPORTNETWORK($B, \mathbf{D}$)
repeat
　　$\Delta\boldsymbol{\lambda} \leftarrow 0$
　　set associated conditional probability distribution of N according to $\boldsymbol{\lambda}$
　　foreach $D_l \in \mathbf{D}$ **do**
　　　　set the evidence in N from D_l
　　　　foreach *Bayesian clause* $c \in B$ **do**
　　　　　　foreach *ground instance* $c\theta$ s.t. $\{\text{head}(c\theta)\} \cup \text{body}(c\theta) \subset N$ **do**
　　　　　　　　foreach *single parameter* $\text{cpd}(c\theta)_{jk}$ **do**
　　　　　　　　　| $\Delta\text{cpd}(c)_{jk} \leftarrow \Delta\text{cpd}(c)_{jk} + (\partial \log P_N(D_l)/\partial \text{cpd}(c\theta)_{jk})$

　　$\Delta\boldsymbol{\lambda} \leftarrow$ PROJECTIONONTOCONSTRAINTSURFACE($\Delta\boldsymbol{\lambda}$)
　　$\boldsymbol{\lambda} \leftarrow \boldsymbol{\lambda} + \alpha \cdot \Delta\boldsymbol{\lambda}$
until $\Delta\boldsymbol{\lambda} \approx 0$
return B

Equation (10) can be viewed as the first-order logical equivalent of the Bayesian network formula. A simplified skeleton of a gradient-based algorithm employing (10) is shown in Algorithm 2.

Before showing how to adapt the EM algorithm, we have to explain two points, which we have left out so far for the sake of simplicity: Constraint satisfaction and decomposable combining rules.

In the problem at hand, the gradient ascent has to be modified to take into account the constraint that the parameter vector $\boldsymbol{\lambda}$ consists of probability values, i.e. $\text{cpd}(c_i)_{jk} \in [0, 1]$ and $\sum_j \text{cpd}(c_i)_{jk} = 1$. Following [3], there are two ways to enforce this:

1. Projecting the gradient onto the constraint surface (as used to formulate the Algorithm 2), and
2. Reparameterizing the problem.

In the experiments, we chose the reparameterization approach because the new parameters automatically respect the constraints on $\text{cpd}(c_i)_{jk}$ no matter what their values are. More precisely, we define the parameters $\boldsymbol{\beta}$ with $\beta_{ijk} \in \mathbb{R}$ such that

$$\text{cpd}(c_i)_{jk} = \frac{e^{\beta_{ijk}}}{\sum_l e^{\beta_{ilk}}} \tag{11}$$

where the β_{ijk} are indexed like $\text{cpd}(c_i)_{jk}$. This enforces the constraints given above, and a local maximum with respect to the $\boldsymbol{\beta}$ is also a local maximum with respect to λ, and vice versa. The gradient with respect to β can be found by computing the gradient with respect to λ and then deriving the gradient with respect to β using the chain rule of derivatives. More precisely, the chain rule of derivatives yields

$$\frac{\partial LL(\mathbf{D}, \lambda)}{\partial \beta_{ijk}} = \sum_{i'j'k'} \frac{\partial LL(\mathbf{D}, \lambda)}{\partial \operatorname{cpd}(c_{i'})_{j'k'}} \cdot \frac{\partial \operatorname{cpd}(c_{i'})_{j'k'}}{\partial \beta_{ijk}} \qquad (12)$$

Since $\partial \operatorname{cpd}(c_{i'})_{j'k'}/\partial \beta_{ijk} = 0$ for all $i \neq i'$, and $k \neq k'$, (12) simplifies to

$$\frac{\partial LL(\mathbf{D}, \lambda)}{\partial \beta_{ijk}} = \sum_{j'} \frac{\partial LL(\mathbf{D}, \lambda)}{\partial \operatorname{cpd}(c_i)_{j'k}} \cdot \frac{\partial \operatorname{cpd}(c_i)_{j'k}}{\partial \beta_{ijk}}$$

The quotient rule yields

$$\frac{\partial LL(\mathbf{D}, \lambda)}{\partial \beta_{ijk}} = \sum_{j'} \left\{ \frac{\partial LL(\mathbf{D}, \lambda)}{\partial \operatorname{cpd}(c_i)_{j'k}} \cdot \frac{\left(\frac{\partial e^{\beta_{ij'k}}}{\partial \beta_{ijk}} \cdot \sum_l e^{\beta_{ilk}} \right) - \left(e^{\beta_{ij'k}} \cdot \frac{\partial \sum_l e^{\beta_{ilk}}}{\partial \beta_{ijk}} \right)}{\left(\sum_l e^{\beta_{ilk}} \right)^2} \right\}$$

$$= \frac{\left\{ \sum_{j'} \left(\frac{\partial LL(\mathbf{D}, \lambda)}{\partial \operatorname{cpd}(c_i)_{j'k}} \cdot \frac{\partial e^{\beta_{ij'k}}}{\partial \beta_{ijk}} \cdot \sum_l e^{\beta_{ilk}} \right) \right\}}{\left(\sum_l e^{\beta_{ilk}} \right)^2} - \frac{\left\{ \sum_{j'} \left(\frac{\partial LL(\mathbf{D}, \lambda)}{\partial \operatorname{cpd}(c_i)_{j'k}} \cdot e^{\beta_{ij'k}} \cdot \frac{\partial \sum_l e^{\beta_{ilk}}}{\partial \beta_{ijk}} \right) \right\}}{\left(\sum_l e^{\beta_{ilk}} \right)^2}$$

Because $\partial e^{\beta_{ij'k}}/\partial \beta_{ijk} = 0$ for $j' \neq j$ and $\partial e^{\beta_{ijk}}/\partial \beta_{ijk} = e^{\beta_{ijk}}$, this simplifies to

$$\frac{\partial LL(\mathbf{D}, \lambda)}{\partial \beta_{ijk}} = \frac{\left(\frac{\partial LL(\mathbf{D}, \lambda)}{\partial \operatorname{cpd}(c_i)_{jk}} \cdot e^{\beta_{ijk}} \cdot \sum_l e^{\beta_{ilk}} \right)}{\left(\sum_l e^{\beta_{ilk}} \right)^2} - \frac{\sum_{j'} \left(\frac{\partial LL(\mathbf{D}, \lambda)}{\partial \operatorname{cpd}(c_i)_{j'k}} \cdot e^{\beta_{ij'k}} \cdot e^{\beta_{ilk}} \right)}{\left(\sum_l e^{\beta_{ilk}} \right)^2}$$

$$= \frac{e^{\beta_{ijk}}}{\left(\sum_l e^{\beta_{ilk}} \right)^2} \cdot \left\{ \frac{\partial LL(\mathbf{D}, \lambda)}{\partial \operatorname{cpd}(c_i)_{jk}} \cdot \left(\sum_l e^{\beta_{ilk}} \right) - \sum_{j'} \frac{\partial LL(\mathbf{D}, \lambda)}{\partial \operatorname{cpd}(c_i)_{j'k}} \cdot e^{\beta_{ij'k}} \right\} \qquad (13)$$

To further simplify the partial derivative, we note that $\partial LL(\mathbf{D}, \lambda)/\partial \operatorname{cpd}(c_i)_{jk}$ can be rewritten as

$$\frac{\partial LL(\mathbf{D}, \lambda)}{\partial \operatorname{cpd}(c_i)_{jk}} = \frac{\operatorname{en}(c_{ijk} \mid \mathbf{D})}{\operatorname{cpd}(c_i)_{jk}} = \frac{\operatorname{en}(c_{ijk} \mid \mathbf{D})}{\frac{e^{\beta_{ijk}}}{\sum_l e^{\beta_{ijk}}}} = \frac{\operatorname{en}(c_{ijk} \mid \mathbf{D})}{e^{\beta_{ijk}}} \cdot \left(\sum_l e^{\beta_{ijk}} \right)$$

by substituting (11) in (9). Using the last equation, (13) simplifies to

$$\frac{\partial LL(\mathbf{D}, \lambda)}{\partial \beta_{ijk}}$$

$$= \frac{e^{\beta_{ijk}}}{\left(\sum_l e^{\beta_{ilk}} \right)^2} \cdot \left\{ \frac{\operatorname{en}(c_{ijk} \mid \mathbf{D})}{e^{\beta_{ijk}}} \cdot \left(\sum_l e^{\beta_{ilk}} \right)^2 - \sum_{j'} \frac{\operatorname{en}(c_{ij'k} \mid \mathbf{D})}{e^{\beta_{ij'k}}} \cdot \left(\sum_l e^{\beta_{ilk}} \right) \cdot e^{\beta_{ij'k}} \right\}$$

$$= \operatorname{en}(c_{ijk} \mid \mathbf{D}) - \frac{e^{\beta_{ijk}}}{\sum_l e^{\beta_{ilk}}} \cdot \sum_{j'} \operatorname{en}(c_{ij'k} \mid \mathbf{D}).$$

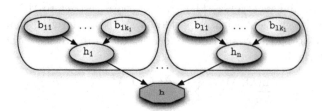

Fig. 5. The scheme of decomposable combining rules. Each rectangle corresponds to a ground instance $c\theta \equiv h_i|b_{1i}, \ldots, b_{ki}$ of a Bayesian clause $c \equiv h|b_1, \ldots, b_k$. The node h is a deterministic node, i.e., its state is deterministic function of the parents joint state.

Using once more (11), the following proposition is proven:

Proposition 2 (Partial Derivative of Log-Likelihood of an Reparameterized BLP). *Let B be a Bayesian logic program reparameterized according to (11). The partial derivative of the log-likelihood of B with respect to β_{ijk} for a given data set $\mathbf{D}$ is*

$$\frac{\partial LL(\mathbf{D}, \lambda)}{\partial \beta_{ijk}} = \mathrm{en}(c_{ijk} \mid \mathbf{D}) - \mathrm{cpd}(c_i)_{jk} \sum_{j'} \mathrm{en}(c_{ij'k} \mid \mathbf{D}) . \tag{14}$$

Equation (14) shows that the partial derivative can be expressed solely in terms of expected counts and original parameters. Consequently, its computational complexity is linear in (10).

We assumed *decomposable* combining rules.

Definition 9 (Decomposable Combining Rule). *Decomposable combining rules can be expressed using a set of separate, deterministic nodes in the support network such that the family of every non-deterministic node uniquely corresponds to a ground Bayesian clause, as shown in Figure 5.*

Most combining rules commonly employed in Bayesian networks such as *noisy_or* or linear regression are decomposable (cp. [23]). The definition of decomposable combining rules directly imply the following proposition.

Proposition 3. *For each node x in the support network n there exist at most one clause c and a substitution θ such that* $\mathrm{body}(c\theta) \subset \mathrm{LH}(B)$ *and* $\mathrm{head}(c\theta) = x$.

Thus, while the same clause c can induce more than one node in N, all of these nodes have identical local structure: the associated conditional probability distributions (and so the parameters) have to be identical, i.e.,

$$\forall \text{ substitutions } \theta : \ \mathrm{cpd}(c\theta) = \mathrm{cpd}(c) .$$

Example 1. Consider the nodes bt(ann), mc(ann), pc(ann) and bt(brain), mc(brain), pc(brian). Both families contribute to the conditional probability distribution associated with the clause defining bt(X).

This is the same situation as for dynamic Bayesian networks where the parameters that encode the stochastic model of state evolution appear many times in the network. However, gradient methods might be applied to non-decomposable combining functions as well. In the general case, the partial derivatives of an inner function has to be computed. For instance, [3] derive the gradient for *noisy_or* when it is not expressed in the structure. This seems to be more difficult in the case of the EM algorithm, which we will now devise.

5.3 Expectation-Maximization (EM)

The Expectation-Maximization (EM) algorithm [15] is another classical approach to maximum likelihood parameter estimation in the presence of missing values. The basic observation of the Expectation-Maximization algorithm is as that *if the states of all random variables are observed, then learning would be easy*. Assuming that no value is missing, Lauritzen [36] showed that maximum likelihood estimation of Bayesian network parameters simply corresponds to frequency counting in the following way. Let $n(\mathbf{a} \mid \mathbf{D})$ denote the *counts* for a particular joint state $\mathbf{a}$ of variables $\mathbf{A}$ in the data, i.e. the number of cases in which the variables in $\mathbf{A}$ are assigned the evidence $\mathbf{a}$. Then the maximum likelihood value for the conditional probability value $P(X = x \mid \mathbf{Pa}(X) = \mathbf{u})$ is the ratio

$$\frac{n(X = x, \mathbf{Pa}(X) = \mathbf{u}_k \mid D_l)}{n(\mathbf{Pa}(X) = \mathbf{u}_k \mid D_l)} . \tag{15}$$

However, in the presence of missing values, the maximum likelihood estimates typically cannot be written in closed form. Therefore, the Expectation-Maximization algorithm iteratively performs the following two steps:

(E-Step) Based on the current parameters $\boldsymbol{\lambda}$ and the observed data $\mathbf{D}$ the algorithm computes a distribution over all possible completions of each partially observed data case. Each completion is then treated as a fully-observed data case weighted by its probability.

(M-Step) A new set of parameters is then computed based on Equation (15) taking the weights into accounts.

[36] showed that this idea leads to a modified Equation (15) where the *expected counts*

$$\mathrm{en}(\mathbf{a}|\mathbf{D}) := \sum_{l=1}^{m} P_N(\mathbf{a} \mid D_l) \tag{16}$$

are used instead of *counts*. Again, essentially any Bayesian network engine can be used to compute $P(\mathbf{a}|D_l)$.

To apply the EM algorithm to parameter estimation of Bayesian logic programs, we assume decomposable combining rules. Thus,

- Each node in the support network was "produced" by exactly one Bayesian clause c, and
- Each node derived from c can be seen as a separate "experiment" for the conditional probability distribution cpd(c).

Formally, due to the reduction of our problem at hand to parameter estimation within the support network N, the update rule becomes

$$cpd(c_i)_{jk} \leftarrow \frac{\text{en}(c_i|\mathbf{D})}{\text{en}(\text{body}(c_i)|\mathbf{D})} = \frac{\text{en}(\text{head}(c_i), \text{body}(c_i)|\mathbf{D})}{\text{en}(\text{body}(c_i)|\mathbf{D})} \qquad (17)$$

where $\text{en}(\cdot|\mathbf{D})$ refers to the first order expected counts as defined in Equation (9). Note that the summation over data cases and ground instances is hidden in $\text{en}(\cdot|\mathbf{D})$. Equation (17) is similar to the one already encountered in Equation (10) for computing the gradient.

5.4   Gradient vs. EM

As one can see, the EM update rule in equation (17) and the corresponding equation (7) for the gradient ascent are very similar. Both rely on computing expected counts. The comparison between EM and (advanced) gradient techniques like conjugate gradient is not yet well understood. Both methods perform a greedy local search, which is guaranteed to converge to stationary points. They both exploit expected counts, i.e., sufficient statistics as their primary computation step. However, there are important differences.

The EM is easier to implement because it does not have to enforce the constraint that the parameters are probability distributions. It converges much faster (at least initially) than simple gradient, and is somewhat less sensitive to starting points. (Conjugate) gradients estimate the step size with a line search involving several additional Bayesian network inferences compared to EM. On the other hand, gradients are more flexible than EM, as they allow one to learn non-multinomial parameterizations using the chain rule for derivatives [3] or to choose other scoring functions than the likelihood [26]. Furthermore, although the EM algorithm is quite successful in practice due to its simplicity and fast initial progress, it has been argued (see e.g. [25,38] and references in there) that the EM convergence can be extremely slow in particular close to the solution, and that more advanced second-order methods should in general be favored to EM or one should switch to gradient-based method after a small number of initial iterations.

Finally, though we focused here on parameter estimation, methods for computing the gradient of the log-likelihood with respect to the parameters of a probabilistic model can also be used to employ generative models within discriminative learners such as SVMs. In the context of probabilistic ILP, this yields relational kernel methods [31,16].

6   Experiments

The presented learning algorithm for Bayesian logic programs is mainly meant as an overall and general framework. Indeed, it leaves several aspects open such as scoring functions. Nevertheless in this section, we report on experiments that show that the algorithm and its underlying principles work.

We implemented the score-based algorithm in Sicstus Prolog 3.9.0 on a Pentium-III 700 MHz Linux machine. The implementation has an interface to the Netica API (http://www.norsys.com) for Bayesian network inference and maximum likelihood estimation. To do the maximum likelihood estimation, we adapted the scaled conjugate gradient (SCG) as implemented in Bishop and Nabney's Netlab library (http://www.ncrg.aston.ac.uk/netlab/, see also [4]) with an upper bound on the scale parameter of $2 \cdot 10^6$. Parameters were initialized randomly. To avoid zero entries in the conditional probability tables, m-estimates were used.

6.1 Genetic Domain

The goal was to learn a global, descriptive model for our genetic domain, i.e. to learn the program *bloodtype*. We considered two totally independent families using the predicates given by *bloodtype* having 12 respectively 15 family members. For each least Herbrand model 1000 data cases from the induced Bayesian network were sampled with a fraction of 0.4 of missing at random values of the observed nodes making in total 2000 data cases.

Therefore, we first had a look at the (logical) hypotheses space. The space could be seen as the first order equivalent of the space for learning the structure of Bayesian networks (see Figure 3). The generating hypothesis is a member of it. In a further experiment, we fixed the definitions for m/2 and f/2. The hypothesis scored best included bt(X) | mc(X), pc(X), i.e. the algorithm re-discovered the intensional definition which was originally used to build the data cases. However, the definitions of mc/1 and pc/1 considered genetic information of the grandparents to be important. It failed to re-discover the original definitions for reasons explained above. The predicates m/2 and f/2 were not part of the learned model rendering them to be extensionally defined. Nevertheless, the founded global model had a slightly better likelihood than the original one.

6.2 Bongard Domain

The Bongard problems (due to the Russian scientist M. Bongard) are well-known problems within inductive logic programming. Consider Figure 6. Each example or scene consists of

- A variable number of geometrical objects such as triangles, rectangles and circles etc (predicate obj/2 with $\mathbf{S}(obj) = \{triangle, circle\}$)., each having a number of different properties such as color, orientation, size etc., and
- A variable number of relations between objects such as in (predicate in/3 having states *true, false*), leftof, above etc.

The task is to find a set of rules which *discriminates* positive from negative examples (represented by class/1 over the states *pos, neg*) by looking at the kind of objects they consists of. Though the Bongard problems are toy problems, they are very similar to real-world problems in e.g. the field of molecular biology where essentially the same representational problems arise. Data consists of

Fig. 6. A Bongard problem consisting of 12 scenes, six positive ones and six negative ones. The goal is to discriminate between the two classes.

molecules, each of which is composed of several atoms with specific properties such as charge. There exists a number of relations between atoms like e.g. bonds, structure etc.

In most real-world applications, the data is *noisy*. Class labels or object properties might be wrong or missing in the data cases. One extreme case concerns clustering where no class labels are given. Furthermore, we might be uncertain about relations among objects. Some positive examples might state that a triangle is not in a circle due to noise. In such cases, Bayesian logic programs are a natural method of choice. We conducted the following experiments.

First, we generated 20 positive and 20 negative examples of the concept *"there is a triangle in a circle."* The number of objects varied from 2 to 8. We conducted three different experiments. We assumed the *in* relation to be deterministic and given as background knowledge in(Example,Obj1,Obj2), i.e. we assumed that there is no uncertainty about it. Due to that, no conditional probability distribution has to take in/3 into account. Because each scene is independent of the other, we represented the whole training data as one data case

$$\{class(e1) = pos, obj(e1, o1) = triangle, obj(e1, o2) = circle,$$
$$class(e2) = neg, obj(e2, o1) = triangle, size(e2, o1) = large,$$
$$obj(e2, o2) = \text{`?'}, \ldots\}$$

with the background knowledge $in(e1, o1, o2), \ldots$ where $e1, e2, \ldots$ are identifiers for examples and $o1, o2, \ldots$ for objects. A fraction of 0.2 of the random variables were not observed. Our algorithm scored the hypothesis

```
class(Ex) | obj(Ex,O1),in(Ex,O1,O2),obj(Ex,O2).
```

best after specifying obj(Ex,O2) as a lookahead for in(Ex,O1,O2). The conditional probability distribution assigned *pos* a probability higher than 0.6 only if object O1 was a triangle and O2 a circle. Without the lookahead, adding in(Ex,O1,O2) yield no improvement in score, and the correct hypothesis was not considered. The hypothesis is not a well-defined Bayesian networks, but it says that ground atoms over `obj/2` extensional defined. Therefore, we estimated the maximum likelihood parameters of

```
obj(Ex,O) | dom(Ex,O).
class(Ex) | obj(Ex,O1),in(Ex,O1,O2),obj(Ex,O2).
```

where `dom/2` ensured range-restriction and was part of the deterministic background knowledge. Using 0.6 as threshold, the learned Bayesian logic program had accuracy 1.0 on the training set and on an independently generated validation set consisting of 10 positive and 10 negative examples.

In a second experiments, we fixed the structure of the program learned in the first experiment, and estimated its parameters on a data set consisting of 20 positive and 20 negative examples of the disjunctive concept *"there is a (triangle or a circle) in a circle."* The estimated conditional probability distribution gave almost equal probability for the object `O1` to be a triangle or circle.

In third experiment, we assumed uncertainty about the *in* relation. We enriched the data case used for the first experiment in the following way

$$\{ class(e1) = pos, obj(e1, o1) = triangle, obj(e1, o2) = circle,$$
$$in(e1, o1, o2) = true, class(e2) = neg, obj(e2, o1) = triangle,$$
$$size(e2, o1) = large, obj(e2, o2) = \text{'?'}, in(e2, o1, o2) = false, \dots \},$$

i.e. for each pair of objects that could be related by *in* a ground atom over *in* was included. Note that the state need not to be observed. Here, the algorithm did not re-discovered the correct rule but

```
class(X) | obj(X,Y)
obj(X,Y) | in(X,Y,Z), obj(Z).
```

This is interesting, because when these rules are used to classify examples, only the first rule is needed. The class is independent of any information about `in/3` given full knowledge about the objects. The likelihood of the founded solution was close to the one of `class(Ex) | obj(Ex,O1),in(Ex,O1,O2),obj(Ex,O2)` on the data (absolute difference less than 0.001). However, the accuracy decreased (about 0.6 on an independently generated training set (10 pos / 10 neg)) for the reasons explained above: We learned a global model not focusing on the classification error. [18] showed for Bayesian network classifier that maximizing the conditional likelihood of the class variable comes close to minimizing the classification error. In all experiments we assumed *noisy or* as combining rule.

Finally, we conducted a simple clustering experiments. We generated 20 positive and 20 negative examples of the disjunctive concept *"there is a triangle."* There were triangles, circles and squares. The number of objects varied from 2 to 8. All class labels were said to be observed, and 20% of the remaining stated were missing at random. The learned hypothesis was `class(X) | obj(X,Y)` totally separating the two classes.

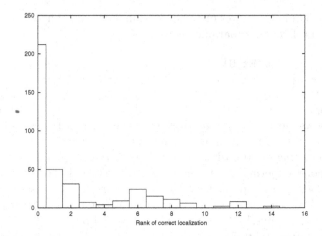

Fig. 7. An histogram of the ranks of the correct localization computed by an Bayesian logic program on the KDD Cup 2001 test set

6.3   KDD Cup 2001

We also started to conduct experiments on *large scale* data sets namely the KDD Cup 2001[4] data sets, cf. [8]. Task 3 is to predict the localizations local(G) of proteins encoded by the genes G. This is a multi-class problem because there are 16 possible localizations. The training set consisted of 862 genes, the test set of 381 genes. The information we used included whether organisms with an mutation in this gene can survive ess(G) (3 states), the class class(G) of a gene/protein G (24 states), the complex compl(G) (56 states), and the other proteins G with which each protein G is known to interact encoded by inter(G1,G2). To avoid a large interaction table, we considered only those interactions with a correlation higher than 0.85. Furthermore, we introduced a hidden predicate hidden/1 with domain 0,1,2 to compress the representation size because e.g. the conditional probability table of local(G1) | inter(G1,G2),local(G2) would consists of 225 entries (instead of 45 using hidden(G1)). The ground atoms over hidden/1 were never observed in the data. Nevertheless, the naive Prolog representation of the support networks induced by some hypothesis (more than 4.400 random variables with more than 60.000 parameters) in our current implementation broke the memory capabilities of Sicstus Prolog. Due to that, we can only report on preliminary results. We only considered maximum likelihood parameter estimation on the training set. The (logical) structure is based on naive Bayes taking relational information into account:

```
local(G1)   | gene(G1).
hidden(G1)  | local(G1).
hidden(G1)  | inter(G1,G2),local(G2).
```

---

[4] For details see http://www.cs.wisc.edu/~dpage/kddcup2001/

```
class(G1)   | hidden(G1).
compl(G1)   | local(G1).
ess(G1)     | local(G1).
```

As combining rule, we used for all predicates *average*. The given ground atoms over `inter/2` were used as pure logical background knowledge. Therefore, the conditional probability distribution associated to `hidden(G1) | inter(G1,G2)`, `local(G2)` had not to take it into account. The parameters were randomly initialized. Again, the training set was represented as one data case, so that no artificial independencies among genes were postulated. Estimate the parameters took 12 iteration (about 30 min). The learned Bayesian logic program achieved an accuracy of 0.57 (top 50% level of submitted models was 0.61, best predictive accuracy was 0.72). A learner predicting always the majority class would achieve an predictive accuracy of 0.44. Furthermore, when we rank for each test gene its possible localizations according to the probability computed by the program, then the correct localization was among the three highest ranked localizations in 293 out of 381 cases (77%) (cf. Figure 7). Not that it took 40 iterations to learn the corresponding grounded Bayesian logic program.

7   Related Work

The learning of Bayesian networks has been thoroughly investigated in the Uncertainty in AI community, see e.g. [22,7]. Binder *et al.* [3], whose approach we have adapted, present results for a gradient-based method. But so far – to the best of our knowledge – there has not been much work on learning within first order extensions of Bayesian networks. Koller and Pfeffer [33] adapt the EM algorithm for probabilistic logic programs [40], a framework which in contrast to Bayesian logic programs sees ground atoms as states of random variables. Although the framework seems to theoretically allow for continuous random variables there exists no (practical) query-answering procedure for this case; to the best of our knowledge, Ngo and Haddawy [40] give only a procedure for variables having finite domains. Furthermore, Koller and Pfeffer's approach utilizes support networks, too, but requires the intersection of the support networks of the data cases to be empty. This could be in our opinion in some cases too restrictive, e.g. in the case of dynamic Bayesian networks. Friedman *et al.* [19,20] adapted the Structural-EM to learn the structure of probabilistic relational models. It applies the idea of the standard EM algorithm for maximum likelihood parameter estimation to the problem of learning the structure. If we know the values for all random variables, then the maximum likelihood estimate can be written in closed from. Based on the current hypothesis a distribution over all possible completions of each partially observed data case is computed. Then, new hypotheses are computed using a score-based method. However, the algorithm does not consider logical constraints on the space of hypotheses. Indeed, the considered clauses need not be logically valid on the data. Therefore, combining our approach with the structural EM seems to be reasonable and straightforward. Finally, there is work on learning object-oriented Bayesian networks [35,1].

There exist also methods for learning within first order probabilistic frameworks which do not build on Bayesian networks. Sato and Kameya [43] introduce an EM method for parameter estimation of PRISM programs, see also Chapter 5. Cussens [11] investigates EM like methods for estimating the parameters of stochastic logic programs (SLPs). As a reminder, SLPs lift probabilistic context-free grammars to the first order case by replacing production rules with probability values with clauses labeled with probability values. In turn, they define probability distributions over proofs. As discussed in Chapter 1, this is quite different from Bayesian logic programs, which lift Bayesian networks by defining probability distributions over an interpretation. Nevertheless, mappings between the two approaches exist as shown by Muggleton and Chen in Chapter 12. Cussens' EM approach for SLPs has been successfully applied to protein fold discovery by Chen *et. al* as reported in Chapter 9. Muggleton [39] uses ILP techniques to learn the logical structure/program of stochastic logic programs. The used ILP setting is different to *learning from interpretations*, it is not based on learning Bayesian networks, and so far considers only for single predicates definitions.

To summarize, the related work on learning probabilistic relational models mainly differs in three points from ours:

- The underlying (logical) frameworks lack important knowledge representational features which Bayesian logic programs have.
- They adapt the EM algorithm to do parameter estimation which is particularly easy to implement. However, there are problematic issues both regarding speed of convergence as well as convergence towards a local (sub-optimal) maximum of the likelihood function. Different accelerations based on the gradient are discussed in [38]. Also, the EM algorithm is difficult to apply in the case of general probability density functions because it relies on computing the sufficient statistics (cf. [22]).
- No probabilistic extension of the *learning from interpretations* is established.

8   Conclusions

A new link between ILP and learning of Bayesian networks was established. We have proposed a scheme for learning both the probabilities and the structure of Bayesian logic programs. We addressed the question "where do the numbers come from?" by showing how to compute the gradient of the likelihood based on ideas known for (dynamic) Bayesian networks. The intensional representation of Bayesian logic programs, i.e. their compact representation should speed up learning and provide good generalization. The general learning setting built on the ILP setting *learning from interpretations*. We have argued that by adapting this setting score-based methods for structural learning of Bayesian networks could be updated to the first order case. The ILP setting is used to define and traverse the space of (logical) hypotheses.

The experiments proved the principle of the algorithm. Their results highlight that future work on improved scoring functions is needed. We plan to conduct experiments on real-world scale problems. The use of refinement operators adding or deleting non constant-free atoms should be explored. Furthermore, it would be interesting to weaken the assumption that a data case corresponds to a complete interpretation. Not assuming all relevant random variables are known would be interesting for learning intensional rules like nat(s(X)) | nat(X). Ideas for handling this within inductive logic programming might be adapted [14,6]. Furthermore, instead of traditional score-based greedy algorithm more advanced UAI methods such as Friedman's Structural-EM or structure search among equivalence classes of Bayesian logic programs may be adapted taking advantage of the logical constraints implied by the data cases. In any case, we believe that the proposed approach is a good point of departure for further research. The link established between ILP and Bayesian networks seems to be bi-directional. Can ideas developed in the UAI community be carried over to ILP?

Acknowledgements

The authors would like to thank Manfred Jaeger, Stefan Kramer and David Page for helpful discussions on the ideas of the paper. Furthermore, the authors would like to thank Jan Ramon and Hendrik Blockeel for making available their Bongard problems generators. This research was partly supported by the European Union IST programme under contract number IST-2001-33053 (Application of Probabilistic Inductive Logic Programming – APRIL).

References

1. Bangsø, O., Langseth, H., Nielsen, T.D.: Structural learning in object oriented domains. In: Russell, I., Kolen, J. (eds.) Proceedings of the Fourteenth International Florida Artificial Intelligence Research Society Conference (FLAIRS 2001), Key West, Florida, USA, pp. 340–344. AAAI Press, Menlo Park (2001)
2. Bauer, H.: Wahrscheinlichkeitstheorie, 4th edn., Walter de Gruyter, Berlin, New York (1991)
3. Binder, J., Koller, D., Russell, S., Kanazawa, K.: Adaptive Probabilistic Networks with Hidden Variables. Machine Learning 29(2–3), 213–244 (1997)
4. Bishop, C.M.: Neural Networks for Pattern Recognition. Oxford University Press, Oxford (1995)
5. Blockeel, H., De Raedt, L.: Lookahead and discretization in ilp. In: Džeroski, S., Lavrač, N. (eds.) ILP 1997. LNCS, vol. 1297, pp. 77–85. Springer, Heidelberg (1997)
6. Blockeel, H., De Raedt, L.: ISIDD: An Interactive System for Inductive Database Design. Applied Artificial Intelligence 12(5), 385 (1998)
7. Buntine, W.: A guide to the literature on learning probabilistic networks from data. IEEE Transaction on Knowledge and Data Engineering 8, 195–210 (1996)
8. Cheng, J., Hatzis, C., Krogel, M.–A., Morishita, S., Page, D., Sese, J.: KDD Cup 2002 Report. SIGKDD Explorations 3(2), 47–64 (2002)
9. Cooper, G.F.: The computational complexity of probabilistic inference using Bayesian belief networks. Artificial Intelligence 42, 393–405 (1990)
10. Cowell, R.G., Dawid, A.P., Lauritzen, S.L., Spiegelhalter, D.J.: Probabilistic networks and expert systems. In: Statistics for engineering and information, Springer, Heidelberg (1999)

11. Cussens, J.: Parameter estimation in stochastic logic programs. Machine Learning 44(3), 245–271 (2001)
12. De Raedt, L.: Logical settings for concept-learning. Artificial Intelligence 95(1), 197–201 (1997)
13. De Raedt, L., Bruynooghe, M.: A theory of clausal discovery. In: Bajcsy, R. (ed.) Proceedings of the Thirteenth International Joint Conference on Artificial Intelligence (IJCAI 1993), Chambery, France, pp. 1058–1063. Morgan Kaufmann, San Francisco (1993)
14. De Raedt, L., Dehaspe, L.: Clausal discovery. Machine Learning 26(2-3), 99–146 (1997)
15. Dempster, A.P., Laird, N.M., Rubin, D.B.: Maximum likelihood from incomplete data via the EM algorithm. J. Royal Stat. Soc. B 39, 1–39 (1977)
16. Dick, U., Kersting, K.: Fisher Kernels for relational data. In: Fürnkranz, J., Scheffer, T., Spiliopoulou, M. (eds.) Proceedings of the 17th European Conference on Machine Learning (ECML 2006), Berlin, Germany, pp. 112–125 (2006)
17. Flach, P.A., Lachiche, N.: 1BC: A first-order Bayesian classifier. In: Džeroski, S., Flach, P.A. (eds.) ILP 1999. LNCS (LNAI), vol. 1634, pp. 92–103. Springer, Heidelberg (1999)
18. Friedman, N., Geiger, D., Goldszmidt, M.: Bayesian network classifiers. Machine Learning 29, 131–163 (1997)
19. Friedman, N., Getoor, L., Koller, D., Pfeffer, A.: Learning probabilistic relational models. In: Dean, T. (ed.) Proceedings of the Sixteenth International Joint Conferences on Artificial Intelligence (IJCAI 1999), Stockholm, Sweden, pp. 1300–1309. Morgan Kaufmann, San Francisco (1999)
20. Getoor, L., Koller, D., Taskar, B., Friedman, N.: Learning probabilistic relational models with structural uncertainty. In: Getoor, L., Jensen, D. (eds.) Proceedings of the AAAI-2000 Workshop on Learning Statistical Models from Relational Data, AAAI Press, Menlo Park (2000)
21. Gilks, W.R., Thomas, A., Spiegelhalter, D.J.: A language and program for complex bayesian modelling. The Statistician 43 (1994)
22. Heckerman, D.: A Tutorial on Learning with Bayesian Networks. Technical Report MSR-TR-95-06, Microsoft Research (1995)
23. Heckerman, D., Breese, J.: Causal Independence for Probability Assessment and Inference Using Bayesian Networks. Technical Report MSR-TR-94-08, Microsoft Research (1994)
24. Jaeger, M.: Relational Bayesian networks. In: Geiger, D., Shenoy, P.P. (eds.) Proceedings of the Thirteenth Annual Conference on Uncertainty in Artificial Intelligence (UAI 1997), Providence, Rhode Island, USA, pp. 266–273. Morgan Kaufmann, San Francisco (1997)
25. Jamshidian, M., Jennrich, R.I.: Accleration of the EM Algorithm by using Quasi-Newton Methods. Journal of the Royal Statistical Society B 59(3), 569–587 (1997)
26. Jensen, F.V.: Gradient descent training of bayesian networks. In: Hunter, A., Parsons, S. (eds.) ECSQARU 1999. LNCS (LNAI), vol. 1638, pp. 190–200. Springer, Heidelberg (1999)
27. Jensen, F.V.: Bayesian networks and decision graphs. Springer, Heidelberg (2001)
28. Kersting, K., De Raedt, L.: Bayesian logic programs. In: Cussens, J., Frisch, A. (eds.) Work-in-Progress Reports of the Tenth International Conference on Inductive Logic Programming (ILP 2000) (2000),
http://SunSITE.Informatik.RWTH-Aachen.DE/Publications/CEUR-WS/Vol-35/
29. Kersting, K., De Raedt, L.: Bayesian logic programs. Technical Report 151, University of Freiburg, Institute for Computer Science (submitted) (April 2001)

30. Kersting, K., De Raedt, L., Kramer, S.: Interpreting Bayesian Logic Programs. In: Getoor, L., Jensen, D. (eds.) Working Notes of the AAAI-2000 Workshop on Learning Statistical Models from Relational Data (SRL), Austin, Texas, AAAI Press, Menlo Park (2000)
31. Kersting, K., Gärtner, T.: Fisher Kernels for Logical Sequences. In: Boulicaut, J.-F., Esposito, F., Giannotti, F., Pedreschi, D. (eds.) ECML 2004. LNCS (LNAI), vol. 3201, p. 205. Springer, Heidelberg (2004)
32. Koller, D.: Probabilistic relational models. In: Džeroski, S., Flach, P.A. (eds.) ILP 1999. LNCS (LNAI), vol. 1634, pp. 3–13. Springer, Heidelberg (1999)
33. Koller, D., Pfeffer, A.: Learning probabilities for noisy first-order rules. In: Proceedings of the Fifteenth Joint Conference on Artificial Intelligence (IJCAI 1997), Nagoya, Japan, pp. 1316–1321 (1997)
34. Lam, W., Bacchus, F.: Learning Bayesian belief networks: An approach based on the MDL principle. Computational Intelligence 10(4) (1994)
35. Langseth, H., Bangsø, O.: Parameter learning in object oriented Bayesian networks. Annals of Mathematics and Artificial Intelligence 32(1-2), 221–243 (2001)
36. Lauritzen, S.L.: The EM algorithm for graphical association models with missing data. Computational Statistics and Data Analysis 19, 191–201 (1995)
37. Lloyd, J.W.: Foundations of Logic Programming, 2nd edn. Springer, Berlin (1989)
38. McKachlan, G.J., Krishnan, T.: The EM Algorithm and Extensions. John Eiley & Sons, Inc. (1997)
39. Muggleton, S.H.: Learning stochastic logic programs. In: Getoor, L., Jensen, D. (eds.) Working Notes of the AAAI-2000 Workshop on Learning Statistical Models from Relational Data (SRL), Austin, Texas, AAAI Press, Menlo Park (2000)
40. Ngo, L., Haddawy, P.: Answering queries from context-sensitive probabilistic knowledge bases. Theoretical Computer Science 171, 147–177 (1997)
41. Pearl, J.: Reasoning in Intelligent Systems: Networks of Plausible Inference, 2nd edn. Morgan Kaufmann, San Francisco (1991)
42. Poole, D.: Probabilistic Horn abduction and Bayesian networks. Artificial Intelligence 64, 81–129 (1993)
43. Sato, T., Kameya, Y.: Parameter learning of logic programs for symbolic-statistical modeling. Journal of Artificial Intelligence Research 15, 391–454 (2001)
44. Srinivasan, A., Muggleton, S., Bain, M.: The justification of logical theories based on data compression. In: Furukawa, K., Michie, D., Muggleton, S. (eds.) Machine Intelligence, vol. 13, Oxford University Press, Oxford (1994)
45. Sterling, L., Shapiro, E.: The Art of Prolog: Advanced Programming Techniques. MIT Press, Cambridge (1986)
46. Taskar, B., Segal, E., Koller, D.: Probabilistic clustering in relational data. In: Nebel, B. (ed.) Seventeenth International Joint Conference on Artificial Intelligence (IJCAI 2001), Seattle, Washington, USA, pp. 870–887. Morgan Kaufmann, San Francisco (2001)
47. Xiang, Y., Wong, S.K.M., Cercone, N.: Critical remarks on single link search in learning belief networks. In: Horvitz, E., Jensen, F.V. (eds.) Proceedings of the Twelfth Annual Conference on Uncertainty in Artificial Intelligence (UAI 1996), Portland, Oregon, USA, pp. 564–571. Morgan Kaufmann, San Francisco (1996)

The Independent Choice Logic and Beyond

David Poole

Department of Computer Science
University of British Columbia
2366 Main Mall
Vancouver, B.C., Canada V6T 1Z4
poole@cs.ubc.ca
http://www.cs.ubc.ca/spider/poole/

Abstract. The Independent Choice Logic began in the early 90's as a way to combine logic programming and probability into a coherent framework. The idea of the Independent Choice Logic is straightforward: there is a set of independent choices with a probability distribution over each choice, and a logic program that gives the consequences of the choices. There is a measure over possible worlds that is defined by the probabilities of the independent choices, and what is true in each possible world is given by choices made in that world and the logic program. ICL is interesting because it is a simple, natural and expressive representation of rich probabilistic models. This paper gives an overview of the work done over the last decade and half, and points towards the considerable work ahead, particularly in the areas of lifted inference and the problems of existence and identity.

1 Introduction

There are good normative arguments for using logic to represent knowledge [Nilsson, 1991; Poole et al., 1998]. These arguments are usually based on reasoning with symbols with an explicit denotation, allowing relations amongst individuals, and permitting quantification over individuals. This is often translated as needing (at least) the first-order predicate calculus.

There are also good normative reasons for using Bayesian decision theory for decision making under uncertainty [Neumann and Morgenstern, 1953; Savage, 1972]. These arguments can be intuitively interpreted as seeing decision making as a form of gambling, and that probability and utility are the appropriate calculi for gambling. These arguments lead to the assignment of a single probability to a proposition; thus leading to the notion of probability as a measure of subjective belief.

These two normative arguments are not in conflict with each other. Together they suggest having probability measures over rich structures. How this can be done in a simple, straightforward manner is the motivation behind a large body of research over the last 20 years.

The independent choice logic (ICL) started off as Probabilistic Horn Abduction [Poole, 1991a,b, 1993a,b] (the first three of these papers had a slightly different language), which allowed for probabilistically independent choices and a logic program to give the consequences of the choices. The independent choice logic extends probabilistic Horn abduction in allowing for multiple agents each making their own choices

L. De Raedt et al. (Eds.): Probabilistic ILP 2007, LNAI 4911, pp. 222–243, 2008.

[Poole, 1997b] (where nature is a special agent who makes choices probabilistically) and in allowing negation as failure in the logic [Poole, 2000b].

The ICL is still one of the simplest and most powerful representations available. It is simple to define, straightforward to represent knowledge in and powerful in that it is a Turing-complete language that can represent arbitrary finite[1] probability distributions at least as compactly as can Bayesian belief networks. It can also represent infinite structures such as Markov chains.

In this paper we overview the base logic, and give some representation, inference and learning challenges that still remain.

2   Background

2.1   Logic Programming

The independent choice logic builds on a number of traditions. The first is the idea of logic programs [Lloyd, 1987].

A logic program is built from constants (that denote particular individuals), variables (that are universally quantified over the set of individuals), function symbols (that are used to indirectly describe individuals), predicate symbols (that denote relations). A term is either a constant, a variable or of the form $f(t_1, \ldots, t_k)$ where f is a function symbol and the t_i are terms. A predicate symbol applied to a set of terms is an atomic formula (or just an atom). A **clause** is either an atom or is of the form

$$h \leftarrow a_1 \wedge \cdots \wedge a_k$$

where h is an atom and each a_i is an atom or the negation of an atom. We write the negation of atom a as $\neg a$. A logic program is a set of clauses. We use the Prolog convention of having variables in upper case, and constants, predicate symbols and function symbols in lower case. We also allow for Prolog notation for lists and allow the standard Prolog infix operators.

A ground atom is one that does not contain a variable. The grounding of a program is obtained by replacing the variables in the clauses by the ground terms. Note that if there are function symbols, the grounding contains countably infinitely many clauses.

Logic programs are important because they have:

- A logical interpretation in terms of truth values of clauses (or their completion [Clark, 1978]). A logic program is a logical sentence from which one can ask for logical consequences.
- A procedural semantics (or fixed-point semantics). A logic program is a non-deterministic pattern-matching language where predicate symbols are procedures and function symbols give data structures.
- A database semantics in terms of operations on the relations denoted by predicate symbols. As a database language, logic programs are more general than the relational algebra as logic programs allow recursion and infinite relations using function symbols.

---

[1] It can represent infinite distributions, just not arbitrarily complex ones. A simple counting argument says that no finite language can represent arbitrary probability distributions over countable structures.

Thus a logic program can be interpreted as logical statements, procedurally or as a database and query language.

Logic programming research has gone in two general directions. In the first, are those frameworks, such as acyclic logic programs [Apt and Bezem, 1991], that ensure there is a single model for any logic program. Acyclic logic programs assume that all recursions for variable-free queries eventually halt. In particular, a program is acyclic if there is assignment of an natural number to each ground atom so that the natural number assigned to the head of each clause in the grounding of the program is greater than the atoms in the body. The acyclicity disallows programs such as $\{a \leftarrow \neg a\}$ and $\{a \leftarrow \neg b, b \leftarrow \neg a\}$. Intuitively, acyclic programs are those that don't send top-down interpreters, like Prolog, into infinite loops. Under this view, cyclic programs are buggy as recursions are not well founded.

The other direction, exemplified by Answer Set Programming [Lifschitz, 2002], allows for cyclic theories, and considers having multiple models as a virtue. These multiple models can correspond to multiple ways the world can be. Baral et al. [2004] have investigated having probability distributions over answer sets.

The stable model semantics [Gelfond and Lifschitz, 1988] provides a semantics for logic programs where the clauses contain negations in the body (i.e., for "negation as failure"). The stable model semantics is particularly simple with acyclic logic programs [Apt and Bezem, 1991]. A model specifies the truth value for each ground atom. A stable model M is a model where a ground atom a is true in M if and only if there is a clause in the grounding of the logic program with a as the head where the body is true in M. An acyclic logic program has a unique stable model.

Clark [1978] gives an alternative definition of negation as failure in terms of completions of logic programs. Under Clark's completion, a logic program means a set of if and only if definitions of predicates. Acyclic logic programs have the nice property that what is true in the unique stable model corresponds to what logically follows from Clark's completion.

2.2 Belief Networks

A belief network or Bayesian network [Pearl, 1988] is a representation of independence amongst a finite set of random variables. In this paper we will write random variables starting with a *lower case* letter, so they are not confused with logical variables.

In particular, a belief network uses an acyclic directed graph to represent the dependence amongst a finite set of random variables: the nodes in the graph are the variables, and the network represents the independence assumption that a node is independent of its non-descendants given its parents. That is, if x is a random variable, and $par(x)$ is the set of parents of in the graph, then $P(x|par(x), u) = P(x|par(x))$. It follows from the chain rule of probability that if $x_1, \ldots, x_n$ are all of the variables, then

$$P(x_1, \ldots, x_n) = \prod_{i=1}^{n} P(x_i|par(x_i))$$

The probability distribution $P(x_1, \ldots, x_n)$ can be used to define any conditional probability.

Bayesian networks have become very popular over the last 20 years, essentially because:

- The independence assumption upon which belief networks are based is useful in practice. In particular, causality would be expected to obey the independence assumption if the direct causes of some event are the parents. The notion of locality of influence is a good modelling assumption for many domains.
- There is a natural specification of the parameters. Humans can understand and debug the structure and the numbers. The specification in terms of conditional probability means that the probabilities can be learned independently; adding and removing variables from the models only have a local effect.
- There are algorithms that can exploit the structure for efficient inference.
- The probabilities and the structure can be learned effectively.

Note that the first two properties are not true of undirected models such as Markov networks (see e.g., Pearl [1988], pages 107–108). Markov Logic Networks [Richardson and Domingos, 2006] inherit all of the problems of undirected models.

Example 1. Consider the problem of diagnosing errors that students make on simple multi-digit addition problems [Brown and Burton, 1978]:

$$
\begin{array}{r}
x_2 \; x_1 \\
+ \quad y_2 \; y_1 \\
\hline
z_3 \; z_2 \; z_1
\end{array}
$$

The students are presented with the digits x_1, x_2, y_1 and y_2 and are expected to provide the digits z_1, z_2 and z_3. From observing their behaviour, we want to infer whether they understand how to do arithmetic, and if not, what they do not understand so that they can be taught appropriately.

For simplicity, let's assume that the student can make systematic and random errors on the addition of digits and on the carrying. This problem can be represented as a belief network as depicted in Figure 1. The tabular representation of the conditional probabilities often used in belief networks makes the representation cumbersome, but it can be done.

In this figure, x_1 is a random variable with domain $\{0, \ldots, 9\}$ that represents the digit in the top-right. Similarly for the variables $x_2, y_1, y_2, z_1, z_2, z_3$. The *carry$_i$* variables represent the carry into digit i, and have domain $\{0, 1\}$. The Boolean variable *knows_addition* represents whether the student knows basic addition of digits. (A Boolean variable has domain $\{true, false\}$). The variable *knows_carry* has domain $\{knowCarry, carryZero, carryRandom\}$ representing whether they know how to carry, whether they always carry zero, or whether they carry randomly (i.e., we are modelling more than one error state). By conditioning on the observed values of the x_i, y_i and z_i variables, inference can be used to compute the probability that a student knows how to carry or knows addition.

The belief network representation for the domain of diagnosing students with addition problems becomes impractical when there are multiple problems each with multiple digits, multiple students, who take the same or different problems at different times (and change their skills through time).

One way to represent this is to duplicate nodes for the different digits, problems, students and times. The resulting network can be depicted using plates [Buntine, 1994], as

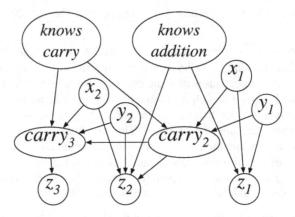

Fig. 1. A belief network for the simple addition example

in Figure 2. The way to view this representation is that there are copies of the variables in the Student/Time plate for each student-time pair. There are copies of the variables in the Digit/Problem plate for each digit-problem pair. Think of the plates as coming out of the diagram. Thus, there are different instances of variables x and y for each digit and each problem; different instances of *carry* and z for each digit, problem, student and time; and different instances of *knowsCarry* and *knowsAddition* for each student and time. Note that there can be cycles in this graph when the dependence is on different instances of variables. For example, there is a cycle on the node *carry* as the carry for one digit depends on the carry from the previous digit. The cycle on *knowsCarry* is because a student's knowledge of carrying depends on the student's knowledge at the previous time.

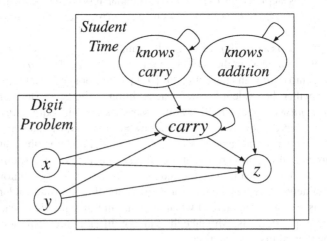

Fig. 2. A belief network with plates addition example

The plate notation is very convenient and natural for many problems and leads to what could be called parametrized belief networks that are networks that are built from templates [Horsch and Poole, 1990]. Note that it is difficult to use plates when one variable depends on different instances of the same relation. For example, if whether two authors collaborate depends on whether they have coauthored a paper (collaborate depends on two different instances of an author relationship). Heckerman et al. [2004] show how plates relate to PRMs and discuss limitations of both models.

2.3   The Independent Choice Logic

The ICL can either be seen as adding independent stochastic inputs to a logic program, or as a way to have rule-based specification of Bayesian networks with logical variables (or plates).

We assume that we have atomic formulae as logic programming. We allow function symbols.

An **atomic choice** is an atom that does not unify with the head of any clause. An **alternative** is a set of atomic choices. A **choice space** is a set of alternatives such that the atomic choices in different alternatives do not unify.

An ICL theory consists of:

F: The facts, an acyclic logic program.

C: A choice space, which is a set of sets of atoms. The elements of the choice space are called alternatives. The elements of the alternatives are called atomic choices. Atomic choices in the same or different alternatives cannot unify with each other. Atomic choices cannot unify with the head of any clause in F.

P_0: A probability distribution over the alternatives in C. That is $P_0 : \cup \mathbf{C} \rightarrow [0, 1]$ such that

$$\forall \chi \in \mathbf{C} \sum_{\alpha \in \chi} P_0(\alpha) = 1$$

The restrictions on the unification of atomic choices are there to enable a free choice of an atomic choice from each alternative.

Example 2. Here is a meaningless example (that is simple enough to show the semantics below):

$$\mathbf{C} = \{\{c_1, c_2, c_3\}, \{b_1, b_2\}\}$$

$$\mathbf{F} = \{f \leftarrow c_1 \wedge b_1, f \leftarrow c_3 \wedge b_2,$$
$$d \leftarrow c_1, \qquad d \leftarrow \neg c_2 \wedge b_1,$$
$$e \leftarrow f, \qquad e \leftarrow \neg d\}$$

$$P_0(c_1) = 0.5 \ P_0(c_2) = 0.3 \ P_0(c_3) = 0.2$$
$$P_0(b_1) = 0.9 \ P_0(b_2) = 0.1$$

2.4 Semantics

The semantics is defined in terms of possible worlds.

A **total choice** for choice space C is a selection of exactly one atomic choice from each grounding of each alternative in C. Note that while choice spaces can contain free variables, a total choice does not.

There is a **possible world** for each total choice. What is **true** in a possible world is defined by the atoms chosen by the total choice together with the logic program. In particular an atom is true in a possible world if it is in the (unique) stable model of the total choice corresponding to the possible world together with the logic program [Poole, 2000b]. The acyclicity of the logic program and the restriction that atomic choices don't unify with the head of clauses guarantees that there is a single model for each possible world.

For the case where there are only a finite number of possible worlds (when there are no function symbols), the probability for a possible world is the product of the probabilities of the atomic choices that make up the possible world. The probability of any proposition is the sum of the probabilities of the possible worlds in which the proposition is true.

Example 3. In the ICL theory of example 2, there are six possible worlds, and each possible world can be given a probability:

$$
\begin{aligned}
w_1 &\models c_1 \; b_1 \quad f \quad d \quad e \qquad P(w_1) = 0.45 \\
w_2 &\models c_2 \; b_1 \; \neg f \; \neg d \quad e \qquad P(w_2) = 0.27 \\
w_3 &\models c_3 \; b_1 \; \neg f \quad d \; \neg e \qquad P(w_3) = 0.18 \\
w_4 &\models c_1 \; b_2 \; \neg f \quad d \; \neg e \qquad P(w_4) = 0.05 \\
w_5 &\models c_2 \; b_2 \; \neg f \; \neg d \quad e \qquad P(w_5) = 0.03 \\
w_6 &\models c_3 \; b_2 \quad f \; \neg d \quad e \qquad P(w_6) = 0.02
\end{aligned}
$$

The probability of any proposition can be computed by summing the measures of the worlds in which the proposition is true. For example

$$P(e) = \mu(\{w_1, w_2, w_5, w_6\}) = 0.45 + 0.27 + 0.03 + 0.02 = 0.77$$

where μ is the measure on sets of possible worlds.

When there are function symbols, there are infinitely many possible worlds, and there needs to be a more sophisticated definition of probabilities. To define the probabilities, we put a measure over sets of possible worlds. In particular, we define the sigma algebra of sets of possible worlds that can be described by finite logical formulae of ground atomic choices. We define a probability measure over this algebra as follows. Any formula of ground atomic choices can be put into disjunctive normal form, such that the disjuncts are mutually exclusive. That is, it can be put in the form

$$(a_{11} \wedge \ldots \wedge a_{1k_1}) \vee \ldots \vee (a_{m1} \wedge \ldots \wedge a_{mk_m})$$

such that for each $i \neq j$ there is some such s, t that a_{is} and a_{jt} are in the grounding of an alternative. The set of all possible worlds where this formula is true, has measure

$$\sum_i \prod_j P_0(a_{ij})$$

Note that this is assuming that different ground instances of alternatives are probabilistically independent.

The probability of a proposition is the measure of the set of possible worlds in which the proposition is true.

2.5   ICL and Belief Networks

It may seem that, with independent alternatives, that the ICL is restricted in what it can represent. This is not the case; the ICL can represent anything that is representable by a belief network. Moreover the translation is local, and there is the same number of alternatives as there are free parameters in the belief network.

A random variable having a value is a proposition and is represented in ICL as a logical atom. Random variable x having value v can be represented as the atom $x(v)$, with the intended interpretation of $x(v)$ being that x has value v. We can also use the infix relation $=$ and write $x = v$. Note that there is nothing special about $=$ (the only thing built-in is that it is an infix operator). As long as you use a consistent notation for each random variable, any syntax can be used, except that names that start with an upper case represent logical variables (following the Prolog tradition), not random variables.

A Boolean random variable x can be optionally directly represented as an ICL atom x; i.e., $x = true$ is x and $x = false$ is $\neg x$.

Conditional probabilities can be locally translated into rules, as in the following example.

Example 4. Suppose a, b and c are variables, where b and c are the parents of a. We could write the conditional probability as a single rule:

$$a(A) \leftarrow b(B) \wedge c(C) \wedge na(A, B, C)$$

where there is an alternative for $na(A, B, C)$ for each value of B and C. For example, if a, b and c were Boolean (have domain $\{true, false\}$), one alternative is:

$$\{na(true, true, false), na(false, true, false)\}$$

where $P_0(ca(true, true, false))$ has the same value as $P(a|b, \neg c)$ in the belief network.

As a syntactic variant, if you had decided to use the $=$ notation, this clause could be written as:

$$a = A \leftarrow b = B \wedge c = C \wedge aifbnc(A, B, C)$$

If a, b and c are Boolean variables, replacing the $P(a|b, \neg c)$ in the belief network, we could have rules such as

$$a \leftarrow b \wedge \neg c \wedge aifbnc$$

where $aifbnc$ is an atomic choice where $P_0(aifbnc)$ has the same value as the conditional probability as $P(a|b, \neg c)$ in the belief network.

As a syntactic variant, this can be abbreviated as

$$a \leftarrow b \wedge \neg c : p_0$$

where p_0 is $P(a|b, \neg c)$. In general

$$H \leftarrow b : p$$

is an abbreviation for $H \leftarrow b \wedge n$, where n is an atom that contains a new predicate symbol and contains all of the free variables of the clause. This notation, however, tends to confuse people (students who are familiar with logic programming), and they make more mistakes than when they have a clear separation of the logical and probabilistic parts. It is not as useful when there are logical variables, as there tends to be many numbers (as in the table for $aifbnc(A, B, C)$), often we don't want a fully parametrized atomic choice, and you often want to say what happens when the atomic choice is false.

The ICL representation lets us naturally specify context-specific independence [Boutilier et al., 1996; Poole, 1997a], where, for example, a may be independent of c when b is false but be dependent when b is true. Context-specific independence is often specified in terms of a tree for each variable; the tree has probabilities at the leaves and parents of the variable on the internal nodes. It is straightforward to translate these into the ICL. The logic program also naturally represents the "noisy or", when the bodies are not disjoint which is a form of causal independence [Zhang and Poole, 1996]. Standard algorithms such as clique-tree propagation are not good at reasoning with these representations, but there are ways to exploit noisy-or and context specific independence using modifications of variable elimination [Zhang and Poole, 1996; Díez and Galán, 2002; Poole and Zhang, 2003] or recursive conditioning [Allen and Darwiche, 2003].

Example 5. Continuing our addition example, it is difficult to specify a Bayesian network even for the simple case of adding two digits case as shown in Figure 1, as specifying how, say z_2 depends on its parents is non-trivial, let alone for arbitrary digits, problems, students and times.

In this example, we will show the complete axiomatization for z that runs in our CILog2 interpreter[2]. The plates of Figure 2 correspond to logical variables. For example, $z(D, P, S, T)$ gives the z value for the digit D, problem P, student S, at time T. $x(D, P)$ is the x value for digit D and problem P. $knowsCarry(S, T)$ is true when student S knows how to carry at time T.

The rules for z are as follows. Note that the symbol "=" here is just a syntactic sugar; it just as easily could be made the last argument of the predicate.

If the student knows addition and didn't make a mistake on this case, they get the right answer:

$$z(D, P, S, T) = V \leftarrow$$
$$x(D, P) = Vx \wedge$$
$$y(D, P) = Vy \wedge$$
$$carry(D, P, S, T) = Vc \wedge$$
$$knowsAddition(S, T) \wedge$$
$$\neg mistake(D, P, S, T) \wedge$$
$$V \text{ is } (Vx + Vy + Vc) \text{ div } 10.$$

---

[2] The complete code can be found at http://www.cs.ubc.ca/spider/poole/ci2/ code/cilog/CILog2.html

If the student knows addition but made a mistake, they pick a digit at random:

$z(D, P, S, T) = V \leftarrow$
$knowsAddition(S, T) \wedge$
$mistake(D, P, S, T) \wedge$
$selectDig(D, P, S, T) = V.$

If the student doesn't know addition, they pick a digit at random:

$z(D, P, S, T) = V \leftarrow$
$\neg knowsAddition(S, T) \wedge$
$selectDig(D, P, S, T) = V.$

The alternatives are:

$\{noMistake(D, P, S, T), mistake(D, P, S, T)\}$
$\{selectDig(D, P, S, T) = V \mid V \in \{0..9\}\}$

There are similar rules for $carry(D, P, S, T)$ that depend on $x(D, P)$, $y(D, P)$, $knowsCarry(S, T)$ and $carry(D1, P, S, T)$ where $D1$ is the previous digit. And similar rules for $knowsAddition(S, T)$ that depends on the previous time.

By observing x, y and z for a student on various problems, we can query on the probability the student knows addition and knows how to carry.

2.6   Unknown Objects

BLOG [Milch et al., 2005] claims to deal with unknown objects. In this section we will show how to write one of the BLOG example in ICL. First note that BLOG has many built-in procedures, and ICL (as presented) has none. I will simplify the example only to make concrete the distributions used which are not specified in the BLOG papers.

I will do the aircraft example, which is Example 3 of Milch et al. [2005]. As I will make the low levels explicit I will assume that the locations are a 10x10 grid and there are 8 directions of aircraft.

First we can generate a geometric distribution of the number of aircraft[3]. We can do this by generating the number by repeatedly asking whether there are any more; the resulting number follows a geometric distribution with parameter P (which must be specified when $numObj$ is called). $numObj(T, N, N1, P)$ is true if there are $N1$ objects of type T given there are N. Note that the type is provided because we want the number of planes to be independent of the number of other objects (even though it isn't strictly needed for this example):

$numObj(T, N, N, P) \leftarrow$
$\neg more(T, N, P).$

---

[3] I would prefer to have the initial number of aircraft and to allow for new aircraft to enter, but as the BLOG paper did not do this, I will not either. I will assume that all of the aircraft are in the grid at the start, these can leave, and no new aircraft can arrive.

$$numObj(T, N, N2, P) \leftarrow$$
$$\quad more(T, N, P) \wedge$$
$$\quad N1 \text{ is } N + 1 \wedge$$
$$\quad numObj(T, N + 1, N2, P).$$

Where the alternative is (in CILog syntax):

prob $more(T, N, P) : P.$

which means that $\{more(T, N, P), \neg more(T, N, P)\}$ is an alternative with $P(more(T, N, P)) = P$. Note that we could have equivalently used $noMore(T, N, P)$ instead of $\neg more(T, N, P)$.

Note also that if the probability changes with the number of objects (as in a Poisson distribution), that can be easily represented too.

We can define the number of aircraft using $numObj$ with $P = 0.8$:

$$numAircraft(N) \leftarrow$$
$$\quad numObj(aircraft, 0, N, 0.8).$$

The aircraft will just be described in terms of the index from 1 to the number of aircraft:

$$aircraft(I) \leftarrow$$
$$\quad numAircraft(N) \wedge$$
$$\quad between(1, N, I).$$

where $between(L, H, V)$ is true if $L \leq V \leq H$ and is written just as it would be in Prolog. Note that $aircraft(I)$ is true in the worlds where there are at least I aircraft.

We can now define the state of the aircraft. The state of the aircraft with consist of an x-coordinate (from 0 to 9), a y-coordinate (from 0 to 9), a direction (one of the 8 directions) and a predicate to say whether the aircraft is inside the grid. Let $xpos(I, T, V)$ mean the x-position of aircraft I at time T is V if it is in the grid, and the value V can be arbitrary if the aircraft I is outside the grid.

We will assume the initial states are independent, so they can be stated just in choices. One alternative defines the initial x-coordinate.

$$\{xpos(I, 0, 0), xpos(I, 0, 1), \ldots, xpos(I, 0, 9)\}$$

with $P_0(xpos(I, 0, V)) = 0.1$.

We can axiomatize the dynamics similarly to the examples of Poole [1997b], with rules such as:

$$xpos(I, next(T), P) \leftarrow$$
$$\quad xpos(I, T, PrevPos) \wedge$$
$$\quad direction(I, T, Dir) \wedge$$
$$\quad xDer(I, T, Dir, Deriv) \wedge$$
$$\quad P \text{ is } PrevPos + Deriv.$$

Where $xDer(I, T, Dir, Deriv)$ gives the change in the x-position depending on the direction. It can be defined using CILog syntax using alternatives such as:

prob $xDer(I, T, east, 0) : 0.2, xDer(I, T, east, 1) : 0.7, xDer(I, T, east, 2) : 0.1$

We can now define how blips occur. Let $blip(X, Y, T)$ be true if a blip occurs at position X-Y at time T. Blips can occur at random or because of aircraft. We could either have a distribution over the number of blips, as we did for aircraft, and have a random variable for the location of each blip. Alternatively, we could have a random variable for the existence of a blip at each location. To do the latter, we would have:

$blip(X, Y, T) \leftarrow$
$blipRandomlyOccurs(X, Y, T).$

Suppose that blips can randomly occur at any position with probability 0.02, independently for each position and time. This can be stated as:

prob $blipRandomlyOccurs(X, Y, T) : 0.1.$

To axiomatize how blips can be produced by aircraft we can write[4]:

$blip(X, Y, T) \leftarrow$
$aircraft(I) \wedge$
$inGrid(I, T) \wedge$
$xpos(I, T, X) \wedge$
$ypos(I, T, Y) \wedge$
$producesBlip(I, T).$

Aircraft produce blips where they are with probability 0.9 can be stated as:

prob $producesBlip(I, T) : 0.1.$

Observing blips over time, you can ask the posterior distributions of the number of aircraft, their positions, etc. [Poole, 1997b] gives some similar examples (but without uncertainty over the number).

Ultimately, you want to axiomatize the actual dynamics of blip production; how are blips actually produced? Presumably the experts in radars have a good intuition (which can presumably be combined with data).

The complete axiomatization is available at the CILog2 web site (Footnote 2). Note that the CILog implementation can solve this as it generates only some of the proofs and bounds the error [Poole, 1993a] (although not as efficiently as possible; see Section 3).

BLOG lets you build libraries of distributions in Java. ICL, lets you build them in (pure) Prolog. The main difference is that a pure Prolog program is an ICL program (and so the libraries will be in ICL), but a Java program is not a BLOG program.

---

[4] I will leave it as an exercise to the reader to show how to represent the case where there can be noise in the location of the blip.

2.7   ICL as a Programming Language

The procedural interpretation of logic programs gives another way to look at ICL. It turns out that any belief network can be represented as a deterministic system with (independent) probabilistic exogenous inputs [Pearl, 2000, p. 30]. One technique for making a probabilistic programming language is to use a standard programming language to define the deterministic system and to allow for random inputs. This is the basis for a number of languages which differ in the language used to specify the deterministic system:

- ICL uses acyclic logic programs (they can even have negation as failure) to specify the deterministic system
- IBAL [Pfeffer, 2001] uses an ML-like functional language to specify the deterministic system
- A-Lisp [Andre and Russell, 2002] uses Lisp to specify the deterministic system
- CES [Thrun, 2000] uses C++ to specify the deterministic system.

While each of these have their advantages, the main advantage if ICL is the declarative semantics and the relational view (it is also an extension of Datalog).

All of these can be implemented stochastically, where the inputs are chosen using a random number generator and the programming language then gives the consequences. You can do rejection sampling by just running the program in the randomly generated inputs, but the real challenge is to do more sophisticated inference which has been pursued by for all of these.

2.8   ICL, Abduction and Logical Argumentation

Abduction is a powerful reasoning framework. The basic idea of abduction is to make assumptions to explain observations. This is usually carried out by collecting the assumptions needed in a proof and ensuring they are consistent.

The ICL can also be seen as a language for abduction. In particular, if all of the atomic choices are assumable (they are abducibles or possible hypotheses). An **explanation**[5] for g is a consistent set of assumables that implies g. A set of atomic choices is consistent if there is at most one element in any alternative.

Recall that the semantic of ICL is defined in terms of a measure over set of atomic choices. Abduction can be used to derive those atomic choices over which the measure is defined.

Each of the explanations has an associated probability obtained by computing the product of the probabilities of the atomic choices that make up the explanation. If the rules for each atom are mutually exclusive, the probability of g can be computed by summing the probabilities of the explanations for g [Poole, 1993b, 2000b]. We need to do something a bit more sophisticated if the rules are not disjoint or contain negation as failure [Poole, 2000b]. In these cases we can make the set of explanations disjoint (in a similar way to build a decision tree from rules) or find the duals of the set of explanations of g to find the explanations of $\neg g$.

---

[5] We need to extend the definition of explanation to account for negation as failure. The explanation of $\neg a$ are the duals of the explanations of a [Poole, 2000b].

If we want to do evidential reasoning and observe *obs*, we compute

$$P(g|obs) = \frac{P(g \wedge obs)}{P(obs)}$$

In terms of explanations, we can first find the explanations for *obs* (which would give us $P(obs)$) and then try to extend these explanations to also explain g (this will give us $P(g \wedge obs)$). Intuitively, we explain all of the observations and see what these explanations also predict. This is similar to proposals in the non-monotonic reasoning community to mix abduction and default reasoning [Poole, 1989; Shanahan, 1989; Poole, 1990].

We can also bound the prior and posterior probabilities by generating only a few of the most plausible explanations (either top-down [Poole, 1993a] or bottom-up [Poole, 1996]). Thus we can use inference to find the best explanations to do sound (approximate) probabilistic reasoning.

2.9   Other Formalisms

There are other formalisms that combine logic programming and probabilities.

Some of them such as Bayesian Logic Programs [Kersting and De Raedt, 2007] use some of the theory of logic programming, but mean different things by their logic programs. For example, the *allHappy* program (where *allHappy(L)* is true if all elements of list L are happy):

allHappy([]).
allHappy([X|A]) ← happy(X) ∧ allHappy(A).

is a perfectly legal ICL program and means exactly the same as it does in Prolog. The meaning does not depend on whether there is uncertainty about whether someone is happy, or even uncertainty in the elements of the list. However, this is not a Bayesian logic program (even if *allHappy* was a Boolean random variable).

Bayesian logic program do not mean the same things by atoms in logic programs. In normal logic programs, atoms represent propositions (and so can be combined with logical expressions). In Bayesian logic programs they mean random variables. The rules in ICL mean implication, whereas in BLP, the rules mean probabilistic dependency. BLP does not allow negation as failure, but ICL does.

In ICL, familiar theorems of logic programs are also theorems of ICL. Any theorem about acyclic logic programs is also a theorem of ICL (as it is true in all possible worlds). In particular, Clark's completion [Clark, 1978] is a theorem of ICL. For example, the completion of *allHappy* is true in the ICL when the above definition of *allHappy* is given in the facts. This is true even if *happy(X)* is stochastic: *happy(a)* may have different values in different possible worlds, and so might *allHappy([a, b, c])*, but the completion is true in all of the possible worlds.

Kersting and De Raedt [2007] claim that "Bayesian logic programs [compared to ICL theories] have not as many constraints on the representation language, ..., have a richer representation language and their independence assumptions represent the causality of the domain". All these claims are false. ICL can represent arbitrary logic programs with no restrictions except the acyclicity restriction that recursions have to be well-founded (none of the textbooks on logic programming I know of contains any logic programs that are not acyclic, although they do have logic programs that contains cuts and input/output that ICL does not handle.). ICL is Turing-complete. Pearl [2000] *defines* causality in terms of structural equation models with exogenous noise. The ICL represents this causality directly with the structural equation models represented as logic programs.

Note that the logical reasoning in Bayesian logic programming constructs a belief network, whereas logical reasoning in ICL *is* probabilistic reasoning. Finding the proofs of the observation and query is enough to determine the posterior probability of the query [Poole, 1993a]. Reasoning in ICL has inspired many of the algorithms for Bayesian network inference [Zhang and Poole, 1996; Poole, 1996; Poole and Zhang, 2003]. We should not assume that we can just pass of the probabilistic inference problem to a general-purpose solver and expect it to work (although [Chavira et al., 2006] comes close to this goal) as there is much more structure in the high-level representations.

Stochastic logic programs [Muggleton, 1996] are quite different in their goal to the other frameworks presented here. Stochastic logic programs give probability distributions over proofs, rather than defining the probability that some proposition is true.

3   Ongoing Research

This section outlines some challenges that we are currently working on. These correspond to representation, inference and learning.

3.1   Continuous Variables

There is nothing in the ICL semantics that precludes having continuous random variables, or even continuously many random variables. In the example of aircraft and blips, we could have a continuous random variable that is the position of the aircraft. We could also have continuously many random variables, about whether there is a blip at each of continuously many x-y positions.

The semantics of ICL mimics the standard definition of the limiting process of a continuous variable. The logical formulae could describe finer partitions of the continuous space, and we get the standard definition.

However, reasoning in terms of finer partitions is not computationally satisfactory. It is better to define continuous variables in terms of a mixture of kernel functions, such as mixtures of Gaussian distributions, truncated Gaussians [Cozman and Krotkov, 1994] or truncated exponential distributions [Cobba and Shenoy, 2006]. This can be done by having Gaussian alternatives. Allowing Gaussian alternatives and conditions in the logic programs, means that the program has to deal with truncated Gaussians; but it also means that it is easy to represent truncated Gaussians in terms of logic programs with Gaussian alternatives.

There are still many problems to be solved to get this to work satisfactorily. It needs to be expressive enough to state what we need, but it also needs to be able to be reasoned with efficiently.

3.2 Reasoning in the ICL

To reason in the ICL we can either do

- Variable elimination (marginalization) to simplify the model [Poole, 1997a]. We sum out variables to reduce the detail of the representation. This is similar to partial evaluation in logic programs.
- Generating some of the explanations to bound the probabilities [Poole, 1993a, 1996]. If we generated all of the explanations we could compute the probabilities exactly, but there are combinatorially many explanations. It should be possible to combine this with recursive conditioning [Darwiche, 2001] to get the best of both worlds.
- Stochastic simulation; generating the needed atomic choices stochastically, and estimating the probabilities by counting the resulting proportions.

One of the things that we would like in a logical language like the ICL to allow lifted inference, where we don't ground out the theory, but reason at the first-order level, with unification. There have been a number of attempts at doing this for various simple languages [Poole, 1997a, 2003; de Salvo Braz et al., 2005], but the final solution remains elusive. The general idea of [Poole, 2003] is that we can do lifted reasoning as in theorem proving or as in Prolog, using unification for matching, but instead of applying a substitution such as $\{X/c\}$, we need to split on $X = c$, giving the equality case, $X = c$, and the inequality case, $X \neq c$. Lifted inference gets complicated when we have aggregation, such as the "or" in the logic program, as shown in the following example:

Example 6. Consider the simple ICL theory with a single clause

$$f \leftarrow e(Y).$$

where the alternative is $\{e(Y), not\_e(Y)\}$, where $P(e(X)) = \alpha$. In this case, the probability of f depends on the number of individuals. In particular,

$$P(f) = 1 - (1 - \alpha)^n$$

where n is the population size (the number of individuals). It is of this form as we assume that the choices are independent. The probability of f can be computed in $O(\log n)$ time. The challenge is to define a general algorithm to compute probabilities in time less than linear in the population size.

This example has shown why we need to type existential variables, and a population size for each type.

There are more complicated cases where how to solve it in time less than linear in the population size is part of ongoing research:

Example 7. Consider the rules:

$$f \leftarrow e(Y).$$
$$e(Y) \leftarrow d(Y) \wedge n_1(Y).$$
$$e(Y) \leftarrow \neg d(Y) \wedge n_2(Y).$$

$$d(Y) \leftarrow c(X, Y).$$
$$c(X, Y) \leftarrow b(X) \wedge n_3(X, Y).$$
$$c(X, Y) \leftarrow \neg b(X) \wedge n_4(X, Y).$$
$$b(X) \leftarrow a \wedge n_5(X)$$
$$b(X) \leftarrow \neg a \wedge n_6(X)$$
$$a \leftarrow n_7$$

Where the n_i are atomic choices. Suppose $P(n_i(\cdot)) = \alpha_i$. There are thus, 7 numbers to specify, but the interdependence is very complicated. Suppose X has domain size n and the Y has domain size m, and the grounding of e is $e_1, \ldots, e_m$, and similarly for the other variables. The grounding can be seen as the belief network of Figure 3.

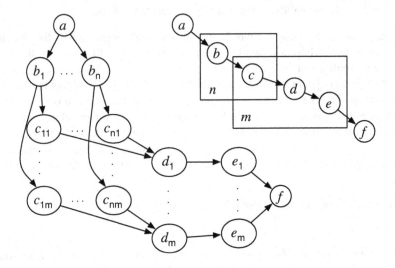

Fig. 3. A belief network and plate representation from Example 7

Note that the parents of f are all interdependent. Not only do they depend on a, but on each of the b variables. It is still an open problem to be able to solve networks such as this in time that is less than linear in n and m.

3.3   ICL and Learning

There is a large body of work on learning and belief networks. This means either:

- Using the belief network as a representation for the problem of Bayesian learning of models [Buntine, 1994]. In Bayesian learning, we want the posterior distribution of hypotheses (models) given the data. To handle multiple cases, Buntine uses the notion of plates that corresponds to the use of logical variables in the ICL [Poole, 2000a]. Poole [2000a] shows the tight integration of abduction and induction. These papers use belief networks and the ICL to learn various representations including decision trees and neural networks, as well us unsupervised learning.

- Learning the structure and probabilities of belief networks [Heckerman, 1995]. There has been much work on learning parameters for the related system called PRISM [Sato and Kameya, 2001].

There are a number of reasons that the ICL makes a good target language for learning:

- Being based on logic programming, it can build on the successes of inductive logic programming [Muggleton and De Raedt, 1994; Quinlan and Cameron-Jones, 1995; Muggleton, 1995]. The fact that parts of ICL theories are logic programs should aid in this effort.
- There is much local structure that naturally can be expressed in the ICL that can be exploited. One of the most successful methods for learning Bayesian networks is to learn a decision tree for each variable given its predecessors in a total ordering [Friedman and Goldszmidt, 1996; Chickering et al., 1997]. These decision trees correspond to a particular form of ICL rules.
- Unlike many representations such as Probabilistic Relational Models [Getoor et al., 2001], the ICL is not restricted to a fixed number of parameters to learn; it is possible to have representations where each individual has associated parameters. This should allow for richer representations and so better fit to the data. That is, it lets you learn about particular individuals.

3.4   Existence and Identity

The first-order aspects of theorem proving and logic programming are based on Skolemization and Herbrand's theorem. See for example [Chang and Lee, 1973].

Skolemization is giving a name to something that is said to exist. For example, consider the formula $\forall X \exists Y p(X, Y)$. When Skolemizing, we name the Y that exists for each X, say $f(X)$. The Skolemized form of this formula is then $p(X, f(X))$. After Skolemization, there are only universally quantified variables.

Herbrand's theorem [1930] says:

- If a logical theory has a model it has a model where the domain is made of ground terms, and each term denotes itself.
- If a logical theory T is unsatisfiable, there is a finite set of ground instances of formulae of T which is unsatisfiable.

This theorem is the basis for the treatment of variables in theorem proving and logic programming. We may as well reason in terms of the grounding. It also implies that if a logical theory does not contain equality as a built-on predicate, that we can reason as though different ground terms denote different objects.

As soon as we have negation as failure, the implicit assumption that ground terms denote different objects needs to be made explicit.

Languages such as the ICL make two very strong assumptions:

- You know all of the individuals in the domain
- Different constants denote different individuals

The first assumption isn't a problem in logic programming and theorem proving, due to Herbrand's theorem. The second assumption, the unique names assumption, is very common in logic programming. Lifting the second assumption is important when we want to reason about identity uncertainty.

To consider the first assumption, suppose that you want to have a naive Bayesian model of what apartment a person likes. Suppose you want to say that if a person likes an apartment, then it's very likely there exists a bedroom in the apartment that is large and green. To do this, it is common to create a constant, say c, for the room that exists. There are two main problems with this. First is semantic, you need to reason about the existence of the room. A common solution is to have existence as a predicate [Pasula et al., 2003; Laskey and da Costa, 2005]. Unfortunately this doesn't work in this situation, as it has to be clear exactly what doesn't exist when the predicate is false, and the properties of the room do not make sense when this room doesn't exist. The second reason is more pragmatic: it isn't obvious how to condition on observations, as you may not know which room that you have observed corresponds to the room c. For example, consider the case where you have observed a green bedroom, but you haven't observed its size, and a large bedroom, but you haven't observed its colour. It isn't well defined (or obvious) how to condition on the observation of c. A solution proposed in [Poole, 2007] is to only have probabilities over well-defined propositions, and for the theory to only refer to closed formulae; this avoids the need to do correspondence between objects in the model and individuals in the world when conditioning. We are currently integrating this idea with the independence assumptions of the ICL.

4   Conclusion

This paper has provided a brief overview of the Independent Choice Logic as well as some issues that arise from representing and reasoning about first-order probabilistic theories. There are some fundamental open issues that cut across representations that we have touched on this paper. Which representations will prove to be most effective remains to be seen.

Acknowledgements

This work was supported by the Natural Sciences and Engineering Research Council of Canada Operating Grant OGPOO44121. Thanks to Michael Chiang, Mark Crowley, Jacek Kisyński, Brian Milch, Kristian Kersting and the anonymous reviewers for valuable feedback.

References

Allen, D., Darwiche, A.: New advances in inference by recursive conditioning. In: Proceedings of the 18th International Joint Conference on Artificial Intelligence (IJCAI 2003), pp. 2–10. Morgan Kaufmann, San Francisco (2003)

Andre, D., Russell, S.: State abstraction for programmable reinforcement learning agents. In: Proc. AAAI 2002 (2002)

Apt, K.R., Bezem, M.: Acyclic programs. New Generation Computing 9(3-4), 335–363 (1991)

Baral, C., Gelfond, M., Rushton, N.: Probabilistic reasoning with answer sets. In: Proceedings of LPNMR7, pp. 21–33 (2004)

Boutilier, C., Friedman, N., Goldszmidt, M., Koller, D.: Context-specific independence in Bayesian networks. In: Horvitz, E., Jensen, F. (eds.) UAI 1996, Portland, OR, pp. 115–123 (1996)

Brown, J.S., Burton, R.R.: Diagnostic models for procedural bugs in basic mathematical skills. Cognitive Science 2, 155–191 (1978)

Buntine, W.L.: Operations for learning with graphical models. Journal of Artificial Intelligence Research 2, 159–225 (1994)

Chang, C.L., Lee, R.C.T.: Symbolic Logical and Mechanical Theorem Proving. Academic Press, New York (1973)

Chavira, M., Darwiche, A., Jaeger, M.: Compiling relational bayesian networks for exact inference. International Journal of Approximate Reasoning (IJAR) 42, 4–20 (2006)

Chickering, D.M., Heckerman, D., Meek, C.: A Bayesian approach to learning Bayesian networks with local structure. In: UAI 1997, pp. 80–89 (1997)

Clark, K.L.: Negation as failure. In: Gallaire, H., Minker, J. (eds.) Logic and Databases, pp. 293–322. Plenum Press, New York (1978)

Cobba, B.R., Shenoy, P.P.: Inference in hybrid bayesian networks with mixtures of truncated exponentials. International Journal of Approximate Reasoning 41(3), 257–286 (2006)

Cozman, F., Krotkov, E.: Truncated gaussians as tolerance sets. Technical Report CMU-RI-TR-94-35, Robotics Institute, Carnegie Mellon University, Pittsburgh, PA (September 1994)

Darwiche, A.: Recursive conditioning. Artificial Intelligence 126(1-2), 5–41 (2001)

de Salvo Braz, R., Amir, E., Roth, D.: Lifted first-order probabilistic inference. In: IJCAI 2005, Edinburgh (2005),
http://www.cs.uiuc.edu/~eyal/papers/fopl-res-ijcai05.pdf

Díez, F.J., Galán, S.F.: Efficient computation for the noisy max. International Journal of Intelligent Systems (to appear, 2002)

Friedman, N., Goldszmidt, M.: Learning Bayesian networks with local structure. In: UAI 1996, pp. 252–262 (1996), http://www2.sis.pitt.edu/~dsl/UAI/UAI96/Friedman1.UAI96.html

Gelfond, M., Lifschitz, V.: The stable model semantics for logic programming. In: Kowalski, R., Bowen, K. (eds.) Proceedings of the Fifth Logic Programming Symposium, Cambridge, MA, pp. 1070–1080 (1988)

Getoor, L., Friedman, N., Koller, D., Pfeffer, A.: Learning probabilistic relational models. In: Dzeroski, S., Lavrac, N. (eds.) Relational Data Mining, pp. 307–337. Springer, Heidelberg (2001)

Heckerman, D.: A tutorial on learning with Bayesian networks. Technical Report MSR-TR-95-06, Microsoft Research, March 1995. URL Revised (November 1996), http://www.research.microsoft.com/research/dtg/heckerma/heckerma.html

Heckerman, D., Meek, C., Koller, D.: Probabilistic models for relational data. Technical Report MSR-TR-2004-30, Microsoft Research (March 2004)

Horsch, M., Poole, D.: A dynamic approach to probabilistic inference using Bayesian networks. In: Proc. Sixth Conference on Uncertainty in AI, Boston, July 1990, pp. 155–161 (1990)

Kersting, K., De Raedt, L.: Bayesian logic programming: Theory and tool. In: Getoor, L., Taskar, B. (eds.) An Introduction to Statistical Relational Learning, MIT Press, Cambridge (2007)

Laskey, K.B., da Costa, P.G.C.: Of klingons and starships: Bayesian logic for the 23rd century. In: Uncertainty in Artificial Intelligence: Proceedings of the Twenty-First Conference (2005)

Lifschitz, V.: Answer set programming and plan generation. Artificial Intelligence 138(1–2), 39–54 (2002)

Lloyd, J.W.: Foundations of Logic Programming, 2nd edn. Symbolic Computation Series. Springer, Berlin (1987)

Milch, B., Marthi, B., Russell, S., Sontag, D., Ong, D.L., Kolobov, A.: BLOG: Probabilistic models with unknown objects. In: IJCAI 2005, Edinburgh (2005)

Muggleton, S.: Stochastic logic programs. In: De Raedt, L. (ed.) Advances in Inductive Logic Programming, pp. 254–264. IOS Press, Amsterdam (1996)

Muggleton, S.: Inverse entailment and Progol. New Generation Computing 13(3-4), 245–286 (1995)

Muggleton, S., De Raedt, L.: Inductive logic programming: Theory and methods. Journal of Logic Programming 19(20), 629–679 (1994)

Neumann, J.V., Morgenstern, O.: Theory of Games and Economic Behavior, 3rd edn. Princeton University Press, Princeton (1953)

Nilsson, N.J.: Logic and artificial intelligence. Artificial Intelligence 47, 31–56 (1991)

Pasula, H., Marthi, B., Milch, B., Russell, S., Shpitser, I.: Identity uncertainty and citation matching. In: NIPS, vol. 15 (2003)

Pearl, J.: Causality: Models, Reasoning and Inference. Cambridge University Press, Cambridge (2000)

Pearl, J.: Probabilistic Reasoning in Intelligent Systems: Networks of Plausible Inference. Morgan Kaufmann, San Mateo (1988)

Pfeffer, A.: IBAL: A probabilistic rational programming language. In: IJCAI 2001 (2001), http://www.eecs.harvard.edu/~avi/Papers/ibal.ijcai01.ps

Poole, D.: Logical generative models for probabilistic reasoning about existence, roles and identity. In: 22nd AAAI Conference on AI (AAAI 2007) (2007)

Poole, D.: Learning, Bayesian probability, graphical models, and abduction. In: Flach, P., Kakas, A. (eds.) Abduction and Induction: Essays on their relation and integration, Kluwer, Dordrecht (2000a)

Poole, D.: First-order probabilistic inference. In: Proc. Eighteenth International Joint Conference on Artificial Intelligence (IJCAI 2003), Acapulco, Mexico, pp. 985–991 (2003)

Poole, D.: Explanation and prediction: An architecture for default and abductive reasoning. Computational Intelligence 5(2), 97–110 (1989)

Poole, D.: A methodology for using a default and abductive reasoning system. International Journal of Intelligent Systems 5(5), 521–548 (1990)

Poole, D.: Representing diagnostic knowledge for probabilistic Horn abduction. In: IJCAI 1991, Sydney, pp. 1129–1135 (August 1991a)

Poole, D.: Representing Bayesian networks within probabilistic Horn abduction. In: UAI 1991, Los Angeles, July 1991, pp. 271–278 (1991b)

Poole, D.: Logic programming, abduction and probability: A top-down anytime algorithm for computing prior and posterior probabilities. New Generation Computing 11(3–4), 377–400 (1993a)

Poole, D.: Probabilistic Horn abduction and Bayesian networks. Artificial Intelligence 64(1), 81–129 (1993b)

Poole, D.: Probabilistic conflicts in a search algorithm for estimating posterior probabilities in Bayesian networks. Artificial Intelligence 88, 69–100 (1996)

Poole, D.: Probabilistic partial evaluation: Exploiting rule structure in probabilistic inference. In: IJCAI 1997, Nagoya, Japan, pp. 1284–1291 (1997a), http://www.cs.ubc.ca/spider/poole/abstracts/pro-pa.html

Poole, D.: The independent choice logic for modelling multiple agents under uncertainty. Artificial Intelligence 94, 7–56 (special issue on economic principles of multi-agent systems) (1997b), http://www.cs.ubc.ca/spider/poole/abstracts/icl.html

Poole, D.: Abducing through negation as failure: stable models in the Independent Choice Logic. Journal of Logic Programming 44(1–3), 5–35 (2000), http://www.cs.ubc.ca/spider/poole/abstracts/abnaf.html

Poole, D., Zhang, N.L.: Exploiting contextual independence in probabilistic inference. Journal of Artificial Intelligence Research 18, 263–313 (2003)

Poole, D., Mackworth, A., Goebel, R.: Computational Intelligence: A Logical Approach. Oxford University Press, New York (1998)

Quinlan, J.R., Cameron-Jones, R.M.: Induction of logic programs: FOIL and related systems. New Generation Computing 13(3-4), 287–312 (1995)

Richardson, M., Domingos, P.: Markov logic networks. Machine Learning 62, 107–136 (2006)

Sato, T., Kameya, Y.: Parameter learning of logic programs for symbolic-statistical modeling. Journal of Artificial Intelligence Research (JAIR) 15, 391–454 (2001)

Savage, L.J.: The Foundation of Statistics, 2nd edn. Dover, New York (1972)

Shanahan, M.: Prediction is deduction, but explanation is abduction. In: IJCAI-1989, Detroit, MI, pp. 1055–1060 (August 1989)

Thrun, S.: Towards programming tools for robots that integrate probabilistic computation and learning. In: Proceedings of the IEEE International Conference on Robotics and Automation (ICRA), San Francisco, CA, IEEE, Los Alamitos (2000)

Zhang, N.L., Poole, D.: Exploiting causal independence in Bayesian network inference. Journal of Artificial Intelligence Research 5, 301–328 (1996)

Protein Fold Discovery Using Stochastic Logic Programs

Jianzhong Chen[1], Lawrence Kelley[2], Stephen Muggleton[1], and Michael Sternberg[2]

[1] Department of Computing, Imperial College London, London SW7 2AZ, UK
{cjz,shm}@doc.ic.ac.uk
[2] Department of Biological Sciences, Imperial College London, London SW7 2AZ, UK
{l.a.kelley,m.sternberg}@imperial.ac.uk

Abstract. This chapter starts with a general introduction to *protein folding*. We then present a probabilistic method of dealing with multi-class classification, in particular multi-class protein fold prediction, using Stochastic Logic Programs (SLPs). Multi-class prediction attempts to classify an observed datum or example into its proper classification given that it has been tested to have multiple predictions. We apply an SLP parameter estimation algorithm to a previous study in the protein fold prediction area, in which logic programs have been learned by Inductive Logic Programming (ILP) and a large number of multiple predictions have been detected. On the basis of several experiments, we demonstrate that PILP approaches (eg. SLPs) have advantages for solving multi-class (protein fold) prediction problems with the help of learned probabilities. In addition, we show that SLPs outperform ILP plus majority class predictor in both predictive accuracy and result interpretability.

1 Introduction to Protein Folding

1.1 The Importance of Proteins in Biology

Proteins are the molecular machines that drive and control virtually all of the features of biological organisms at the molecular level. The primary function of DNA and the genes contained therein is to store the instructions for creating proteins. Many proteins are enzymes that catalyze biochemical reactions, and are vital to metabolism. Other proteins have structural or mechanical functions, such as the proteins in the cytoskeleton, which forms a system of scaffolding that maintains cell shape. Proteins are also central in cell signalling, immune responses, cell adhesion, and cell division for growth and reproduction. An understanding of how proteins perform this diverse range of functions allows a deep insight into how organisms function as a whole, how they evolved, and how to treat disease.

1.2 The Importance of Protein Structure

A protein is a biological macromolecule composed of a series of smaller, building-block molecules called amino acids, linked together in a linear chain, like beads on a string. Amino acids are small molecules containing nitrogen, hydrogen, carbon, oxygen and sometimes sulphur. There are 20 different types of amino acid found in biological systems and they range in size from 6 to 25 atoms. A protein can be thought of as a

L. De Raedt et al. (Eds.): Probabilistic ILP 2007, LNAI 4911, pp. 244–262, 2008.

linear sequence of letters taken from the alphabet of 20 amino acids. Proteins in biology range from between 30 amino acids in length up to several thousand, with a typical size of 200.

The actual protein sequences found in nature are derived from the genetic sequences of DNA that we call genes. A gene is essentially a stretch of DNA composed of the four letter alphabet A,T,G,C. Within the cell, molecular machinery (itself composed of an amalgam of protein and nucleic acid) converts such stretches of DNA (genes) into stretches of amino acids linked in a chain (a protein) through a straightforward mapping where triplets of DNA bases (A,T,G,C) map to specific amino acid types. In this way, genes are the instructions for making proteins: a mapping of strings from an alphabet of 4 to strings from an alphabet of 20.

Thus a bewildering diversity of protein sequences are possible. Currently over 4 million different protein sequences have been elucidated by gene sequencing projects around the world. Human beings are currently estimated to have approximately 30,000 genes. Each of these genes 'codes' for a unique protein. How do proteins carry out such a diverse range of complex biological functions?

Protein sequences, that is, linear polymers of amino acids do not simply remain in a loose extended, or even random shape. The power of proteins hinges on the fact that protein sequences found in nature fold up into compact, complex three-dimensional shapes. These complex shapes, or folds, are generally stable and the same sequence will reliably fold to the same shape. However, different sequences fold into different shapes. It is the highly complex, nonlinear interaction of regions of the chain with itself via a variety of physical and chemical forces that means the sequence determines the final folded structure. It is the specific three-dimensional shape of a folded protein that provides it with the power to carry out highly specific molecular functions, such as specifically binding to other proteins or DNA. Or specifically binding a small drug molecule, breaking specific bonds in the drug, and releasing the pieces. The complex folds of proteins can be likened to specifically shaped molecular 'hands' with the power to make and break chemical bonds, transport electrons, and interact with one another. The central principle of structural biology is that "**sequence determines structure, and structure determines function**". This principle has been called "the other half of the genetic code" (Fig. 1).

Given the three-dimensional structure of a protein one can gain deep insights into how the protein carries out its function. In addition, it permits researchers to design methods to intervene in its function by rationally designing drug molecules to bind to specific three-dimensional features of the protein molecule. This is of particular importance in the case of proteins involved in disease states.

The biological community has now gathered vast amounts of data regarding the sequences of genes and their corresponding proteins across hundreds of species. However, elucidating the three-dimensional shape of a folded protein by experimental methods is an extremely costly, time-consuming and sometimes impossible task. As a result only a tiny fraction (42,000) of the proteins found in nature (4 million so far) have had their three-dimensional structure experimentally determined (Data as of March 2007). Naturally, since the three-dimensional structure of a protein is determined by the amino acid sequence, and given that we know the amino acid sequence, one would imagine it to

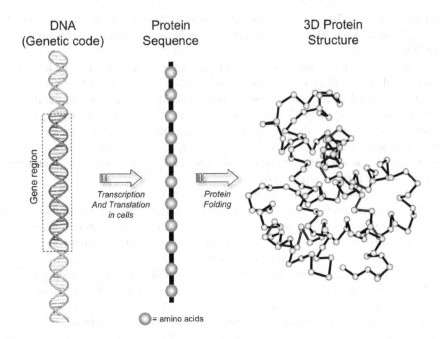

Fig. 1. Schematic illustration of the flow of biological information from a gene sequence in DNA, to a protein sequence, which then folds to its unique three-dimensional shape. This three-dimensional fold can then perform its specific biological function. This process is known as the sequence, structure, function paradigm.

be a straightforward task to model the folding process given the sequence and to thus computationally determine the folded structure.

1.3   Computational Protein Folding Is Extremely Difficult

Unfortunately, understanding the relationship between the amino acid sequence of a protein and its resulting three-dimensional structure is one of the major long-standing problems in molecular biology. This is known as the *protein folding* problem and has been under intensive study by research groups around the world for almost 30 years. The protein folding problem is so difficult primarily for two reasons: 1) The computational resources needed to run a physics (or quantum physics) simulation are so vast that even taking Moore's law into account, it would be at least 50 years before we could simulate the folding of a very small protein and hundreds of years before we could model a typical protein. 2) Heuristic approaches do not work because of the size of the protein conformational space, inadequacies in the approximations, and the nonlinearity of the problem. It is a conservative approximation to say each amino acid along the protein chain has five degrees of freedom. Thus a 100 amino acid protein has approximately 5100 possible conformations, which is more than the number of quarks in the observable universe. Even when reasonable heuristics are used to significantly restrict this search space, there is still a problem of determining whether a given conformation

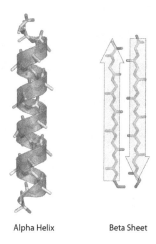

Alpha Helix   Beta Sheet

Fig. 2. Most proteins are composed of combinations of simple substructures known as secondary structure elements. This figure illustrates the basic connectivity of the chain (with many atoms omitted for clarity) of the alpha helix and the beta-strand. Cartoon representations are superposed on the protein chain for ease of visualisation.

is correct. Dozens of papers are written each year with new attempts at designing an 'energy function' that can pick out correctly folded computer models from incorrect ones, but with little success.

1.4 Fold Space

Fortunately however, the repertoire of three-dimensional folds found in nature appears to be severely limited. Of the 42,000 experimentally determined protein structures, there are less than 1000 fundamentally different 'folds'. That is, many of the protein structures seen in nature bear strong resemblances to one another when one considers the general path of the protein chain in space. In addition, almost all protein structures seen in nature are composed of packed arrangements of more regular, more primitive substructures, known as alpha-helices and beta-strands (Fig. 2). These substructure types are termed the secondary structure elements (in contrast with the primary structure which is the 1 dimensional sequence of amino acids). Multiple helices and/or strands pack together in space to form the tertiary structure of the protein (Fig. 3). This hierarchical organisation has been taken further in databases such as SCOP [1] and CATH [2]. In SCOP (Structural Classification Of Proteins), tertiary protein structures are grouped into families, superfamilies, folds, and classes. This classification has been designed and maintained by world experts on protein structure and is based on a combination of evolutionary and structural/topological considerations. Understanding or discovering rules that govern the structure of fold space, i.e. how patterns of secondary structure elements partition the space of folds, is an important step in deepening our understanding of the sequence, structure, function paradigm.

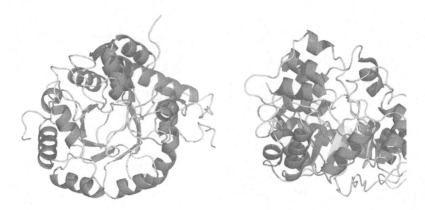

Fig. 3. Two views of the tertiary structure of a typical protein fold. It can be seen that the protein is a complex combination of alpha helices and beta strands connected by variable loop regions. In this particular protein, there is a central barrel structure surrounded by helices. This general fold is known as a TIM-Barrel after the first such structure solved with this fold - Triose phosphate Isomerase.

2   Problems to Be Addressed

2.1   Multi-class Prediction/Classification

Multi-class classification is a central problem in machine learning, as applications that require a discrimination among several classes are ubiquitous [3]. We consider the problem of multi-class prediction/classification[1] using Probabilistic Inductive Logic Programming (PILP) approaches [4]. A conventional Inductive Logic Programming (ILP) program is given with a training data set consisting of examples belonging to $N > 2$ different classes, and the goal is to construct a method that, given a new unlabeled datum, will correctly predict the class to which the datum belongs[2].

From machine learning point of view, solving the above problem requires us to deal with *multi-class prediction* rather than the *binary classification* approach[3] used in the original study. Generally speaking, binary classification can be used to predict whether an example belong to a class or not, whereas multi-class prediction can classify an example into one class from multiple ambiguous predictions based on some given 'ranking' or 'leveraging' mechanism. Precisely, a *binary predictor* defines a function

---

[1] It is also called multiclass prediction, multiple classification or multi-classification in some references.

[2] We distinguish the case where each datum is required to belong to a single class from the other case where a given example is allowed to be a member of more than one class simultaneously. The former case of requirement is assumed in our framework, where multiple predictions have been detected due to some reasons in practice and the goal is to solve the uncertainty from the observations so that a single prediction could be made correctly for each datum.

[3] When there are multiple classes and the class labels are assumed to be independent, a conventional ILP classifier actually provides a set of binary classifiers, each of which is used to distinguish whether or not an example is in a certain class.

f that maps an example e and a class label cl to a binary set, ie. $f : (e, cl) \mapsto \{yes, no\}$; and a *multi-class predictor* defines a function g that maps an example e and a set of class labels $\{cl_1, \ldots, cl_m\}$ to one class label $cl_i \in \{cl_1, \ldots, cl_m\}$ with some ranking mechanism r_i, ie. $g : \{(e, cl_1), \ldots, (e, cl_m)\} \mapsto (e, cl_i, r_i), m > 1, 1 \leq i \leq m$. Multi-class predictor is more useful for unlabeled/unseen data classification. In majority voting, the ranking mechanism is the class size, ie. the number of predicted examples of a class. In PILP approaches, the class-conditional probabilities are computed for each example e as the ranking mechanism, which specify a distribution of the prediction probabilities of e over multiple classes.

While binary classification [6,7], which classifies the members of a given set of objects into two groups on the basis of whether they have some property or not, is well understood, multi-class classification requires extra techneques. Most of the current multi-class classification techniques are developed in the discriminative classification methods, including decision trees, kernel methods, support vector machine and neural networks. Standard one uses measures that go back to CN2 and ID3. Some of them extend the binary classification algorithms to handle multi-class problems directly, such as decision trees, regression, discriminant analysis, etc [7,8]. The others build multi-class methods on the basic binary classification methods, such as one-versus-others, pairwise classification, all-versus-all, error-correcting output coding, etc [3,6,9,10]. There has also been some work on the combination of these methods with probabilistic modeling [10,11]. The above approaches have limited relevance to ILP-based classifiers, as most of them are based on regularization, modeling the decision boundaries or evaluating several binary classification methods. In logic-based classification methods, such as ILP, majority voting is often used to solve the multiple prediction problems, however the performance depends on the empirical distribution and the (im)balance feature of data.

To solve the multiple prediction uncertainty that naturally exists in the ILP classifiers, we use PILP techniques, which aim at integrating three underlying constituents: statistical learning and probabilistic reasoning within logical or relational knowledge representations [4]. There have been increasing number of attempts to use PILP methods in practical settings recently [12,13]. In this chapter, we present applications of Stochastic Logic Programs (SLPs) [14], one of the existing PILP frameworks, to learn probabilistic logic programs that help to solve the multi-class prediction problem detected in a protein fold prediction study and a working example. We apply a comparative experimental strategy to demonstrate our method in which SLPs are learned from the existing ILP programs and training data, and then the results, including the predictive accuracy and interpretability, are compared between SLP predictors against ILP plus majority class predictors.

2.2 Multi-class Protein Fold Prediction

Protein fold prediction is one of the major unsolved problems in modern molecular biology. Given the amino acid sequence of a protein, the aim is to predict the corresponding three-dimensional structure or local fold [15]. It has been proved that determining actual structure of a protein is hard. It is a good idea to predict the structure and machine learning methods are useful. A major event in the area is the well-known Comparative Assessment of protein Structure Prediction (CASP) competition and CAFASP2 [15].

A variety of machine learning approaches have been successful, such as decision trees [9], support vector machines [6,9] and kernel methods [16,17], neural networks [6], hidden Markov models [12,13], ILP [5,18,19], etc. ILP is useful as it can learn explainable logic rules from examples with the help of relational background knowledge. Multi-class protein fold prediction has been investigated in [6,9,13].

An experimental study of applying ILP to automatically and systematically discover the structural signatures of protein folds and functions has been explored in [5]. The rules derived by ILP from observation and encoded principles are readily interpreted in terms of concepts used by biology experts. For 20 populated folds in SCOP database [20], 59 logical rules were found by ILP system Progol [21]. With the same experiments, the effect of relational background knowledge on learning protein three-dimensional fold signatures has also been addressed in [18]. However, there exists a problem of multiple predictions unsolved in the previous study, ie. a number of protein domains have been predicted to belong to more than one of 20 protein folds or can be explained by rules across multiple folds. In fact, only one protein fold prediction is expected for each protein. We have investigated that, in the previous study, about 40% of the examples have been involved in the problem (Table 1). For example, the worst case we have found is where protein domain 'd1xyzb_' is given to be in fold 'β/α (TIM)-barrel', however it has been tested to have up to four fold predictions - 'β/α (TIM)-barrel', 'NAD(P)-binding Rossmann-fold domains', 'α/β-Hydrolases' and 'Periplasmic binding protein-like II'. This is called the 'False Positive' problem in binary classification [6], where in practice many examples show positive on more than one class which leads to ambiguous prediction results.

Table 1. Ratio of examples with multiple predictions in the previous study

all-α class	all-β class	α/β class	$\alpha + \beta$ class	overall
30/77=38.96%	34/116=29.31%	67/115=58.26%	23/73=31.51%	154/381=40.42%

One of the main reasons for the false positive problem is that the decision boundary between ILP rules can be naturally overlapped due to the complex nature of protein folding, the quality and noise of acquired background knowledge and data, etc. From biology point of view, our study is motivated by finding ways to solve the multiple prediction problem so that, deriving the ILP program and data from the previous study, only one unique fold prediction can be discovered for each protein domain.

In order to make comparison, we inherit the same classification scheme in our study from the previous one [5], which predicts protein folds from the knowledge of protein domains based on the famous SCOP classification scheme [20]. The scheme is a classification done manually by the experts on protein structure and facilitates the understanding of protein structure which can be served as a starting point for machine learning experiments. Table 2 illustrates the hierarchy of protein structures we are using. A *domain* is the building block of the classification; a *fold* represents a classification for a group of protein domains. At the top level, a *class* is used to group folds based on the overall distribution of their secondary structure elements.

Table 2. Protein structure classification scheme

Level	Description	Examples
CLASS	folds are grouped into classes based on the overall distribution of their secondary structure elements.	all-α α/β
FOLD	proteins that share the same core secondary structures and the same interconnections.	Globins Cytokines
superfamily	a group of families.	
family	a group of domains.	
DOMAIN	a structure or substructure that is considered to be folded independently; small proteins have a single domain, and for larger ones, a domain is a substructure.	d1scta_ d1xyzb_

In our study, we have been using the above classification scheme to design the experiments. The data are a set of (positive) examples in protein domain level associated with known protein fold classification for training and test purpose. The background knowledge are a set of domain knowledge for representing the structural and inter-relational information of the domains. The learned ILP rules derived from the previous study stand for the prediction knowledge discovered from the data with the help of background knowledge in protein fold level. The learned SLP probabilities associated with rules and background knowledge represent the probabilistic distributions or statistical frequencies of the protein fold predictions that can be used as the ranking mechanism for solving multiple prediction problem.

3 Multi-class Prediction Using SLPs

3.1 Stochastic Logic Programs

Stochastic logic programs (SLPs) [14] have been chosen as the PILP framework in the study as SLPs provide a natural way in associating probabilities with logical rules. SLPs were introduced originally as a way of lifting stochastic grammars to the level of first-order logic programs. SLPs were considered as a generalization of hidden Markov models and stochastic context-free grammars. SLPs have later been used to define distributions for sampling within inductive logic programming (ILP). It is clear that SLPs provide a way of probabilistic logic representations and make ILP become better at inducing models that represent uncertainty. Please see chapter 2 for a tutorial of SLPs.

Syntactically, an SLP S is a definite logic program, where each clause C is a first-order range-restricted definite clause[4] and some of the definite clauses are labelled/parameterised with non-negative numbers, $l : C$. S is said to be a *pure* SLP if all clauses have parameters, as opposed to an *impure* SLP if not all clauses have labels. The subset S_q of clauses in S whose head share the same predicate symbol q is called the definition of q. For each definition S_q, we use π_q to denote the sum of the labels of the clauses

---

[4] A definite logical clause C is range-restricted if every variable in C^+, the head of C, is found in C^-, the body of C.

in S_q. S is *normalised* if $\pi_q = 1$ for each q and *unnormalised* otherwise. Till now, the definition does not show SLPs represent probability distributions, as each label can be any non-negative number and there is no constraints for the parameters of unnormalised SLPs. For our interest, SLPs are restricted to define probability distributions over logic clauses, where each l is set to be a number in the interval [0,1] and, for each S_q, π_q must be at most 1. In this case, a normalised SLP is also called a *complete* SLP, as opposed to a *incomplete* SLP for unnormalised one. In a pure normalised/complete SLP, each choice for a clause C has a parameter attached and the parameters sum to one, so they can therefore be interpreted as probabilities. Pure normalised/complete SLPs are defined such that each parameter l denotes the probability that C is the next clause used in a derivation given that its head C^+ has the correct predicate symbol. Impure SLPs are useful to define logic programs containing both probabilistic and deterministic rules, as shown in this paper. Unnormalised SLPs can conveniently be used to represent other existing probabilistic models, such as Bayesian nets.

Semantically, SLPs have been used to define probability distributions for sampling within ILP [14]. Generally speaking, an SLP S has a *distributional semantics* [22], that is one which assigns a probability distribution to the atoms of each predicate in the Herbrand base of the clauses in S. The probabilities are assigned to ground atoms in terms of their proofs according to a stochastic SLD-resolution process which employs a stochastic selection rule based on the values of the probability labels. Furthermore, some quantitative results are shown in [23], in which an SLP S with parameter $\lambda = \log l$ together with a goal G defines up to three related distributions in the stochastic SLD-tree of G: $\psi_{\lambda,S,G}(x)$, $f_{\lambda,S,G}(r)$ and $p_{\lambda,S,G}(y)$, defined over derivations $\{x\}$, refutations $\{r\}$ and atoms $\{y\}$, respectively. An example is illustrated in Fig. 4, in which the example SLP S defines a distribution $\{0.1875, 0.8125\}$ over the sample space $\{s(a), s(b)\}$. It is important to understand that SLPs do not define distributions over possible worlds, i.e., $p_{\lambda,S,G}(y)$ defines a distribution over atoms, not over the truth values of atoms.

3.2   Failure-Adjusted Maximization Algorithm

There are two tasks for learning SLPs. Parameter estimation aims to learn the parameters from observations assuming that the underlying logic program is fixed. *Failure-Adjusted Maximization* (FAM) [23] is a parameter estimation algorithm for pure normalised SLPs. Structure learning tries to learn both logic program and parameters from data. Although some fundamental work have been done for SLP structure learning [22,24], it is still an open hard problem in the area which requires one to solve almost all the existing difficulties in ILP learning. In this paper, we apply the two-phase SLP learning method developed in [22] to solve multi-class protein fold predication problem, in which SLP structure has been learned by some ILP learning system and SLP parameters will then be estimated by playing with FAM to the learned ILP program.

FAM is designed to deal with SLP parameter learning from incomplete or ambiguous data in which the atoms in the data have more than one refutation that can yield them. It is an adjustment to the standard EM algorithm where the adjustment is explicitly expressed in terms of failure derivation. The key step in the algorithm is the computation

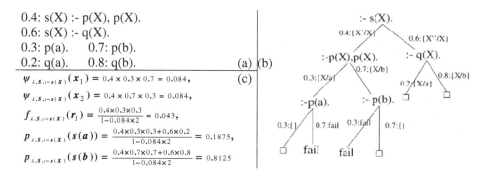

Fig. 4. (a) An example of SLP S (adapted from [23]); (b) A stochastic SLD-tree for S with goal:-
s(X), including 6 derivations in which 4 are refutations (end with □) and 2 are fail derivations;
(c) Probability distributions defined in S for the two fail derivations x_1 and x_2, for the leftmost
refutation r_1, and for the two atoms $s(a)$ and $s(b)$, respectively

of $\psi_{\lambda^h}[v_i|y]$, the expected frequency for clause C_i given the observed data y and the
current parameter estimate λ^h

$$\psi_{\lambda^h}[v_i|y] = \sum_{k=1}^{t-1} N_k \psi_{\lambda^h}[v_i|y_k] + N(Z_{\lambda^h}^{-1} - 1)\psi_{\lambda^h}[v_i|fail],$$

where v_i counts times C_i appeared in some derivation, N_k is the number of times datum
y_k occurred in the observed data, $N = \sum_k N_k$ is the number of observed data, $\psi_{\lambda^h}[v_i|y_k]$
is the expected number of times C_i was used in refutations yielding y_k, $\psi_{\lambda^h}[v_i|fail]$
denotes the expected contribution of C_i to failed derivations, and Z_{λ^h} is the probability
of success. Therefore, the first part corresponds to refutations while the second term to
failed derivations. Broadly speaking, the equation gathers together the contributions of
a particular clause C_i to derivations against the program, the current parameters and the
data. The counts are used to estimate the probabilities for the parameterised clauses in
each FAM iteration. FAM can be used to estimate the parameters for normalized impure
SLP in which some rules are set to be probabilistic and others are pure logical rules.

3.3   An SLP Example for Multi-class Prediction

An artificially generated working example is defined and processed in order to clarify
the problem of multi-class prediction and to demonstrate our method. It also shows that
multiple prediction problem may naturally happen in logic-based classifiers due to the
overlapping among logic clauses.

The so-called *multi-class animal classification* example starts from a logic program
illustrated in Table 3, which contains a set of logic rules learned using ILP and can be
used to classify animals into three classes, ie. mammal, bird or fish. The program was
learned from an artificial data set with 10% of noise and 30% of multiple prediction
examples. An example of multiple predictions can be gained by testing that a bat belongs to both mammal and bird classes, or a dolphin is predicted to be in both mammal

Table 3. A working example: multi-class animal classification program

Prob.	Logic rules	Comments
0.195:	class(mammal,A) :- has_milk(A).	%classification rules
0.205:	class(mammal,A) :- animal_running(A).	%A is in 'mammal' class
0.222:	class(bird,A) :- animal_flying(A).	%A is in 'bird' class
0.189:	class(fish,A) :- has_gills(A).	%A is in 'fish' class
0.189:	class(fish,A) :- habitat(A,water),has_covering(A,none),has_legs(A,0).	
0.433:	animal_running(A) :- hemeothermic(A),habitat(A,land), has_legs(A,4).	*% extensional* background knowledge
0.567:	animal_running(A) :- hemeothermic(A),habitat(A,caves).	
0.6:	animal_flying(A) :- hemeothermic(A),habitat(A,air), has_covering(A,feathers),has_legs(A,2).	
0.4:	animal_flying(A) :- hemeothermic(A),habitat(A,air), has_covering(A,hair),has_legs(A,2).	
	animal(bat).has_milk(bat).hemeothermic(bat).habitat(bat,air). habitat(bat,caves).has_covering(bat,hair).has_legs(bat,2).	*% intensional* background knowledge
	animal(dolphin).has_milk(dolphin).hemeothermic(dolphin). habitat(dolphin,water).has_covering(dolphin,none).has_legs(dolphin,0).······	
	class(mammal,bat).class(mammal,dolphin).······	% data,examples

and fish classes. An example of SLP is also listed in Table 3 in which probabilities are estimated for some rules from data.

3.4   Multi-class Prediction Algorithm Using SLPs

Our method of multi-class prediction using SLPs is illustrated as an algorithm in Table 4. As shown in Table 3, an SLP for multi-class prediction is an impure SLP that has a hierarchical structure, consisting of a set of probabilistic *classification/prediction rules*, a set of probabilistic clauses for *extensional background knowledge*, and a set of non-probabilistic clauses for *intensional background knowledge*. Probabilities are parameter-estimated by FAM algorithm (step 2.1).

Given an example e and a set of predictions $\{cl_1, \ldots, cl_N\}$ for e, a FAM-learned SLP defines a distribution over the predictions, ie. $\{p(e \mid cl_1, \lambda, S), \ldots, p(e \mid cl_N, \lambda, S)\}$. Each $p(e \mid cl_n, \lambda, S)$ denotes a class-conditional prediction probability of e in class cl_n and can be computed (step 2.3.3) as

$$p(e \mid cl_n, \lambda, S) = p(\text{class}(cl_n, e)) = \frac{\sum_{i=1}^{M_n} \prod_{j=1}^{M_i} l_j^{v_j(r_i)}}{\sum_{k=1}^{N} p(e \mid cl_k, \lambda, S)}$$

in its stochastic SLD-tree given S and goal :-class(cl_n, e), where r_i is the i-th refutation that satisfies e, M_n denotes the total number of refutations of e in cl_n, l_j is the probability of clause C_j, $v_j(r_i)$ is the number of times C_j has been used in r_i, and M_i denotes the number of clauses occurred in r_i. Because impure SLPs are allowed, some clauses are unparameterised in a derivation. We apply the 'equivalence class' feature developed in [23] to deal with the case where an unparameterised clause, with probability 1, either succeeds or fails in a derivation (exclusively). Two stochastic

Table 4. The algorithm of multi-class prediction using SLPs

1. Initialize matrix M^{ILP} and M^{SLP} to be zero matrix;
2. Apply n-fold cross validation or leave-one-out test to the data set that are thus divided into n (training,test) subsets; for each subset repeat
   - 2.1. Learn SLP from training data by playing FAM algorithm, which associates probabilities to the probabilistic rules;
   - 2.2. for each class cl count the number of predicted examples in the training set $d(cl)$;
   - 2.3. for each labeled example (e, cl_i) in the test set do
     - 2.3.1. if e has only one prediction cl_j then set M^{ILP}_{ij} + + and M^{SLP}_{ij} + +; else
     - 2.3.2. in all possible class predictions, apply majority class voting to choose cl_j that has the maximum value of $d(cl)$; (in the case when equivalent values happen, cl_j is randomly chosen from the set)
     - 2.3.3. for each possible class prediction cl, apply the learned SLP to compute the prediction probability of e in cl, $p(e \mid cl, \lambda, S)$;
     - 2.3.4. choose cl_k that has the maximum value of $p(e \mid cl_k, \lambda, S)$;
     - 2.3.5. set M^{ILP}_{ij} + + and M^{SLP}_{ik} + +;
3. Compute predictive accuracies pa^{ILP} and pa^{SLP} bases on M^{ILP} and M^{SLP};
4. Learn the final SLP from the whole data set.

SLD-trees for the animal classification working example are illustrated in Fig. 5, from which we have $p(\text{bat} \mid \text{mammal}, \lambda, S) = \frac{0.195+0.205\times0.567}{0.3112+0.0888} = 0.778$ and $p(\text{bat} \mid \text{bird}, \lambda, S) = \frac{0.222\times0.4}{0.3112+0.0888} = 0.222$ given the SLP presented in Table 3. They thus define a distribution over {class(mammal,bat),class(bird,bat)}, the two predictions of bat.

In the algorithm, two multi-class confusion matrixes are built in a n-fold cross validation or leave-one-out test in order to evaluate the corresponding predictive accuracies. We informally define a multi-class confusion matrix to be an integer square matrix $M_{(m+1)\times(m+1)}$ for m known classes[5], in which an arbitrary element M_{ij}, $1 \le i, j \le (m+1)$,

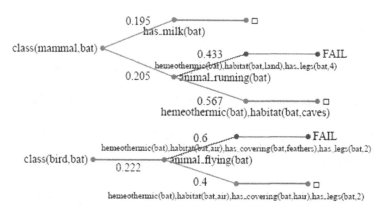

Fig. 5. Stochastic SLD-trees for goals class(mammal,bat) and class(bird,bat)

---

[5] The $(m + 1)$-th column is set to be an 'unknown' class, where an example in some cases fails to be predicted in any of m known classes.

will be increased by 1 if a labeled example taken from class cl_i is predicted to be in class cl_j (or in class cl_{m+1} if no prediction). The overall predictive accuracy based on the multi-class confusion matrix can then be computed by $pa = \frac{\sum_{i=1}^{m+1} M_{ii}}{\sum_{i,j=1}^{m+1} M_{ij}}$ (step 3). Two matrixes with the predictive accuracies for the working example are shown in Fig. 6.

$$M_{4\times4}^{\text{ILP}} = \begin{pmatrix} 16 & 0 & 2 & 0 \\ 2 & 10 & 0 & 0 \\ 2 & 0 & 17 & 1 \\ 0 & 0 & 0 & 0 \end{pmatrix}, pa^{\text{ILP}} = 86 \pm 4.9\%; \quad M_{4\times4}^{\text{SLP}} = \begin{pmatrix} 18 & 0 & 0 & 0 \\ 2 & 10 & 0 & 0 \\ 2 & 0 & 17 & 1 \\ 0 & 0 & 0 & 0 \end{pmatrix}, pa^{\text{SLP}} = 90 \pm 4.2\%.$$

Fig. 6. Confusion matrixes with the predictive accuracies for the animal working example

4 Experiments

A set of scientific experiments[6] are designed to demonstrate and evaluate our methods.

4.1 Hypotheses to Be Tested

The **null hypotheses** to be empirically investigated in the study are as follows,

- PILP approaches based on highly expressive probabilistic logic learning frameworks, eg. SLPs, **do not** outperform any conventional ILP methods on the multiclass prediction showcase.
- For a given logic program with multiple predictions/classifications and a corresponding data set, provision of probabilities **does not** increase predictive accuracy compared with non-probabilistic approaches such as majority class predictor.
- Probabilistic knowledge learned by PILP approaches **does not** produce improved explanatory insight.

4.2 Materials and Inputs

In terms of ILP, the input materials consist of an ILP logic program that has multiple prediction problem and a corresponding data set. An example can be found in Table 3 for the multi-class animal classification and 50 artificial examples are provided. In protein fold prediction, a data set of 381 protein domains together with known protein folds, based on SCOP classification, is provided in Prolog format, for example,

    fold('Globin-like',d1scta_).   fold('beta/alpha (TIM)-barrel', d1xyzb_).

Background knowledge are used to represent the three-dimensional structure information of the examples, eg.

    dom_t(d1scta_). len(d1scta_, 150). nb_alpha(d1scta_,6). nb_beta(d1scta_,0).

Three types of domain background knowledge are further distinguished (Table 5) – *relational knowledge* introduce relationships between secondary structure elements and

---

[6] Details of the experiments can be found at http://www.doc.ic.ac.uk/~cjz/research.

Table 5. List of some predicates of background knowledge

Predicates	Description
extensional relational background knowledge, there are two clauses for each predicate	
adjacent(Dom, S1,S2,Loop, TypeS1,TypeS2)	it returns true if the length of the loop separating two secondary structures S1 of TypeS1 and S2 of TypeS2 is Loop; otherwise, S1 and S2 are bound to two consecutive secondary structure elements.
coil(S1,S2,Len)	bound Len to the length of the loop between secondary structure S1 and S2 or is true if the length of the loop is Len ± 50%.
extensional global background knowledge, there are two clauses for each predicate	
len_interval (Lo=<Dom=<Hi)	is true if the length of the domain Dom is in [Lo,Hi]; otherwise, Lo (Hi) is bound to the length of the smallest (longest) positive example.
nb_alpha_interval (Lo=<Dom=<Hi)	similar to len_interval but process the number of alpha helices.
nb_beta_interval (Lo=<Dom=<Hi)	similar to len_interval but process the number of beta helices.
intensional local background knowledge, there is one clause for each predicate	
unit_len(S,Cst)	is true if the length of the secondary structure S is Cst, the values for Cst are very_lo, lo, hi and very_hi.
unit_aveh(S,Cst)	similar to unit_len but process the average hydrophobicity.
unit_hmom(S,Cst)	similar to unit_len but process the hydrophobic moment.
has_pro(S)	is true if S contains a proline amino acid.

their properties; *global knowledge* encode global characteristics of protein folds, specifically, the number of residues and the number of secondary structures; and *local knowledge* state local information of a single protein element. Some predicates are designed to be intensional, while others are extensional that are generated from intensional knowledge. In addition, 59 prediction *rules* learned by ILP system Progol [21] over 20 populated protein folds have been derived from the original study, eg.

fold('Globin-like',A) :- adjacent(A,B,C,1,h,h), has_pro(C).
fold('beta/alpha (TIM)-barrel',A) :- adjacent(A,B,C,4,h,e), unit_len(B,hi).

4.3   Methods and Results

The method presented in Table 4 has been applied to both multi-class protein fold prediction with a 5-fold cross validation test and multi-class animal classification working example with a leave-one-out test. In order to empirically test the pre-set hypotheses, five sub-experiments (Table 6) are designed and evaluated, each of which has an SLP predictor as well as a majority class predictor. The first two experiments are used to test the convergence property of FAM, while the other three are designed to investigate the influence of empirical data distribution on the performance of the two predictors. Main results of the predictive accuracy for the experiments are shown in Table 7. In summary, SLP predictors outperform majority class predictors in predictive accuracy in all five experiments. The result of protein fold prediction experiment 1 shows a promising improvement in the overall predictive accuracy that is 71.39% achieved by SLP predictor against 64.57% by non-probabilistic majority class predictor, and the difference

Table 6. Description of the experiments

Experiment	Data set and description
1	protein fold prediction, 59 learned ILP rules and 381 protein domains; learning SLP from uniform initial parameters, ie. each parameterised clause with definition S_q is initially set to have a probability $\frac{1}{|S_q|}$
2	protein fold prediction, 59 learned ILP rules and 381 protein domains; learning SLP from random initial parameters
3	animal classification, 18 examples of mammal class, 12 of bird class and 20 of fish class; predicted class size in order: mammal $\geq$ fish > bird
4	animal classification, 17 examples of mammal class, 17 of bird class and 16 of fish class; predicted class size in order: bird $\geq$ mammal > fish
5	animal classification, 14 examples of mammal class, 18 of bird class and 18 of fish class; predicted class size in order: bird $\geq$ fish > mammal

Table 7. Comparison of predictive accuracies for experiment 1 (overall and by four protein classes), 3, 4 and 5

Experiment	1 (overall)	3	4	5
SLP predictor	71.39±2.32%	90±4.24%	90±4.24%	90±4.24%
majority class predictor	64.57±2.45%	86±4.91%	84±5.18%	68±6.60%
Significance of Difference	0.021	0.269	0.185	0.003

Experiment 1 by protein class	all-α class	all-β class	α/β class	$\alpha+\beta$ class
SLP predictor	76.62±4.82%	81.03±3.64%	51.30±4.66%	82.19±4.48%
majority class predictor	71.43±5.15%	69.83±4.26%	44.35±4.63%	80.82±4.61%

of the predictive accuracies (ie. the probability of the second null hypothesis in section 5.1) is significant at the 0.021 level. Experiment 3, 4 and 5 imply that the majority class predictors are dependent on the predicted class size of each class, ie. the number of examples predicted in that class, which is further dependent on the empirical data distribution, ie. the ratio of the number of examples provided in the training data set (Table 6). We can see that the predicted class size of mammal class plays a key role on the predictive accuracy, eg. the accuracy is 86% when it has the largest class size in experiment 3, whereas the accuracy decreases to 68% in the worst case when it has the smallest class size in experiment 5.

4.4 Interpretability

Probabilities not only increase predictive accuracy but also improve the interpretability of the learned programs. These are demonstrated by interpreting Fig. 7 as follows, in which the probabilities are learned from the whole data set for all five experiments.

Fig. 7(a) – The probabilities demonstrate the ranking or importance information of the prediction rules in each protein fold; the values exactly match the power rules selected in [5] that was determined by recall, however different recall thresholds have to be manually set for different folds, whereas the probabilities can be automatically learned; by comparing the results between experiment 1 and 2, which are shown in the

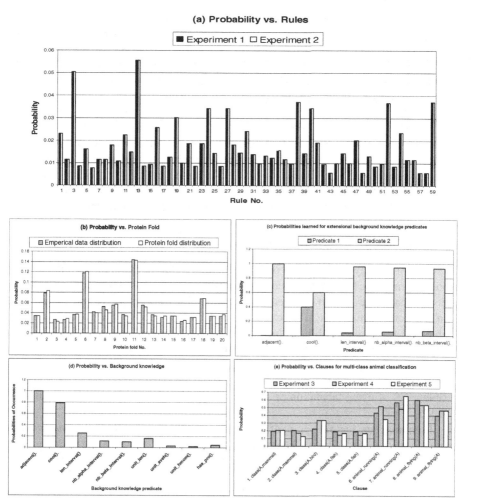

Fig. 7. Probability interpretability

two legends, we claim that the initial parameter settings have no effect on the learning results, ie. the FAM algorithm converges with any normalised initial parameters.

Fig. 7(b) – The fold probabilities, that are computed by summing up corresponding rule probabilities and shown in the first legend, indicate the popularity of different protein folds, which has been agreed by the biologists (the second and fourth authors of the paper); they tend towards empirical data distribution that is shown in the other legend.

Fig. 7(c) – Probabilities have been learned for the extensional background knowledge (Table 5), each of which has two clauses shown in two legends; it is one of the advantages of PILP to learn the probabilities for extensional background knowledge in addition to those for the prediction rules that might be simply estimated from or tend to converge to the empirical data distribution; these probabilities play the key roles in the computation of prediction probabilities for examples (section 4).

Fig. 7(d) – The probabilities shown are summed up from all 59 probabilistic rules by counting the frequencies of use of particular extensional background knowledge: relational > global > local; it is clear to find that relational knowledge are far more frequently used and occurred than the other two; the finding reenforces the conclusions in [5] - different predicates play different roles in defining protein fold signatures, but the predicate coil() has been found to have a higher frequency of use as relational knowledge in our study than that in [5], where it was treated as global knowledge.

Fig. 7(e) – The probabilities of probabilistic clauses in the animal working example (Fig. 3) are illustrated by the three legends for experiments 3, 4 and 5, respectively; the values are slightly changed by providing data sets with different empirical distributions, which result in the SLP predictors having the same predictive accuracy in the three experiments (Table 7); in contrast, the predictive accuracies of majority class predictors are dependent on the empirical data distribution and the predicted class sizes.

5   Discussion and Conclusions

Impure SLPs play a key role in our study, which allow us to model both probabilistic and deterministic knowledge in the probabilistic logic programs. The ability to combine non-probabilistic domain knowledge with probabilities is a central feature of SLPs [23]. In addition, the hierarchical structure of the SLPs improves the interpretability, and the ability of learning probabilities for extensional background knowledge from deterministic intensional background knowledge and ground examples provides SLPs a good representation for solving multi-class prediction problem. On the other hand, efficiency is a problem existed in the current FAM algorithm, especially for large SLPs. Some possible ways of using tabulation or sampling to increase efficiency have been discussed in [23].

From machine learning point of view, it is useful to compare the following two terms used in ILP method [5] with the probability used in SLPs. The measure of compression (the number of positive examples covered – the number of negative examples covered – the length of the rules) was used to seek the specific rules in ILP, but the probability is used to measure the importance of ILP rules with the same definition. While recall (true positive / total number of positive examples) was used to measure the predictive accuracy and to generate the power rules in the original binary classification method, the probability is used to solve multi-class prediction problem. However, ILP can deal with both positive and negative examples, SLPs are learned from positive examples only.

Our method of multi-class prediction using SLPs has significant advantages compared with some existing multi-class classification methods. Firstly, SLPs outperform majority voting in the way that probability has less dependency on the empirical data distribution. Secondly, sample probabilities are learned from data to tackle the uncertainty of multiple predictions naturally existing in logic programs, which are more natural and sound than decision trees [7] and the sequential model [8]. Thirdly, our method does not need to combine or utilize multiple binary classifiers as presented in [6,9,10,11]. Fourthly, SLPs use probabilities to model the decision boundaries among classes, whereas support vector machine and their reduction methods [6,9] use regularization and discriminative methods to evaluate several binary classification methods for stochastic voting and usually result in reduced accuracy and efficiency. Finally, a distingusing feature of SLPs is, in contrast to typical conflict resolution strategies in rule-learning, SLPs attach probability values to all the rules, not just the ones defining the concept.

The same protein folding data set or similar sets have been applied as a benchmark by some other machine learning methods. Improved logic rules have been learned using ILP in [19], in which the multiple predictions have been effectively reduced by rearranging background knowledge. Logical hidden Markov models, another PILP framework, are applied in [13] to deal with multi-class protein fold prediction by representing the secondary structure of protein domains as logical sequences; the work increases predictive efficiency and accuracy by reducing the problem representation complexity. Conditional random fields [12] provide another PILP approach to deal with multi-class protein fold classification using logical sequence method. A novel kernel method on Prolog proof trees for binary protein fold prediction has been studied in [17] which provides higher overall accuracy compared with Progol. Even with the same data set, it is not straightforward to compare the results gained by these methods with those shown in this paper due to our specific research motivation and target, which aims to solve multi-class prediction problem by learning SLPs on the basis of the existing ILP programs and data, while the other methods apply their own binary or multi-class classification solutions to the data without deriving the ILP programs. As our future work, resolving rule conflicts with double induction [25] and using Area Under the Curve (AUC) [26] rather than predictive accuracy for performance evaluation will be considered.

In conclusion, the null hypotheses we have set in experiments were rejected on the basis of the results. Overall we conclude that PILP approaches (eg. SLPs) have demonstrable advantages for solving multi-class prediction problem, in particular multi-class protein fold prediction problem, and SLPs have outperformed ILP plus majority class predictor in both predictive accuracy and result interpretability.

Acknowledgements

The authors would like to acknowledge the support of the EC Sixth Framework Project "Application of Probabilistic Inductive Logic Programming II (APrIL II)" (Grant Ref: FP-508861).

References

1. Andreeva, A., Howorth, D., Brenner, S.E., Hubbard, T.J.P., Chothia, C., Murzin, A.G.: SCOP database in 2004: Refinements integrate structure and sequence family data. Nucl. Acid Res. 32, 226–229 (2004)
2. Pearl, F., Todd, A., Sillitoe, I., Dibley, M., Redfern, O., Lewis, T., Bennett, C., Marsden, R., Grant, A., Lee, D., Akpor, A., Maibaum, M., Harrison, A., Dallman, T., Reeves, G., Diboun, I., Addou, S., Lise, S., Johnston, C., Sillero, A., Thornton, J., Orengo, C.: The CATH Domain Structure Database and related resources Gene3D and DHS provide comprehensive domain family information for genome analysis. Nucleic Acids Research 33(Database Issue), 247–251 (2005)
3. Har-Peled, S., Roth, D., Zimak, D.: Constraint Classification: a New Approach to Multiclass Classification and Ranking. In: Proc. of the Inter. Conf. on Algorithmic Learning Theory, pp. 365–379 (2002)

4. De Raedt, L., Dietterich, T., Getoor, L., Muggleton, S.H.: Probabilistic, Logical and Relational Learning - Towards a Synthesis. In: Dagstuhl Seminar Proceedings 05051 (2006)

5. Turcotte, M., Muggleton, S.H., Sternberg, M.J.E.: Automated Discovery of Structural Signatures of Protein Fold and Function. J. Mol. Biol. 306, 591–605 (2001)

6. Ding, C.H.Q., Dubchak, I.: Multi-class Protein Fold Recognition Using Support Vector Machines and Neural Networks. Bioinformatics 17(4), 349–358 (2001)

7. Mitchell, T.M.: Machine Learning. The McGraw-Hill Companies, Inc, New York (1997)

8. Even-Zohar, Y., Roth, D.: A Sequential Model for Multi Class Classification. In: Proc. of the Conf. on Empirical Methods for Natural Language Processing (EMNLP), pp. 10–19 (2001)

9. Tan, A.C., Giltert D., Deville Y.: Multi-class Protein Fold Classification Using a New Ensemble Machine Learning Approach. In: Inter. Conf. on Genome Informatics, GIW (2003)

10. Wu, T.-F., Lin, C.-J., Weng, R.C.: Probability Estimates for Multi-class Classification by Pairwise Coupling. JMLR 5, 975–1005 (2004)

11. Yukinawa, N., Oba, S., Kato, K., Taniguchi, K., Iwao-Koizumi, K., Tamaki, Y., Noguchi, S., Ishii, S.: A Multi-class Predictor Based on a Probabilistic Model: Application to Gene Expression Profiling-based Diagnosis of Thyroid Tumors. BMC Genomes 7, 190 (2006)

12. Gutmann, B., Kersting, K.: TildeCRF: Conditional Random Fields for Logical Sequences. In: Fürnkranz, J., Scheffer, T., Spiliopoulou, M. (eds.) ECML 2006. LNCS (LNAI), vol. 4212, pp. 18–22. Springer, Heidelberg (2006)

13. Kersting, K., De Raedt, L., Raiko, T.: Logical Hidden Markov Models. JAIR 25, 425–456 (2006)

14. Muggleton, S.H.: Stochastic Logic Programs. In: De Raedt, L. (ed.) Advances in Inductive Logic Programming, pp. 254–264 (1996)

15. Moult, J.: Rigorous Performance Evaluation in Protein Structure Modeling and Implications for Computational Biology. Phil. Trans. R. Soc. B 361, 453–458 (2006)

16. Kersting, K., Gartner, T.: Fisher Kernels for Logical Sequences. In: Boulicaut, J.-F., Esposito, F., Giannotti, F., Pedreschi, D. (eds.) ECML 2004. LNCS (LNAI), vol. 3201, pp. 205–216. Springer, Heidelberg (2004)

17. Passerini, A., Frasconi, P., De Raedt, L.: Kernels on Prolog Proof Trees: Statistical Learning in the ILP Setting. JMLR 7, 307–342 (2006)

18. Turcotte, M., Muggleton, S.H., Sternberg, M.J.E.: The Effect of Relational Background Knowledge on Learning of Protein Three-Dimensional Fold Signature. Machine Learning 43(1-2), 81–95 (2001)

19. Cootes, A.P., Muggleton, S.H., Sternberg, M.J.E.: The Automatic Discovery of Structural Principles Describing Protein Fold Space. J. Mol. Biol. 330, 839–850 (2003)

20. Brenner, S.E., Chothia, C., Hubbard, T.J., Murzin, A.G.: Understanding protein structure: Using SCOP for fold interpretation. Methods in Enzymology 266, 635–643 (1996)

21. Muggleton, S.H., Firth, J.: CProgol4.4: A Tutorial Introduction. In: Džeroski, S., Lavrač, N. (eds.) Relational Data Mining, pp. 160–188 (2001)

22. Muggleton, S.H.: Learning Stochastic Logic Programs. Electronic Transactions in Artificial Intelligence 5(041) (2000)

23. Cussens, J.: Parameter Estimation in Stochastic Logic Programs. Machine Learning 44(3), 245–271 (2001)

24. Muggleton, S.H.: Learning Structure and Parameters of Stochastic Logic Programs. Electronic Transactions in Artificial Intelligence 6 (2002)

25. Lindgren, T., Boström, H.: Resolving Rule Conflicts with Double Induction. Intell. Data Anal. 8(5), 457–468 (2004)

26. Hand, D.J., Till, R.J.: A Simple Generalisation of the Area Under the ROC Curve for Multiple Class Classification Problems. Machine Learning 45(2), 171–186 (2001)

Probabilistic Logic Learning from Haplotype Data

Niels Landwehr[1] and Taneli Mielikäinen[2]

[1] Machine Learning Lab, Institute for Computer Science, University of Freiburg
Georges-Koehler Allee, Building 079, 79110 Freiburg, Germany
`landwehr@informatik.uni-freiburg.de`
[2] Helsinki Institute for Information Technology, University of Helsinki, Finland
`Taneli.Mielikainen@cs.helsinki.fi`

Abstract. The analysis of *haplotype* data of human populations has received much attention recently. For instance, problems such as *Haplotype Reconstruction* are important intermediate steps in gene association studies, which seek to uncover the genetic basis of complex diseases. In this chapter, we explore the application of probabilistic logic learning techniques to haplotype data. More specifically, a new haplotype reconstruction technique based on Logical Hidden Markov Models is presented and experimentally compared against other state-of-the-art haplotyping systems. Furthermore, we explore approaches for combining haplotype reconstructions from different sources, which can increase accuracy and robustness of reconstruction estimates. Finally, techniques for discovering the structure in haplotype data at the level of haplotypes and population are discussed.

1 Introduction

In this chapter, we will look at applications of probabilistic logic learning and related approaches in the area of genetic data analysis. More specifically, we are concerned with analyzing *haplotype* data—a concise representation of the individual genetic make-up of an organism, that is encoded in a set of genetic *markers*. The analysis of haplotype data has become a central theme in modern bioinformatics, and is considered to be a promising approach to many important problems in human biology and medicine. Application areas range from the quest to identify genetic roots of complex diseases to analyzing the evolution history of populations or developing "personalized" medicine based on the individual genetic disposition of the patient.

The rest of the chapter is organized as follows. After starting with a brief introduction to the basic concepts of genetics, such as the genome, chromosomes, and haplotypes, three different haplotype data analysis problems will be discussed. The first problem concerns *haplotype reconstruction*: the problem of resolving the hidden phase information in genotype data obtained from laboratory measurements. For this problem a new statistical method based on Logical Hidden Markov Models is introduced. The second, related, problem is that of *combining*

L. De Raedt et al. (Eds.): Probabilistic ILP 2007, LNAI 4911, pp. 263–286, 2008.

haplotypings, that is, the question how different haplotype reconstructions obtained from different algorithmic methods can be combined and jointly analyzed. The third problem is concerned with discovering the *structure* in haplotype data, at the level of haplotypes and populations of individuals.

1.1   Genomes, Chromosomes and Haplotypes

The *genome* is organized as a set of *chromosomes* [TJHBD97]. A chromosome is a *DNA molecule* consisting of *nucleotides*, small molecules that connect to form the long chain-like DNA molecule. Basically, four different nucleotides occur (Adenine, Cytosine, Guanine, Thymine), and the genetic information is encoded in the sequence of "letters" A,C,G and T. Thus, for our purposes, a DNA molecule is a sequence over the alphabet $\{A, C, G, T\}$, and a genome is then a collection of sequences in $\{A, C, G, T\}^*$.

Most of the genome is invariant between different human individuals. However, the genetic variations that do exist play a crucial role in determining our genetic individuality, they can e.g. contribute to risk factors of complex diseases or influence how an individual patient responds to a certain drug treatment. The analysis of genetic variation in human populations has therefore become a focus of attention in human biology recently [The05]. Most studied differences in the genome are single-nucleotide variations at particular positions in the genome, which are called *single nucleotide polymorphisms* (SNPs). The positions are also called *markers* and the different possible values *alleles*. A *haplotype* is a sequence of SNP alleles along a region of a chromosome, and concisely represents the (variable) genetic information in that region.

The genetic variation in SNPs is mostly due to two causes: *mutation* and *recombination*. A mutation changes a single nucleotide in the chromosome. Mutations are relatively rare, they occur with a frequency of about 10^{-8}. While SNPs are themselves results of ancient mutations, mutations are usually ignored in statistical haplotype models due to their rarity. Recombination introduces variability by breaking up the chromosomes of the two parents and reconnecting the resulting segments to form a new and different chromosome for the offspring. Because the probability of a recombination event between two markers is lower if they are near to each other, there is a statistical correlation (so-called *linkage disequilibrium*) between markers which decreases with increasing marker distance. Statistical approaches to haplotype modeling are based on exploiting such patterns of correlation.

In diploid organisms such as humans there are two *homologous* (i.e., almost identical) copies of each chromosome. Determining haplotype information for an individual therefore means measuring a set of markers along a chromosome for both copies of the chromosome. Current practical laboratory measurement techniques produce a *genotype*—for m markers, a sequence of m unordered pairs of alleles. The genotype reveals which two alleles are present at each marker, but not their respective chromosomal origin. Genotypes, as sequences of unordered pairs, are an example of the way data is *structured* in haplotype analysis, posing challenges to standard propositional data analysis techniques. Using propositional techniques, a genotype could be represented as a sequence of un-ordered pairs, where each unordered pair is considered as a letter in the alphabet. However, such a representation would not take into account the intrinsic structure in each letter as an unordered pair. These limitations can be overcome using a relational representation of the data, as will be shown in the next section.

A similarly challenging task is the representation of a haplotype pair in propositional form, as a haplotype pair consists of two haplotype sequences and there is no natural order for the sequences in the pair. In some cases it might be known which of the haplotypes is inherited from the maternal/paternal genome, but this does not yield a natural ordering: based on the current knowledge of genetics, it does not matter from which parent a particular copy of a chromosome is inherited. Such representational issues will also be discussed in the forthcoming sections. Furthermore, additional relational information could be taken into account. Individuals can be related (e.g., by family relations), and relations between different regions of the marker maps are sometimes known. For example, certain genes might be known to be correlated. Such information is typically probabilistic.

Because of the outlined difficulties with representing haplotype data in propositional form, this domain is an interesting challenge for statistical relational modeling techniques.

Notational Convention. For our purposes, a haplotype h is a sequence of alleles $h[i]$ in markers $i = 1, \ldots, m$. In most cases, only two alternative alleles occur at an SNP marker, so we can assume that $h \in \{0, 1\}^m$. A genotype g is a sequence of unordered pairs $g[i] = \{h_g^1[i], h_g^2[i]\}$ of alleles in markers $i = 1, \ldots, m$. Hence, $g \in \{\{0,0\}, \{1,1\}, \{0,1\}\}^m$. A marker with alleles $\{0,0\}$ or $\{1,1\}$ is *homozygous* whereas a marker with alleles $\{0,1\}$ is *heterozygous*. The number of heterozygous markers is denoted by m' and the number of individuals in the population by n.

2   Haplotype Reconstruction

This section describes and formalizes the haplotype reconstruction (or *haplotyping*) problem, and presents a new method for statistical haplotype reconstruction based on Logical Hidden Markov Models (LOHMMs, see Chapter 3). We will start by defining the problem setting and present a basic LOHMM model for this domain. Two extensions to the basic model will be presented, and finally the method is compared against several state-of-the-art haplotyping techniques on real-world population data.

2.1   The Haplotype Reconstruction Problem

In order to obtain haplotype data for a set of human individuals, their genotypes are measured in the laboratory, and afterwards the haplotypes must be determined from this genotype data. There are two alternative approaches for this reconstruction: One is to use *family trios*, i.e., genotype two parents and the corresponding child. If trios are available, most of the ambiguity in the phase (the order of the alleles in the genotype data) can be resolved analytically, and haplotypes be inferred. If no trios can be obtained, population-based computational methods have to be used to estimate the haplotype pair for each genotype. These approaches exploit statistical correlations between different markers to estimate a distribution over haplotypes for the population sample in question, and use this estimate to infer the most likely haplotype pair for each genotype in the sample. Because trios are more difficult to recruit and more expensive to genotype, population-based approaches are often the only cost-effective method for large-scale studies. Consequently, the study of such techniques has received much attention recently [SWS05, HBE⁺04].

Problem 1 (haplotype reconstruction). Given a multiset $\mathcal{G}$ of genotypes, find for each $g \in \mathcal{G}$ the most likely haplotypes h_g^1 and h_g^2 which are a *consistent* reconstruction of g, i.e., $g[i] = \{h_g^1[i], h_g^2[i]\}$ for each $i = 1, \ldots, m$.

If $\mathcal{H}$ denotes a mapping $\mathcal{G} \rightarrow \{0,1\}^m \times \{0,1\}^m$, associating each genotype $g \in \mathcal{G}$ with a pair $\langle h_g^1, h_g^2 \rangle$ of haplotypes, the goal is to find the $\mathcal{H}$ that maximizes $\mathbb{P}(\mathcal{H} \mid \mathcal{G})$. It is usually assumed that the sample $\mathcal{G}$ is in Hardy-Weinberg equilibrium, i.e., that $\mathbb{P}(\langle h_g^1, h_g^2 \rangle) = \mathbb{P}(h_g^1)\mathbb{P}(h_g^2)$ for all $g \in \mathcal{G}$, and that genotypes are independently sampled from the same distribution. With such assumptions, the likelihood $\mathbb{P}(\mathcal{H} \mid \mathcal{G})$ of the reconstruction $\mathcal{H}$ given $\mathcal{G}$ is proportional to $\prod_{g \in \mathcal{G}} \mathbb{P}(h_g^1)\mathbb{P}(h_g^2)$ if the reconstruction is consistent for all $g \in \mathcal{G}$, and zero otherwise. In population-based haplotyping, a probabilistic model λ for the distribution over haplotypes is estimated from the available genotype information $\mathcal{G}$. The distribution estimate $\mathbb{P}(h \mid \lambda)$ is then used to find the most likely reconstruction $\mathcal{H}$ for $\mathcal{G}$ under Hardy-Weinberg equilibrium.

2.2   A LOHMM Model for Haplotyping

Logical hidden Markov models (LOHMMs, see Chapter 3) upgrade traditional hidden Markov models to deal with sequences of structured symbols, rather than flat characters. The key idea underlying LOHMMs is to employ logical atoms as structured (output and state) symbols. More specifically, LOHMMs define *abstract* states such as $s(A, B)$ where s is the state name and A, B are logical variables. An abstract state represents a set of "ground" states, namely all variable-free logical specializations of the abstract state expression $s(A, B)$ (e.g., $s(1, 0)$). Abstract transitions such as $s(X, Y) \rightarrow s'(1, Y)$ describe how the model transitions between abstract states, and variable unification is used to share information between states, and between states and observations. Variants of the Expectation-Maximization and Viterbi algorithms used with standard HMMs can be derived for learning and inference in LOHMMs.

The basic motivation for using LOHMMs in haplotyping is that it is straightforward to encode genotypes (sequences of unordered pairs) as sequences of logical atoms. This can be done with a predicate $pair(X, Y)$, which can be grounded to $pair(0, 0)$ (homozygous 0), $pair(1, 1)$ (homozygous 1), and $pair(0, 1)$ (heterozygous). Using logical variables and unification, the two individual alleles in the pair can be accessed. This allows to represent biological knowledge such as the assumption of Hardy-Weinberg equilibrium (the fact that a genotype is sampled by sampling two haplotypes independently and from the same distribution) in the LOHMM structure.

As underlying model for the distribution over haplotypes, we use a straightforward left-to-right Markov model λ over the binary marker values at positions $t = 1, \ldots, m$:

$$\mathbb{P}(h) = \prod_{t=1}^{m} \mathbb{P}_t(h[t] \mid h[t-1], \lambda).$$

This is motivated by the observation that linkage disequilibrium is strongest for adjacent markers. Parameters of this model are of the form $\mathbb{P}_t(h[t] \mid h[t-1])$, the probability of sampling the new allele $h[t]$ at position t after observing the allele $h[t-1]$ at position $t-1$. The Markov model on haplotypes can be extended to a LOHMM on genotypes as follows. The LOHMM is organized as a left-to-right model with layers $t = 1, \ldots, m$. At every layer t, one component of the model encodes the distribution $P(h[t+1] \mid h[t])$. This component is traversed twice for sampling the two new alleles $h^1[t+1], h^2[t+1]$ based on their respective histories $h^1[t], h^2[t]$. Afterwards, the unordered pair corresponding to the new allele pair is emitted.

Figure 1 shows a single layer (at marker t) of the LOHMM model. For sampling two new markers $h^1[t+1], h^2[t+1]$ at position $t+1$ based on the markers $h^1[t], h^2[t]$ at position t, we start at state $m_t(X, Y)$ with $h^1[t], h^2[t]$ bound to X and Y. The model then transitions to the state $s_t(X, Y, x)$ to sample the first new marker $h^1[t+1]$. The multiple transitions from state s_t to state s'_t encode the distribution $P(h[t+1] \mid h[t])$. In $s'_t(A', B, x)$, the new marker $h^1[t+1]$ has been sampled and is bound to A'. Afterwards, the same path is traversed again to sample the second marker, with arguments in state s_t swapped. This effectively samples the new marker $h^2[t+1]$ based on $h^2[t]$ *independently* and *from the same distribution*. Finally, the unordered pair corresponding to the two new markers is emitted in the transition from s'_t to m_{t+1}. This is can be easily accomplished using the logical generality ordering on abstract states in LOHMMs: if the more specific abstract states for homozygous markers match the ground state a homozygous pair is emitted, otherwise, an (unordered) heterozygous pair. Note that this model only has 2 free parameters per layer, in contrast to a naive first-order HMM model on the the joint state of the two haplotypes, which would have 12 free parameters per layer.

This kind of model can be directly trained from genotype data using the EM algorithm for LOHMMs, and the most likely haplotype pair for a genotype can be read off the most likely state sequence for that observation returned by the Viterbi algorithm (see [KDR06]). However, initial experiments using the XANTHOS engine for LOHMMs showed that the computational overhead due to the general-purpose framework used in LOHMMs reduced the computational efficiency of the model. Fortunately, it is possible to compile the presented LOHMM model into an equivalent HMM model with parameter tying constraints. While

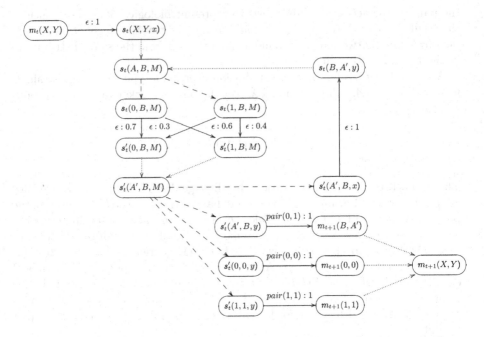

Fig. 1. LOHMM for haplotype reconstruction. One layer at marker position t is shown. The standard syntax for visualizing LOHMMs is used: solid arrows represent abstract transitions, dashed arrows the "more general than" relation, and dotted arrows "must follow" links. For a more detailed description, see Chapter 3.

the details of this transformation are beyond the scope of this article, it generally follows the grounding mechanism for LOHMMs, as described in [KDR06].

2.3   Higher Order Models and Sparse Distributions

The main limitation of the model presented so far is that it only takes into account dependencies between adjacent markers. Expressivity can be increased by using a Markov model of order $k > 1$ for the underlying haplotype distribution [EGT04]:

$$\mathbb{P}(h) = \prod_{t=1}^{m} \mathbb{P}_t(h[t] \mid h[t-k, t-1], \lambda),$$

where $h[j, i]$ is a shorthand for $h[\max\{1, j\}] \dots h[i]$. Unfortunately, the number of parameters in such a model increases exponentially with the history length k. However, observations on real-world data (e.g., [DRS+01]) show that only few conserved haplotype fragments from the set of 2^k possible binary strings of length k actually occur in a particular population. This can be exploited by modeling sparse distributions, where fragment probabilities which are estimated

Algorithm 1. The level-wise SpaMM learning algorithm

Initialize $k := 1$
$\lambda_1 := \text{INITIAL-MODEL}()$
$\lambda_1 := \text{EM-TRAINING}(\lambda_1)$
repeat
    $k := k + 1$
    $\lambda_k := \text{EXTEND-AND-REGULARIZE}(\lambda_{k-1})$
    $\lambda_k := \text{EM-TRAINING}(\lambda_k)$
until $k = k_{max}$

to be very low are set to zero. More precisely, let $p = \mathbb{P}_t(h[t] \mid h[t-k, t-1])$ and define for some small $\epsilon > 0$ a regularized distribution

$$\hat{\mathbb{P}}_t(h[t] \mid h[t-k, t-1]) = \begin{cases} 0 & \text{if } p \leq \epsilon; \\ 1 & \text{if } p > 1 - \epsilon; \\ p & \text{otherwise.} \end{cases}$$

If the underlying distribution is sufficiently sparse, $\hat{\mathbb{P}}$ can be represented using a relatively small number of parameters. The corresponding sparse hidden Markov model structure (in which transitions with probability 0 are removed) will reflect the pattern of conserved haplotype fragments present in the population. How such a sparse model structure can be learned without ever constructing the prohibitively complex distribution $\mathbb{P}$ will be discussed in the next section.

2.4   SpaMM: A Level-Wise Learning Algorithm

To construct the sparse order-k hidden Markov model, we propose a learning algorithm—called **SpaMM** for **Spa**rse **M**arkov **M**odeling—that iteratively refines hidden Markov models of increasing order (Algorithm 1). More specifically, the idea of SpaMM is to identify conserved fragments using a level-wise search, i.e., by extending short fragments (in low-order models) to longer ones (in high-order models), and is inspired by the well-known Apriori data mining algorithm [AMS+96]. The algorithm starts with a first-order Markov model λ_1 on haplotypes where initial transition probabilities are set to $\hat{\mathbb{P}}_t(h[t] \mid h[t-1], \lambda_1) = 0.5$ for all $t \in \{1, \ldots, m\}$, $h[t], h[t-1] \in \{0, 1\}$. For this model, a corresponding LOHMM on genotypes can be constructed as outlined in Section 2.2, which can be compiled into a standard HMM with parameter tying constraints and trained on the available genotype data using EM.

The function EXTEND-AND-REGULARIZE(λ_{k-1}) takes as input a model of order $k-1$ and returns a model λ_k of order k. In λ_k, initial transition probabilities are set to

$$\dot{\mathbb{P}}_t(h[t] \mid h[t-k, t-1], \lambda_{k+1}) = \begin{cases} 0 & \text{if } \mathbb{P}_t(h[t] \mid h[t-k+1, t-1], \lambda_k) \leq \epsilon; \\ 1 & \text{if } \mathbb{P}_t(h[t] \mid h[t-k+1, t-1], \lambda_k) > 1 - \epsilon; \\ 0.5 & \text{otherwise,} \end{cases}$$

i.e., transitions are removed if the probability of the transition conditioned on a shorter history is smaller than ϵ. This procedure of iteratively training, extending

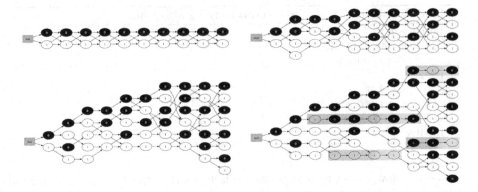

Fig. 2. Visualization of the SpaMM Structure Learning Algorithm. Sparse models $\lambda_1, \ldots, \lambda_4$ of increasing order learned on the Daly dataset are shown. Black/white nodes encode more frequent/less frequent allele in population. Conserved fragments identified in λ_4 are highlighted.

and regularizing Markov models of increasing order is repeated up to a maximum order k_{max}.

Figure 2 visualizes the underlying distribution over haplotypes learned in the first 4 iterations of the SpaMM algorithm on a real-world dataset. The set of paths through the lattice corresponds to the set of haplotypes which have non-zero probability according to the model. Note how some of the possible haplotypes are pruned and conserved fragments are isolated. Accordingly, the number of states and transitions in the final LOHMM/HMM model is significantly smaller than for a full model of that order.

2.5 Experimental Evaluation

The proposed method was implemented in the SpaMM haplotyping system[1]. We compared its accuracy and computational performance to several other state-of-the art haplotype reconstruction systems: PHASE version 2.1.1 [SS05], fastPHASE version 1.1 [SS06], GERBIL as included in GEVALT version 1.0 [KS05], HIT [RKMU05] and HaploRec (variable order Markov model) version 2.0 [EGT06]. All methods were run using their default parameters. The fastPHASE system, which also employs EM for learning a probabilistic model, uses a strategy of averaging results over several random restarts of EM from different initial parameter values. This reduces the variance component of the reconstruction error and alleviates the problem of local minima in EM search. As this is a general technique applicable also to our method, we list results for fastPHASE with averaging (fastPHASE) and without averaging (fastPHASE-NA).

The methods were compared using publicly available real-world datasets, and larger datasets simulated with the Hudson coalescence simulator [Hud02]. As

---

[1] The implementation is available at http://www.informatik.uni-freiburg.de/~landwehr/haplotyping.html

Table 1. Reconstruction Accuracy on Yoruba and Daly Data. Normalized switch error is shown for the Daly dataset, and average normalized switch error over the 100 datasets in the Yoruba-20, Yoruba-100 and Yoruba-500 dataset collections.

Method	Yoruba-20	Yoruba-100	Yoruba-500	Daly
PHASE	**0.027**	**0.025**	*n.a.*	0.038
fastPHASE	0.033	0.031	**0.034**	**0.027**
SpaMM	0.034	0.037	0.040	0.033
HaploRec	0.036	0.038	0.046	0.034
fastPHASE-NA	0.041	0.060	0.069	0.045
HIT	0.042	0.050	0.055	0.031
GERBIL	0.044	0.051	*n.a*	0.034

real-world data, we used a collection of datasets from the Yoruba population in Ibadan, Nigeria [The05], and the well-known dataset of Daly et al [DRS+01], which contains data from a European-derived population. For these datasets, family trios are available, and thus true haplotypes can be inferred analytically.

For the Yoruba population, we sampled 100 sets of 500 markers each from distinct regions on chromosome 1 (**Yoruba-500**), and from these smaller datasets by taking only the first 20 (**Yoruba-20**) or 100 (**Yoruba-100**) markers for every individual. There are 60 individuals in the dataset after preprocessing, with an average fraction of missing values of 3.6%. For the **Daly** dataset, there is information on 103 markers and 174 individuals available after data preprocessing, and the average fraction of missing values is 8%. The number of genotyped individuals in these real-world datasets is rather small. For most disease association studies, sample sizes of at least several hundred individuals are needed [WBCT05], and we are ultimately interested in haplotyping such larger datasets. Unfortunately, we are not aware of any publicly available real-world datasets of this size, so we have to resort to simulated data. We used the well-known Hudson coalescence simulator [Hud02] to generate 50 artificial datasets, each containing 800 individuals (**Hudson** datasets). The simulator uses the standard Wright-Fisher neutral model of genetic variation with recombination. To come as close to the characteristics of real-world data as possible, some alleles were masked (marked as missing) after simulation.

The accuracy of the reconstructed haplotypes produced by the different methods was measured by normalized switch error. The switch error of a reconstruction is the minimum number of recombinations needed to transform the reconstructed haplotype pair into the true haplotype pair. (See Section 3 for more details.) To normalize, switch errors are summed over all individuals in the dataset and divided by the total number of switch errors that could have been made. For more details on the methodology of the experimental study, confer [LME+07].

Table 1 shows the normalized switch error for all methods on the real-world datasets Yoruba and Daly. For the dataset collections Yoruba-20, Yoruba-100 and Yoruba-500 errors are averaged over the 100 datasets. PHASE and Gerbil

Table 2. Average Error for Reconstructing Masked Genotypes on Yoruba-100. From 10% to 40% of all genotypes were masked randomly. Results are averaged over 100 datasets.

Method	10%	20%	30%	40%
fastPHASE	**0.045**	**0.052**	**0.062**	**0.075**
SpaMM	0.058	0.066	0.078	0.096
fastPHASE-NA	0.067	0.075	0.089	0.126
HIT	0.070	0.079	0.087	0.098
GERBIL	0.073	0.091	0.110	0.136

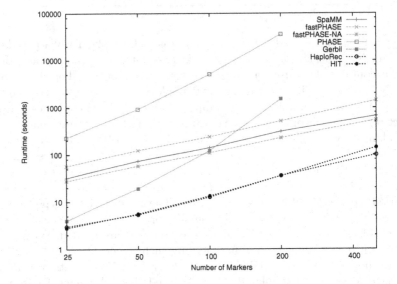

Fig. 3. Runtime as a Function of the Number of Markers. Average runtime per dataset on Yoruba datasets for marker maps of length 25 to 500 for SpaMM, fast-PHASE, fastPHASE-NA, PHASE, Gerbil, HaploRec, and HIT are shown (logarithmic scale). Results are averaged over 10 out of the 100 datasets in the Yoruba collection.

did not complete on Yoruba-500 in two weeks[2]. Overall, the PHASE system achieves highest reconstruction accuracies. After PHASE, fastPHASE with averaging is most accurate, then SpaMM, and then HaploRec. Figure 3 shows the average runtime of the methods for marker maps of different lengths. The most accurate method PHASE is also clearly the slowest. fastPHASE and SpaMM are substantially faster, and HaploRec and HIT very fast. Gerbil is fast for small marker maps but slow for larger ones. For fastPHASE, fastPHASE-NA, HaploRec, SpaMM and HIT, computational costs scale linearly with the length of

---

[2] All experiments were run on standard PC hardware with a 3.2GHz processor and 2GB of main memory.

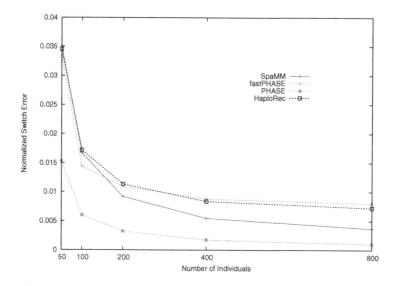

Fig. 4. Reconstruction Accuracy as a Function of the Number of Samples Available. Average normalized switch error on the Hudson datasets as a function of the number of individuals for SpaMM, fastPHASE, PHASE and HaploRec is shown. Results are averaged over 50 datasets.

the marker map, while the increase is superlinear for PHASE and Gerbil, so computational costs quickly become prohibitive for longer maps.

Performance of the systems on larger datasets with up to 800 individuals was evaluated on the 50 simulated Hudson datasets. As for the real-world data, the most accurate methods were PHASE, fastPHASE, SpaMM and HaploRec. Figure 4 shows the normalized switch error of these four methods as a function of the number of individuals (results of Gerbil, fastPHASE-NA, and HIT were significantly worse and are not shown). PHASE was the most accurate method also in this setting, but the relative accuracy of the other three systems depended on the number of individuals in the datasets. While for relatively small numbers of individuals (50–100) fastPHASE outperforms SpaMM and HaploRec, this is reversed for 200 or more individuals.

A problem closely related to haplotype reconstruction is that of genotype imputation. Here, the task is to infer the most likely genotype values (unordered allele pairs) at marker positions where genotype information is missing, based on the observed genotype information. With the exception of HaploRec, all haplotyping systems included in this study can also impute missing genotypes. To test imputation accuracy, between 10% and 40% of all markers were masked randomly, and then the marker values inferred by the systems were compared to the known true marker values. Table 2 shows the accuracy of inferred genotypes for different fractions of masked data on the Yoruba-100 datasets and Table 3 on the simulated Hudson datasets with 400 individuals per dataset. PHASE was

Table 3. Average Error for Reconstructing Masked Genotypes on Hudson.
From 10% to 40% of all genotypes were masked randomly. Results are averaged over 50 datasets.

Method	10%	20%	30%	40%
fastPHASE	0.035	0.041	0.051	0.063
SpaMM	**0.017**	**0.023**	**0.034**	**0.052**
fastPHASE-NA	0.056	0.062	0.074	0.087
HIT	0.081	0.093	0.108	0.127
GERBIL	0.102	0.122	0.148	0.169

too slow to run in this task as its runtime increases significantly in the presence of many missing markers. Evidence from the literature [SS06] suggests that for this task, fastPHASE outperforms PHASE and is indeed the best method available. In our experiments, on Yoruba-100 fastPHASE is most accurate, SpaMM is slightly less accurate than fastPHASE, but more accurate than any other method (including fastPHASE-NA). On the larger Hudson datasets, SpaMM is significantly more accurate than any other method.

To summarize, our experimental results confirm PHASE as the most accurate but also computationally most expensive haplotype reconstruction system [SS06,SS05]. If more computational efficiency is required, fastPHASE yields the most accurate reconstructions on small datasets, and SpaMM is preferable for larger datasets. SpaMM also infers missing genotype values with high accuracy. For small datasets, it is second only to fastPHASE; for large datasets, it is substantially more accurate than any other method in our experiments.

3   Comparing Haplotypings

For haplotype pairs, as structured objects, there is no obvious way of measuring similarity—if two pairs are not identical, their distance could be measured in several ways. At the same time, comparing haplotypings is important for many problems in haplotype analysis, and therefore a *distance* or *similarity* measure on haplotype pairs is needed. The ability to compare haplotypings is useful, for example, for evaluating the quality of haplotype reconstructions, if (at least for part of the data) the correct haplotypings are known. An alternative approach to evaluation would be to have an accurate generative model of haplotype data for the population in question, which could assign probability scores to haplotype reconstructions. However, such a model seems even harder to obtain than known correct haplotype reconstructions (which can be derived from family trios).

Moreover, a distance measure between haplotypes allows to compute *consensus* haplotype reconstructions, which average between different, conflicting reconstructions—for example, by minimizing the sum of distances. This opens up possibilities for the application of ensemble methods in haplotype analysis, which can increase accuracy and robustness of solutions. Finally, comparison operators can be used to study the structure of populations (Section 4.1) or structure of haplotypes (Section 4.2).

Although we could simply represent the haplotype data in a relational form and use standard relational distance measures, distance measures customized to this particular problem will take our knowledge about the domain better into account, and thus yield better results. In the rest of this section we will discuss different approaches to define distances between haplotype pairs and analyze their properties. Afterwards, we discuss algorithms to compute consensus haplotypes based on these distances, and present some computational complexity results.

3.1   Distance Computations

The genetic distance between two haplotype pairs is a complex function, which depends on the information the chromosomes of the two individuals contain (and, in principle, even other chemical properties of the DNA sequences). How-ever, modeling distance functions at this level is rather tedious. Instead, simpler distance functions aiming to capture some aspects of the relevant properties of the genetic similarity have to be used.

In this section we consider distance functions based on markers, i.e., distances between haplotype pairs. These can be grouped into two categories: distances induced by distances between individual haplotypes, and distance functions that work with the pair directly. Pair-wise Hamming distance is the most well-known example for the first category, and switch distance for the second. We will also give a unified view to both of the distance functions by proposing a k-Hamming distance which interpolates between pair-wise Hamming distance and switch distance.

Hamming distance and other distances induced by distances on sequences. The most common distance measure between sequences $s, t \in \Sigma^m$ is the Hamming distance that counts the number of disagreements between s and t, i.e.,

$$d_H(s,t) = |\{i \in \{1, \ldots, m\} : s[i] \neq t[i]\}|. \tag{1}$$

The Hamming distance is not directly applicable for comparing the genetic information of two individuals, as this information consists of a pair of haplotypes. To generalize the Hamming distance to pairs of haplotypes, let us consider hap-lotype pairs $\{h_1^1, h_1^2\}$ and $\{h_2^1, h_2^2\}$. The distance between the pairs should be zero if the sets $\{h_1^1, h_1^2\}$ and $\{h_2^1, h_2^2\}$ are the same. Hence, we should try to pair the haplotypes both ways and take the one with the smaller distance, i.e.,

$$d_H(\{h_1^1, h_1^2\}, \{h_2^1, h_2^2\}) = \min \left\{ d_H(h_1^1, h_2^1) + d_H(h_1^2, h_2^2), d_H(h_1^1, h_2^2) + d_H(h_1^2, h_2^1) \right\}.$$

Note that a similar construction can be used to map any distance function between haplotype sequences to a distance function between pairs of haplotyp-ings. Furthermore, if the distance function between the sequences satisfies the triangle inequality, so does the corresponding distance function for haplotype reconstructions.

Proposition 1. *Let* $d\colon \Sigma^m \times \Sigma^m \to \mathbb{R}_{\geq 0}$ *be a distance function between sequences of length* m *and*

$$d(\{h_1^1, h_1^2\}, \{h_2^1, h_2^2\}) = \min\{d(h_1^1, h_2^1) + d(h_1^2, h_2^2), d(h_1^1, h_2^2) + d(h_1^2, h_2^1)\}$$

for all $h_1^1, h_1^2, h_2^1, h_2^2 \in \Sigma^m$. *If* d *satisfies the triangle inequality for comparing sequences, i.e.,*

$$d(s, t) \leq d(s, u) + d(t, u)$$

for all $s, t, u \in \Sigma^m$, *then* d *satisfies the triangle inequality for comparing unordered pairs of sequences, i.e.,*

$$d(h_1, h_2) \leq d(h_1, h_3) + d(h_2, h_3)$$

for all $h_1^1, h_1^2, h_2^1, h_2^2, h_3^1, h_3^2 \in \Sigma^m$.

Proof. Choose arbitrary sequences $h_1^1, h_1^2, h_2^1, h_2^2, h_3^1, h_3^2 \in \Sigma^m$. We show that the claim holds for them and hence for all sequences of length m over the alphabet Σ. Assume, without loss of generality, that $d(\{h_1^1, h_1^2\}, \{h_2^1, h_2^2\}) = d(h_1^1, h_2^1) + d(h_1^2, h_2^2)$ and $d(\{h_1^1, h_1^2\}, \{h_3^1, h_3^2\}) = d(h_1^1, h_3^1) + d(h_1^2, h_3^2)$. For $d(\{h_2^1, h_2^2\}, \{h_3^1, h_3^2\})$ there are two cases as it is the minimum of $d(h_2^1, h_3^1) + d(h_2^2, h_3^2)$ and $d(h_2^1, h_3^2) + d(h_2^2, h_3^1)$.

If $d(\{h_2^1, h_2^2\}, \{h_3^1, h_3^2\}) = d(h_2^1, h_3^1) + d(h_2^2, h_3^2)$, then

$$d(\{h_1^1, h_1^2\}, \{h_3^1, h_3^2\}) + d(\{h_2^1, h_2^2\}, \{h_3^1, h_3^2\}) =$$
$$d(h_1^1, h_3^1) + d(h_1^2, h_3^2) + d(h_2^1, h_3^1) + d(h_2^2, h_3^2) =$$
$$\left[d(h_1^1, h_3^1) + d(h_2^1, h_3^1)\right] + \left[d(h_1^2, h_3^2) + d(h_2^2, h_3^2)\right] \geq d(h_1^1, h_2^1) + d(h_1^2, h_2^2).$$

If $d(\{h_2^1, h_2^2\}, \{h_3^1, h_3^2\}) = d(h_2^2, h_3^1) + d(h_2^1, h_3^2)$, then

$$d(\{h_1^1, h_1^2\}, \{h_3^1, h_3^2\}) + d(\{h_2^1, h_2^2\}, \{h_3^1, h_3^2\}) =$$
$$d(h_1^1, h_3^1) + d(h_1^2, h_3^2) + d(h_2^2, h_3^1) + d(h_2^1, h_3^2) =$$
$$\left[d(h_1^1, h_3^1) + d(h_2^2, h_3^1)\right] + \left[d(h_1^2, h_3^2) + d(h_2^1, h_3^2)\right] \geq$$
$$d(h_1^1, h_2^2) + d(h_1^2, h_2^1) \geq d(h_1^1, h_2^1) + d(h_1^2, h_2^2).$$

Thus, the claim holds.   □

The approach of defining distance functions between haplotype pairs based on distance functions between haplotypes has some limitations, regardless of the distance function used. This is because much of the variance in haplotypes originates from genetic *cross-over*, which breaks up the chromosomes of the parents and reconnects the resulting segments to form a new chromosome for the offspring. A pair $\{\hat{h}^1, \hat{h}^2\}$ of haplotypes which is the result of a cross-over between two haplotypes h^1, h^2 should be considered similar to the original pair $\{h^1, h^2\}$, even though the resulting sequences can be radically different. This kind of similarity cannot be captured by distance functions on individual haplotypes.

Switch distance. An alternative distance measure for haplotype pairs is to compute the number of *switches* that are needed to transform a haplotype pair to another haplotype pair that corresponds to the same genotype. A switch between markers i and $i + 1$ for a haplotype pair $\{h^1, h^2\}$ transforms the pair $\{h^1, h^2\} = \{h^1[1, i]h^1[i + 1, m], h^2[1, i]h^2[i + 1, m]\}$ into the pair $\{h^1[1, i]h^2[i + 1, m], h^2[1, i]h^1[i + 1, m]\}$. It is easy to see that for any pair of haplotype reconstructions corresponding to the same genotype, there is a sequence of switches transforming one into the other. Thus, this *switch distance* is well defined for the cases we are interested in.

The switch distance, by definition, assigns high similarity to haplotype pairs if one pair can be transformed into the other by a small number of recombination events. It also has the advantage over the Hamming distance that the order of the haplotypes in the haplotype pair does not matter in the distance computation: the haplotype pair can be encoded uniquely as a bit sequence consisting of just the switches between the consecutive heterozygous markers, i.e., as a *switch sequence*:

Definition 1 (Switch sequence). *Let* $h^1, h^2 \in \{0, 1\}^m$ *and let* $i_1 < \ldots < i_{m'}$ *be the heterozygous markers in* $\{h^1, h^2\}$. *The switch sequence of a haplotype pair* $\{h^1, h^2\}$ *is a sequence* $s(h^1, h^2) = s(h^2, h^1) = s \in \{0, 1\}^{m'-1}$ *such that*

$$s[j] = \begin{cases} 0 & \text{if } h^1[i_j] = h^1[i_{j+1}] \text{ and } h^2[i_j] = h^2[i_{j+1}] \\ 1 & \text{if } h^1[i_j] \neq h^1[i_{j+1}] \text{ and } h^2[i_j] \neq h^2[i_{j+1}] \end{cases} \tag{2}$$

The switch distance between haplotype reconstructions can be defined in terms of the Hamming distance between switch sequences as follows.

Definition 2 (Switch distance). *Let* $\{h_1^1, h_1^2\}$ *and* $\{h_2^1, h_2^2\}$ *be haplotype pairs corresponding to the same genotype. The switch distance between the pairs is*

$$d_s(h_1, h_2) = d_s(\{h_1^1, h_1^2\}, \{h_2^1, h_2^2\}) = d_H(s(h_1^1, h_1^2), s(h_2^1, h_2^2))$$

As switch distance is the Hamming distance between the switch sequences, the following proposition is immediate:

Proposition 2. *The switch distance satisfies the triangle inequality.*

k-Hamming distance. Switch distance considers only a very small neighborhood of each marker, namely only the previous and the next heterozygous marker in the haplotype. On the other extreme, the Hamming distance uses the complete neighborhood (via the min operation), i.e., the whole haplotypes for each marker. The intermediate cases are covered by the following k-Hamming distance in which all windows of a chosen length $k \in \{2, \ldots, m\}$ are considered. The intuition behind the definition is that each window of length k is a potential location for a gene, and we want to measure how close the haplotype reconstruction $\{h^1, h^2\}$ gets to the true haplotype $\{h_2^1, h_2^2\}$ in predicting each of these potential genes.

Definition 3 (k-Hamming distance). *Let $\{h_1^1, h_1^2\}$ and $\{h_2^1, h_2^2\}$ be pairs of haplotype sequences corresponding to the same genotype with m' heterozygous markers in positions $i_1, \ldots, i_m$. The k-Hamming distance d_{k-H} between $\{h_1^1, h_1^2\}$ and $\{h_2^1, h_2^2\}$ is defined by*

$$d_{k-H}(h_1, h_2) = \sum_{j=1}^{m'-k+1} d_H(h_1[i_j, \ldots, i_{j+k-1}], h_2[i_j, \ldots, i_{j+k-1}])$$

unless $m' < k$, in which case $d_{k-H}(h_1, h_2) = d_H(h_1, h_2)$.

It is easy to see that $d_{2-H} = 2d_S$, and that for haplotyping pairs with m' heterozygous markers, we have $d_{m'-H} = d_{m-H} = d_H$. Thus, the switch distance and the Hamming distance are the two extreme cases between which d_{k-H} interpolates for $k = 2, \ldots, m' - 1$.

3.2   Consensus Haplotypings

Given a distance function d on haplotype pairs, the problem of finding the *consensus haplotype pair* for a given set of haplotype pairs can be stated as follows:

Problem 2 (Consensus Haplotype). Given haplotype reconstructions $\{h_1^1, h_1^2\}$, $\ldots, \{h_l^1, h_l^2\} \subseteq \{0,1\}^m$, and a distance function $d : \{0,1\}^m \times \{0,1\}^m \to \mathbb{R}_{\geq 0}$, find:

$$\{h^1, h^2\} = \underset{h^1, h^2 \in \{0,1\}^m}{\operatorname{argmin}} \sum_{i=1}^{l} d(\{h_i^1, h_i^2\}, \{h^1, h^2\}).$$

Consensus haplotypings are useful for many purposes. They can be used in ensemble methods to combine haplotype reconstructions from different sources in order to decrease reconstruction errors. They are also applicable when a representative haplotyping is needed, for example for a cluster of haplotypes which has been identified in a haplotype collection.

The complexity of finding the consensus haplotyping depends on the distance function d used. As we will show next, for $d = d_S$ a simple voting scheme gives the solution. The rest of the distances considered in Section 3.1 are more challenging. If $d = d_{k-H}$ and k is small, the solution can be found by dynamic programming. For $d = d_{k-H}$ with large k and $d = d_H$, we are aware of no efficient general solutions. However, we will outline methods that can solve most of the problem instances that one may encounter in practice. For more details, confer [KLLM07].

Switch distance: $d = d_S$. For the switch distance, the consensus haplotyping can be found by the following voting scheme:

(1) Transform the haplotype reconstructions $\{h_i^1, h_i^2\} \subseteq \{0,1\}^m$, $i = 1, \ldots, l$ into switch sequences $s_1, \ldots, s_l \in \{0,1\}^{m'-1}$.
(2) Return the haplotype pair $\{h^1, h^2\}$ that shares the homozygous markers with the reconstructions $\{h_i^1, h_i^2\}$ and whose switch sequence $s \in \{0,1\}^{m'-1}$ is defined by $s[j] = \operatorname{argmax}_{b \in \{0,1\}} |\{j \in \{1, \ldots, m' - 1\} : s_i[j] = b\}|$.

The time complexity of this method is $O(lm)$.

k-Hamming distance: $d = d_{k-H}$. The optimal consensus haplotyping is

$$h_* = \{h_*^1, h_*^2\} = \operatorname*{argmin}_{\{h^1,h^2\}\subseteq\{0,1\}^m} \sum_{i=1}^{l} d_{k-H}(h_i, h).$$

The number of potentially optimal solutions is $2^{m'}$, but the solution can be constructed incrementally based on the following observation:

$$h_* = \operatorname*{argmin}_{\{h^1,h^2\}\subseteq\{0,1\}^m} \sum_{i=1}^{l} d_{k-H}(h_i, h)$$

$$= \operatorname*{argmin}_{\{h^1,h^2\}\subseteq\{0,1\}^m} \sum_{i=1}^{l} \sum_{j=1}^{m'-k+1} d_H(h_i[i_j, \ldots, i_{j+k-1}], h[i_j, \ldots, i_{j+k-1}])$$

Hence, the cost of any solution is a sum of terms

$$D_j(\{x, \bar{x}\}) = \sum_{i=1}^{l} d_H(h_i[i_j, \ldots, i_{j+k-1}], \{x, \bar{x}\}), \quad j = 1, \ldots, m'-k+1, x \in \{0, 1\}^k,$$

where $\bar{x}$ denotes the complement of x. There are $(m' - k + 1)2^{k-1}$ such terms. Furthermore, the cost of the optimal solution can be computed by dynamic programming using the recurrence relation

$$T_j(\{x, \bar{x}\}) = \begin{cases} 0 & \text{if } j = 0 \\ D_j(\{x, \bar{x}\}) + \min_{b \in \{0,1\}} T_{j-1}(\{bx, \overline{bx}\}) & \text{if } j > 0 \end{cases}$$

Namely, the cost of the optimal solution is $\min_{x \in \{0,1\}^k} T_{m'}(\{x, \bar{x}\})$ and the optimal solution itself can be reconstructed by backtracking the path that leads to this position. The total time complexity for finding the optimal solution using dynamic programming is $\mathcal{O}(lm + 2^k kl(m' - k))$: the heterozygous markers can be detected and the data can be projected onto them in time $\mathcal{O}(lm)$, and the optimal haplotype reconstruction for the projected data can be computed in time $\mathcal{O}(2^k kl(m' - k))$. So the problem is fixed-parameter tractable[3] in k.

Hamming distance: $d = d_H$. An ordering (h^1, h^2) of an optimal consensus haplotyping $\{h^1, h^2\}$ with Hamming distance determines an ordering of the unordered input haplotype pairs $\{h_1^1, h_1^2\}, \ldots, \{h_l^1, h_l^2\}$. This ordering can be represented by a binary vector $o = (o_1, \ldots, o_l) \in \{0, 1\}^l$ that states for each $i = 1, \ldots, l$ that the ordering of $\{h_i^1, h_i^2\}$ is $(h_i^{1+o_i}, h_i^{2-o_i})$. Thus, $o_i = \operatorname*{argmin}_{b \in \{0,1\}} d_H(h^1, h_i^{1+b})$, where ties are broken arbitrarily.

---

[3] A problem is called fixed-parameter tractable in a parameter k, if the running time of the algorithm is $f(k)\mathcal{O}(n^c)$ where k is some parameter of the input and c is a constant (and hence not depending on k.) For a good introduction to fixed-parameter tractability and parameterized complexity, see [FG06].

Table 4. The total switch error between true haplotypes and the haplotype reconstructions over all individuals for the baseline methods. For Yoruba and HaploDB, the reported numbers are the averages over the 100 datasets.

Method	Daly	Yoruba	HaploDB
PHASE	145	37.61	108.36
fastPHASE	105	45.87	110.45
SpaMM	127	54.69	120.29
HaploRec	131	56.62	130.28
HIT	121	73.23	123.95
Gerbil	132	75.05	134.22
Ensemble	104	39.86	103.06
Ensemble w/o PHASE	107	43.18	105.68

If the ordering o is known and l is odd, the optimal haplotype reconstruction can be determined in time $\mathcal{O}(lm)$ using the formulae

$$h^1[i] = \operatorname*{argmax}_{b \in \{0,1\}} = \left| \left\{ j \in \{1, \ldots, l\} : h_j^{1+o_j}[i] = b \right\} \right| \tag{3}$$

and

$$h^2[i] = \operatorname*{argmax}_{b \in \{0,1\}} = \left| \left\{ j \in \{1, \ldots, l\} : h_j^{2-o_j}[i] = b \right\} \right|. \tag{4}$$

Hence, finding the consensus haplotyping is polynomial-time equivalent to the task of determining the ordering vector o corresponding to the best haplotype reconstruction $\{h^1, h^2\}$.

The straightforward way to find the optimal ordering is to evaluate the quality of each of the 2^{l-1} non-equivalent orderings. The quality of a single ordering can be evaluated in time $\mathcal{O}(lm)$. Hence, the consensus haplotyping can be found in total time $\mathcal{O}(lm + 2^l lm')$. The runtime can be reduced to $\mathcal{O}(lm + 2^l m')$ by using Gray codes [Sav97] to enumerate all bit vectors o in such order that consecutive bit vectors differ only by one bit. Hence, the problem is fixed-parameter tractable in l (i.e., the number of methods).

3.3   Experiments with Ensemble Methods

Consensus haplotypings can be used to combine haplotypings produced by different systems along the lines of ensemble methods in statistics. In practice, genetics researchers often face the problem that different haplotype reconstruction methods give different results and there is no straightforward way to decide which method to choose. Due to the varying characteristics of haplotyping datasets, it is unlikely that one haplotyping method is generally superior. Instead, different methods have different relative strengths and weaknesses, and will fail in different parts of the reconstruction. The promise of ensemble methods lies in "averaging out" those errors, as far as they are specific to a small subset of methods (rather

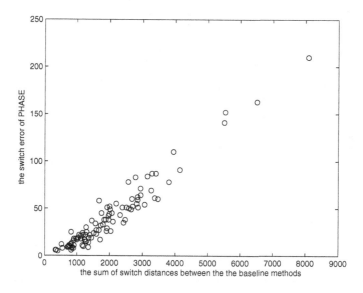

Fig. 5. The switch error of PHASE vs. the sum of the switch distances between the baseline methods for the Yoruba datasets. Each point corresponds to one of the Yoruba datasets, x-coordinate being the sum of distances between the reconstructions obtained by the baseline methods, and y-coordinate corresponding to the switch errors of the reconstructions by PHASE.

than a systematic error affecting all methods). This intuition can be made precise by making probabilistic assumptions about how the reconstruction methods err: If the errors in the reconstructions were small random perturbations of the true haplotype pair, taking a majority vote (in an appropriate sense depending on the type of perturbations) of sufficiently many reconstructions would with high probability correct all the errors.

Table 4 lists the reconstruction results for the haplotyping methods introduced in Section 2 on the Daly, Yoruba and HaploDB [HMK+07] datasets, and results for an ensemble method based on all individual methods (Ensemble) and all individual methods except the slow PHASE system (Ensemble w/o PHASE). The ensemble methods simply return the consensus haplotype pair based on switch distance. For the HaploDB dataset, we sampled 100 distinct marker sets of 100 markers each from chromosome one. The 74 available haplotypes in the data set were paired to form 37 individuals.

It can be observed that the ensemble method generally tracks with the best individual method, which varies for different datasets. Furthermore, if PHASE is left out of the ensemble to reduce computational complexity, results are still close to that of the best method including PHASE (Daly,Yoruba) or even better (HaploDB).

Distance functions on haplotypings can also be used to compute estimates of confidence for the haplotype reconstructions for a particular population. Figure 5 shows that there is a strong correlation between the sum of distances between the individuals methods (their "disagreement") and the actual, normally unknown reconstruction error of the PHASE method (which was chosen as reference method as it was the most accurate method overall in our experiments). This means that the agreement of the different haplotyping methods on a given population is a strong indicator of confidence for the reconstructions obtained for that population.

4   Structure Discovery

The main reason for determining haplotype data for (human) individuals is to relate the genetic information contained in the haplotypes to phenotypic traits of the individual, such as susceptibility to certain diseases. Furthermore, haplotype data yields insight into the organization of the human genome: how individual markers are inherited together, the distribution of variation in the genome, or regions which have been evolutionary conserved (indicating locations of important genes). At the data analysis level, we are therefore interested in analyzing the structure in populations—to determine, for example, the difference in the genetic make-up of a case and a control population—and the structure in haplotypes, e.g. for finding evolutionary conserved regions. In the rest of this section, we will briefly outline approaches to these structure discovery tasks, and in particular discuss representational challenges with haplotype and population data.

4.1   Structure in Populations

The use of haplotype pairs to infer structure in populations is relevant for relating the genetic information to phenotypical properties, and to predict the phenotypical properties based on the genetic information. The main approaches for determining structure in populations are classification and clustering.

As mentioned in the introduction, the main problem with haplotype data is that the data for each individual contains two binary sequences, where each position has a different interpretation. Hence, haplotype data can be considered to consist of unordered pairs of binary feature vectors, with sequential dependencies between nearby positions in the vector (the markers that are close to each other can, for example, be located on the same gene).

A simple way to propositionalize the data is to neglect the latter, i.e., the sequential dependence in the vectors. In that case the unordered pair of binary vectors is transformed into a ternary vector with symbols $\{0,0\}$, $\{0,1\}$, and $\{1,1\}$. However, the dependences between the markers are relevant. Hence, a considerable fraction of the information represented by the haplotypes is then neglected, resulting in less accurate data analysis results.

Another option is to fix the order of the vectors in each pair. The problem in that case is that the haplotype vectors are high-dimensional and hence fixing a total order between them is tedious if not impossible. Alternatively, both ordered pairs could be added to the dataset. However, then the data analysis technique has to take into account that each data vector is in fact a pair of unordered data vectors, which is again non-trivial.

The representational problems can be circumvented considering only the distances/similarities between the haplotype pairs, employing distance functions such as those we defined in the previous section. For example, nearest-neighbor classification can be conducted solely using the class labels and the inter-point distances. Distance information also suffices for hierarchical clustering. Furthermore, K-means clustering is also possible when we are able to compute the consensus haplotype pair for a collection of haplotype pairs. However, the distance functions are unlikely to grasp the fine details of the data, and in genetic data the class label of the haplotype pair (e.g., case/control population in gene mapping) can depend only on a few alleles. Such structure would be learnable e.g. by a rule learner, if the data could be represented accordingly.

Yet another approach is to transform the haplotype data into tabular from by feature extraction. However, that requires some data-specific tailoring and finding a reasonable set of features is a highly non-trivial task, regardless of whether the features are extracted explicitly or implicitly using kernels.

The haplotype data can, however, be represented in a straightforward way using relations. A haplotype pair $\{h^1, h^2\}$ is represented simply by a ternary predicate $m(i, j, h^i[j]), i = 1, 2, j = 1, \ldots, m$. This avoids the problem of fixing an order between the haplotypes, and retains the original representation of the data. Given this representation, probabilistic logical learning techniques could be used for classification and clustering of haplotype data. Some preliminary experiments have indicated that using such a representation probabilistic logic learning methods can in principle be applied to haplotype data, and this seems to be an interesting direction for future work.

4.2   Structure in Haplotypes

There are two main dimensions of structure in haplotype data: horizontal and vertical. The vertical dimension, i.e., structure in populations, has been briefly discussed in the previous section. The horizontal dimension corresponds to linear structure in haplotypes, such as segmentations. In this section, we will briefly discuss approaches for discovering this kind of structure.

Finding segmentation or block structure in haplotypes is considered one of the most important tasks in the search for structure in genomic sequences [DRS+01, GSN+02]. The idea for discovering the underlying block structure in haplotype data is to segment the markers into consecutive blocks in such a way that most of the recombination events occur at the segment boundaries. As a first approximation, one can group the markers into segments with simple (independent) descriptions. Such block structure detection takes the chemical structure of the DNA explicitly into account, assuming certain bonds to be stronger than others, whereas the genetic marker information is handled only implicitly. On the other hand, the genetic information in the haplotype markers could be used in conjunction with the similarity measures on haplotypes described in Section 3 to find haplotype segments, and consensus haplotype fragments for a given segment.

The haplotype block structure hypothesis has been criticized for being overly restrictive. As a refinement of the block model, mosaic models have been suggested. In mosaics there can be different block structures in different parts of the population, which can be modeled as a clustered segmentation [GMT04] where haplotypes are clustered and then a block model is found for each cluster. Furthermore, the model can be further refined by taking into account the sequential dependencies between the consecutive blocks in each block model and the shared blocks in different clusters of haplotypes. This can be modeled conveniently using a Hidden Markov Model [KKM+04]. Finally, the HMM can be extended also to take into account haplotype pairs instead of individual haplotypes.

A global description of the sequential structure is not always necessary, as the relevant sequential structure can concern only a small group of markers. Hence, finding frequent patterns in haplotype data, i.e., finding projections of the haplotype pairs on small sets of markers such that the projections of at least a σ-fraction of the input haplotype pairs agree with the projection for given $\sigma > 0$ is of interest. Such patterns can be discovered by a straightforward modification of the standard level-wise search such as described in [MT97]. For more details on these approaches, please refer to the cited literature.

5   Conclusions

A haplotype can be considered a projection of (a part of) a chromosome to those positions for which there is variation in a population. Haplotypes provide cost-efficient means for studying various questions, ranging from the quest to identify genetic roots of complex diseases to analyzing the evolution history of populations or developing "personalized" medicine based on the individual genetic disposition of the patient. Haplotype data for an individual consists of an unordered pair of haplotypes, as cells carry two copies of each chromosome (maternal and paternal information). This intrinsic structure in haplotype data makes it difficult to apply standard propositional data analysis techniques to this problem. In this chapter, we have studied how (probabilistic) relational/structured data analysis techniques can overcome this representational difficulty, including Logical Hidden Markov Models (Section 2) and methods based on distances between pairs of vectors (Section 3).

In particular, we have proposed the SpaMM system, a new statistical haplotyping method based on Logical Hidden Markov Models, and shown that it yields competitive reconstruction accuracy. Compared to the other haplotyping systems used in the study, the SpaMM system is relatively basic. It is based on a simple Markov model over haplotypes, and uses the logical machinery available in Logical Hidden Markov Models to handle the mapping from propositional haplotype data to intrinsically structured genotype data. A level-wise learning algorithm inspired by the Apriori data mining algorithm is used to construct sparse models which can overcome model complexity and data sparseness problems encountered with high-order Markov chains. We furthermore note that using an embedded implementation LOHMMs can also be competitive with other special-purpose haplotyping systems in terms of computational efficiency.

Finally, we have discussed approaches to discovering structure in haplotype data, and how probabilistic relational learning techniques could be employed in this field.

Acknowledgments. We wish to thank Luc De Raedt, Kristian Kersting, Matti Kääriäinen and Heikki Mannila for helpful discussions and comments, and Sampsa Lappalainen for help with the experimental study.

References

[AMS+96]   Agrawal, R., Mannila, H., Srikant, R., Toivonen, H., Verkamo, A.I.: Fast discovery of association rules. In: Fayyad, U.M., Piatetsky-Shapiro, G., Smyth, P., Uthurusamy, R. (eds.) Advances in Knowledge Discovery and Data Mining, pp. 307–328. AAAI/MIT Press (1996)

[DRS+01]   Daly, M.J., Rioux, J.D., Schaffner, S.F., Hudson, T.J., Lander, E.S.: High-Resolution Haplotype Structure in the Human Genome. Nature Genetics 29, 229–232 (2001)

[EGT04]   Eronen, L., Geerts, F., Toivonen, H.: A Markov Chain Approach to Reconstruction of Long Haplotypes. In: Altman, R.B., Dunker, A.K., Hunter, L., Jung, T.A., Klein, T.E. (eds.) Biocomputing 2004, Proceedings of the Pacific Symposium, Hawaii, USA, 6-10 January 2004, pp. 104–115. World Scientific, Singapore (2004)

[EGT06]   Eronen, L., Geerts, F., Toivonen, H.: HaploRec: efficient and accurate large-scale reconstruction of haplotypes. BMC Bioinformatics 7, 542 (2006)

[FG06]   Flum, J., Grohe, M.: Parameterized Complexity Theory. In: EATCS Texts in Theoretical Computer Science, Springer, Heidelberg (2006)

[GMT04]   Gionis, A., Mannila, H., Terzi, E.: Clustered segmentations. In: 3rd Workshop on Mining Temporal and Sequential Data (TDM) (2004)

[GSN+02]   Gabriel, S.B., Schaffner, S.F., Nguyen, H., Moore, J.M., Roy, J., Blumenstiel, B., Higgins, J., DeFelice, M., Lochner, A., Faggart, M., Liu-Cordero, S.N., Rotimi, C., Adeyemo, A., Cooper, R., Ward, R., Lander, E.S., Daly, M.J., Altshuler, D.: The structure of haplotype blocks in the human genome. Science 296(5576), 2225–2229 (2002)

[HBE+04]   Halldórsson, B.V., Bafna, V., Edwards, N., Lippert, R., Yooseph, S., Istrail, S.: A survey of computational methods for determining haplotypes. In: Istrail, S., Waterman, M.S., Clark, A. (eds.) DIMACS/RECOMB Satellite Workshop 2002. LNCS (LNBI), vol. 2983, pp. 26–47. Springer, Heidelberg (2004)

[HMK+07]   Higasa, K., Miyatake, K., Kukita, Y., Tahira, T., Hayashi, K.: D-HaploDB: A database of definitive haplotypes determined by genotyping complete hydatidiform mole samples. Nucleic Acids Research 35, D685–D689 (2007)

[Hud02]   Hudson, R.R.: Generating samples under a wright-fisher neutral model of genetic variation. Bioinformatics 18, 337–338 (2002)

[KDR06]   Kersting, K., De Raedt, L., Raiko, T.: Logical hidden markov models. Journal for Artificial Intelligence Research 25, 425–456 (2006)

[KKM+04]   Koivisto, M., Kivioja, T., Mannila, H., Rastas, P., Ukkonen, E.: Hidden markov modelling techniques for haplotype analysis. In: Ben-David, S., Case, J., Maruoka, A. (eds.) ALT 2004. LNCS (LNAI), vol. 3244, pp. 37–52. Springer, Heidelberg (2004)

[KLLM07]   Kääriäinen, M., Landwehr, N.: Sampsa Lappalainen, and Taneli Mielikäinen. Combining haplotypers. Technical Report C-2007-57, Department of Computer Science, University of Helsinki (2007)

[KS05]   Kimmel, G., Shamir, R.: A Block-Free Hidden Markov Model for Genotypes and Its Applications to Disease Association. Journal of Computational Biology 12(10), 1243–1259 (2005)

[LME+07]   Landwehr, N., Mielikäinen, T., Eronen, L., Toivonen, H., Mannila, H.: Constrained hidden markov models for population-based haplotyping. BMC Bioinformatics (to appear, 2007)

[MT97]   Mannila, H., Toivonen, H.: Levelwise search and borders of theories in knowledge discovery. Data Mining and Knowledge Discovery 1(3), 241–258 (1997)

[RKMU05]   Rastas, P., Koivisto, M., Mannila, H., Ukkonen, E.: A hidden markov technique for haplotype reconstruction. In: Casadio, R., Myers, G. (eds.) WABI 2005. LNCS (LNBI), vol. 3692, pp. 140–151. Springer, Heidelberg (2005)

[Sav97]   Savage, C.: A survey of combinatorial gray codes. SIAM Review 39(4), 605–629 (1997)

[SS05]   Stephens, M., Scheet, P.: Accounting for Decay of Linkage Disequilibrium in Haplotype Inference and Missing-Data Imputation. The American Journal of Human Genetics 76, 449–462 (2005)

[SS06]   Scheet, P., Stephens, M.: A Fast and Flexible Statistical Model for Large-Scale Population Genotype Data: Applications to Inferring Missing Genotypes and Haplotypic Phase. The American Journal of Human Genetics 78, 629–644 (2006)

[SWS05]   Salem, R., Wessel, J., Schork, N.: A comprehensive literature review of haplotyping software and methods for use with unrelated individuals. Human Genomics 2, 39–66 (2005)

[The05]   The International HapMap Consortium. A Haplotype Map of the Human Genome. Nature, 437, 1299–1320 (2005)

[TJHBD97]   Thompson Jr., J.N., Hellack, J.J., Braver, G., Durica, D.S.: Primer of Genetic Analysis: A Problems Approach, 2nd edn. Cambridge University Press, Cambridge (1997)

[WBCT05]   Wang, W.Y.S., Barratt, B.J., Clayton, D.G., Todd, J.A.: Genome-wide association studies: Theoretical and practical concerns. Nature Reviews Genetics 6, 109–118 (2005)

Model Revision from Temporal Logic Properties in Computational Systems Biology

François Fages and Sylvain Soliman

INRIA Rocquencourt, France

Francois.Fages@inria.fr, Sylvain.Soliman@inria.fr

Abstract. Systems biologists build models of bio-molecular processes from knowledge acquired both at the gene and protein levels, and at the phenotype level through experiments done in wild-life and mutated organisms. In this chapter, we present qualitative and quantitative logic learning tools, and illustrate how they can be useful to the modeler. We focus on biochemical reaction models written in the Systems Biology Markup Language SBML, and interpreted in the Biochemical Abstract Machine BIOCHAM. We first present a model revision algorithm for inferring reaction rules from biological properties expressed in temporal logic. Then we discuss the representations of kinetic models with ordinary differential equations (ODEs) and with stochastic logic programs (SLPs), and describe a parameter search algorithm for finding parameter values satisfying quantitative temporal properties. These methods are illustrated by a simple model of the cell cycle control, and by an application to the modelling of the conditions of synchronization in period of the cell cycle by the circadian cycle.

1   Introduction

One promise of computational systems biology is to model biochemical processes at a sufficiently large scale so that complex system behaviors can be predicted under various conditions. The biochemical reaction systems involved in these processes may contain many cycles and exhibit complex multistationarity and oscillating behaviors. While usually neglected in metabolic networks, these characteristics are preponderant in models of signal transduction and cell control. They thus provide a challenge to representation and inference methods, and the issue of representing complex biochemical systems and their behavior at different levels of abstraction is a central one in systems biology.

The pioneering use in [1] of the π-calculus process algebra for modeling cell signalling pathways, has been the source of inspiration of numerous works in the line of process calculi [2,3,4] and their stochastic extensions [5]. Recently, the question of formalizing the biological properties of the system has also been raised, and formal languages have been proposed for this task, most notably using temporal logics in either boolean [6,7], discrete [8,9,10] or continuous models [11,12].

L. De Raedt et al. (Eds.): Probabilistic ILP 2007, LNAI 4911, pp. 287–304, 2008.

The biochemical abstract machine BIOCHAM[1] [13,14] has been designed as a simplification of the process calculi approach using a logic programming setting and a language of reaction rules compatible with the Systems Biology Markup Language SBML [15] (http://www.sbml.org/). This opens up the whole domain of mathematical biology, through repositories like BioModels.net (http://www.biomodels.net), CMBSlib (http://contraintes.inria.fr/CMBSlib/), PWS (http://jjj.biochem.sun.ac.za/), etc. This rule-based language is used in BIOCHAM for modeling biochemical networks at three abstraction levels:

- The boolean semantics, where one reasons on the presence/absence of molecules,
- The differential semantics, where one reasons on molecular concentrations,
- The stochastic semantics, where one reasons on molecule numbers and reaction probabilities.

A second language is used to formalize the biological properties known from experiments in temporal logic (the *Computation Tree Logic* CTL, *Linear Time Logic* LTL or Probabilistic LTL with constraints, according to the qualitative, quantitative or stochastic nature of the properties). Such a formalization is a first step toward the use of logic learning tools to help the modeler in his tasks [16]. When a model does not satisfy all the expected properties, the purpose of the machine learning system of BIOCHAM is to propose rules or kinetic parameter values in order to curate the model w.r.t. a given specification [12]. This novel approach to biological modeling has been applied to a data set of models about the cell cycle control in different organisms, and signal transduction network (see http://contraintes.inria.fr/APrIL2/).

There has been work on the use of machine learning techniques, such as inductive logic programming (ILP, see Chapter 1 or [17]), to infer gene functions [18], metabolic pathway descriptions [19,20] or gene interactions [8]. However learning biochemical reactions from temporal properties is quite new, both from the machine learning perspective and from the systems biology perspective. A precursor system of this type was the system KARDIO used in drug target discovery [21]. The novelty in our approach is the use of the temporal logic setting to express semi-qualitative semi-quantitative properties of the behaviour of the system to be captured by the model.

In the following, we present successively:

- The boolean semantics of reaction models in Datalog, the representation of biological properties in temporal logic CTL, the application of ILP and model revision from temporal properties,
- The representation of kinetic models, of quantitative properties in temporal logic LTL with constraints, and a parameter search algorithm,
- The evaluation on an application: the modelling of the synchronization in period of the cell cycle by the circadian cycle.

---

[1] BIOCHAM is available for download at http://contraintes.inria.fr/BIOCHAM

2   Reaction Rule Learning from Temporal Properties

2.1   Biochemical Reaction Models in Datalog

From a syntactical point of view, SBML and BIOCHAM models basically consists in a set of reaction rules between molecules, protein complexes and modified proteins such as by phosphorylation. Each reaction rule for synthesis, degradation, complexation, phosphorylation, etc. can be given with a kinetic expression.

Example 1. Here is for instance a simple model of the cell cycle control after Tyson (1991). Each rule is given here with an arithmetic expression (its rate) followed by the keyword `for` and then a list of reactants separated by + on the left side of the reaction arrow => and a list of products on the right side. The notation _ represents the empty list.

```
k1                          for _=>Cyclin.
k2*[Cyclin]                 for Cyclin=>_.
k3*[Cyclin]*[Cdc2~{p1}]     for Cyclin+Cdc2~{p1}=> Cdc2~{p1}-Cyclin~{p1}.
k4p*[Cdc2~{p1}-Cyclin~{p1}] for Cdc2~{p1}-Cyclin~{p1}=> Cdc2-Cyclin~{p1}.
k4*([Cdc2-Cyclin~{p1}])^2*[Cdc2~{p1}-Cyclin~{p1}]
          for Cdc2~{p1}-Cyclin~{p1}=[Cdc2-Cyclin~{p1}]=> Cdc2-Cyclin~{p1}.
k5*[Cdc2-Cyclin~{p1}]       for Cdc2-Cyclin~{p1}=> Cdc2~{p1}-Cyclin~{p1}.
k6*[Cdc2-Cyclin~{p1}]       for Cdc2-Cyclin~{p1}=> Cyclin~{p1}+Cdc2.
k7*[Cyclin~{p1}]            for Cyclin~{p1}=>_.
k8*[Cdc2]                   for Cdc2=> Cdc2~{p1}.
k9*[Cdc2~{p1}]              for Cdc2~{p1}=> Cdc2.
```

The first rule represents the synthesis of a cyclin with a constant rate $k1$. The second rule represents the degradation of the cyclin with a reaction rate proportional to the cyclin concentration. The third rule represents the phosphorylation of the cyclin when it gets complexed with the kinase `Cdc2~{p1}`. The fourth rule is an autocatalyzed dephosphorylation of the complex, etc. For a more complete account of BIOCHAM syntax see for instance [12].

From a semantical point of view, reaction rules can be interpreted under different semantics corresponding to different abstraction levels. The most abstract semantics of BIOCHAM rules is the boolean semantics that associates to each molecule a boolean variable representing its presence or absence in the system, and ignores the kinetic expressions. Reaction rules are then interpreted as an *asynchronous transition system*[2] over states defined by the vector of boolean variables. A rule such as A+B=>C+D defines four possible transitions corresponding to the complete or incomplete consumption of the reactants A and B. Such a rule can only be applied when both A and B are present in the current state. In the next state, C and D are then present, while A and B can either be present (partial consumption) or absent (complete consumption).

---

[2] In this context asynchronous refers to the fact that only one transition is fired at a time, even if several are possible. This choice is justified by the fundamental biochemical phenomena of competition and masking between reaction rules.

The boolean semantics can be straightforwardly represented in Datalog. We use Prolog here for convenience. A state is represented by a Prolog term `state(mol1,...,molN)` where the molecule variable `mol` is 0 if absent, 1 if present, and a variable _ if it can take any value. Transitions are represented by facts `transition(predecessor_state, successor_state)` with variables linking successor and predecessor values.

Example 2. The boolean semantics of the previous cell cycle model can be represented in Prolog as follows:

```
dimension(6).
names('Cyclin','Cdc2~{p1}','Cdc2-Cyclin~{p1,p2}',
      'Cdc2-Cyclin~{p1}','Cdc2','Cyclin~{p1}').
transition(state(_,A,B,C,D,E),state(1,A,B,C,D,E)).
transition(state(1,A,B,C,D,E),state(_,A,B,C,D,E)).
transition(state(1,1,_,A,B,C),state(_,_,1,A,B,C)).
transition(state(A,B,1,_,C,D),state(A,B,_,1,C,D)).
transition(state(A,B,1,1,C,D),state(A,B,_,1,C,D)).
transition(state(A,B,_,1,C,D),state(A,B,1,_,C,D)).
transition(state(A,B,C,1,_,_),state(A,B,C,_,1,1)).
transition(state(A,B,C,D,E,1),state(A,B,C,D,E,_)).
transition(state(A,_,B,C,1,D),state(A,1,B,C,_,D)).
transition(state(A,1,B,C,_,D),state(A,_,B,C,1,D)).
```

Formally, the boolean semantics of a reaction model is a *Kripke structure* (see for instance [22]) $K = (S, R)$ where S is the set of states defined by the vector of boolean variables, and $R \subseteq S \times S$ is the transition relation between states, supposed to be total (i.e. $\forall s \in S, \exists s' \in S$ s.t. $(s, s') \in R$). A path in K, starting from state s_0 is an infinite sequence of states $\pi = s_0, s_1, \cdots$ such that $(s_i, s_{i+1}) \in R$ for all $i \geq 0$. We denote by π^k the path $s_k, s_{k+1}, \cdots$.

2.2 Biological Properties in Temporal Logic CTL

In the boolean semantics of reaction models, the biological properties of interest are *reachability properties*, i.e. whether a particular protein can be produced from an initial state; *checkpoints*, i.e. whether a particular protein or state is compulsory to reach another state; *stability*, i.e. whether the system can (or will) always verify some property; etc.

Such properties can be expressed in the *Computation Tree Logic* CTL* [22] that is an extension of propositional logic for reasoning about an infinite tree of state transitions. CTL* uses operators about branches (non-deterministic choices) and time (state transitions). Two path quantifiers A and E are introduced to handle non-determinism: $A\phi$ meaning that ϕ is true on all branches, and $E\phi$ that it is true on at least one branch. The time operators are F, G, X, U and W; $X\phi$ meaning ϕ is true at the next transition, $G\phi$ that ϕ is always true, $F\phi$ that ϕ is eventually true, $\phi \ U \ \psi$ meaning ϕ is always true until ψ becomes true, and $\phi \ W \ \psi$ meaning ϕ is either always true or until and when ψ becomes true. Table 1 recalls the truth value of a formula in a given Kripke structure.

Table 1. Inductive definition of the truth value of a CTL* formula in a state s or a path π, in a given Kripke structure K

$s \models \alpha$	iff α is a propositional formula true in the state s,
$s \models E\psi$	iff there exists a path π starting from s s.t. $\pi \models \psi$,
$s \models A\psi$	iff for all paths π starting from s, $\pi \models \psi$,
$s \models !\psi$	iff $s \not\models \psi$,
$s \models \psi \ \& \ \psi'$	iff $s \models \psi$ and $s \models \psi'$,
$s \models \psi \mid \psi'$	iff $s \models \psi$ or $s \models \psi'$,
$s \models \psi \Rightarrow \psi'$	iff $s \models \psi'$ or $s \not\models \psi$,
$\pi \models \phi$	iff $s \models \phi$ where s is the first state of π,
$\pi \models X\psi$	iff $\pi^1 \models \psi$,
$\pi \models \psi \ U \ \psi'$	iff there exists $k \geq 0$ s.t. $\pi^k \models \psi'$ and $\pi^j \models \psi$ for all $0 \leq j < k$.
$\pi \models \psi \ W \ \psi'$	iff either for all $k \geq 0$, $\pi^k \models \psi$.
	or there exists $k \geq 0$ s.t. $\pi^k \models \psi \& \psi'$ and for all $0 \leq j < k$, $\pi^j \models \psi$.
$\pi \models !\psi$	iff $\pi \not\models \psi$,
$\pi \models \psi \ \& \ \psi'$	iff $\pi \models \psi$ and $\pi \models \psi'$,
$\pi \models \psi \mid \psi'$	iff $\pi \models \psi$ or $\pi \models \psi'$,
$\pi \models \psi \Rightarrow \psi'$	iff $\pi \models \psi'$ or $\pi \not\models \psi$,

In this logic, $F\phi$ is equivalent to $true \ U \ \phi$, $G\phi$ to $\phi \ W \ false$, and the following duality properties hold: $!(E\phi) = A(!\phi)$, $!(X\phi) = X(!\phi)$, $!(F\phi) = G(!\phi)$, $!(\phi \ U \ \psi) = !\psi \ W \ !\phi$ and $!(\phi \ W \ \psi) = !\psi \ U \ !\phi$, where $!$ denotes negation. The following abbreviation are used in BIOCHAM:

- `reachable(P)` stands for $EF(P)$;
- `steady(P)` stands for $EG(P)$;
- `stable(P)` stands for $AG(P)$;
- `checkpoint(Q,P)` stands for $!E(!Q \ U \ P)$;
- `oscillates(P)` stands for $EG((F \ !P) \wedge (F \ P))$.

These temporal properties can be checked in the Prolog representation of reaction rules, by using a symbolic model-checker written in Prolog. The BIOCHAM model checker in Prolog proceeds by computing both backward and forward frontiers of states, starting from the initial states (resp. the goal states) leading to a goal state (resp. an initial state). These sets of states are represented by Prolog facts with variables. Their cardinalities are reduced by subsumption checks in this representation. In its simplest form, the forward reachability analysis proceeds by computing the transitive closure of the transition relation, starting from the initial state, up to the reaching of a state in the query. The simplest case in such a model checker is thus a standard transitive closure algorithm in Prolog.

For performance reasons in large reaction models however, the symbolic model checker NuSMV [23] based on ordered binary decision diagram (OBDD) is preferred and is used by default in BIOCHAM, through an interface. NuSMV is restricted to the fragment CTL of CTL* in which each time operator must be immediately preceded by a path quantifier. This restriction causes a difficulty for the oscillation properties only, since they cannot be expressed in CTL. In CTL, oscillation properties are thus approximated by the necessary but not sufficient formula $EG((EF\ !P) \wedge (EF\ P))$. We refer to [7,24] for the expressivity and scalability of this approach in reaction models containing several hundreds of variables and rules.

2.3   Model Revision from Temporal Properties

Having the model and the properties defined by a Prolog program, ILP techniques can in principle be used for learning reaction rules from temporal properties, i.e. structure learning of the underlying logic program (see Chapter 1). Here the positive and negative examples are uniformly given as a list of temporal properties to satisfy (expressed in a language closed by negation), instead of by positive and negative *facts*. Because of the relative complexity of the model checker in Prolog, this approach is currently limited to reachability properties. For learning from more general temporal properties, the NuSMV model checker is used in BIOCHAM as a black box, within an enumeration algorithm of all possible rule instances of some given rule pattern.

Furthermore, in the general framework of model revision, one wants to discover deletions as well as additions of reaction rules (of some pattern given as a bias) in order to satisfy a set of CTL formulas given as positive and negative examples. CTL properties can be classified into ECTL and ACTL formulas (i.e. formulas containing only E or A path quantifiers respectively) in order to anticipate whether reaction rules need be added or deleted. Indeed if an ECTL (resp. ACTL) formula is false in a Kripke structure, it remains false in a Kripke structure with less (resp. more) transitions. We refer to [12] for the details of the model revision algorithm implemented in BIOCHAM along these lines.

We show here our results on the model of example 1. For the structure learning phase, some CTL formulae are entered as a specification, expressing here reachability, oscillation and checkpoint properties:

```
add_specs({
        reachable(Cdc2~{p1}),
        reachable(Cdc2),
        reachable(Cyclin),
        reachable(Cyclin~{p1}),
        reachable(Cdc2-Cyclin~{p1}),
        reachable(Cdc2~{p1}-Cyclin~{p1})}).
```

```
add_specs({
        oscil(Cdc2,
        oscil(Cdc2~{p1})),
        oscil(Cdc2~{p1}-Cyclin~{p1}),
        oscil(Cdc2-Cyclin~{p1}),
        oscil(Cyclin),
            checkpoint(Cdc2~{p1}-Cyclin~{p1}, Cdc2-Cyclin~{p1})}).
```

These properties are satisfied by the model and can be automatically checked by the model-checker. The simplest example to illustrate the structural learning

method is to delete one rule in the model and let the learning system revise the
model in order to satisfy the specification.

```
biocham: delete_rules(Cyclin+Cdc2~{p1}=>Cdc2~{p1}-Cyclin~{p1}).
Cyclin+Cdc2~{p1}=>Cdc2~{p1}-Cyclin~{p1}
```

```
biocham: check_all.
The specification is not satisfied.
This formula is the first not verified: Ai(oscil(Cdc2~{p1}-Cyclin~{p1}))
```

```
biocham: revise_model(more_elementary_interaction_rules).
Success
Modifications found:
  Deletion(s):
  Addition(s):
Cyclin+Cdc2~{p1}=[Cdc2]=>Cdc2~{p1}-Cyclin~{p1}.
```

The first solution found is correct, even though it does not correspond to the
deleted rule. In fact, there are four solutions consisting in adding one rule, the
third one corresponds to the original model:

```
biocham: learn_one_addition(elementary_interaction_rules).
(1) Cyclin+Cdc2~{p1}=[Cdc2]=>Cdc2~{p1}-Cyclin~{p1}
(2) Cyclin+Cdc2~{p1}=[Cyclin]=>Cdc2~{p1}-Cyclin~{p1}
(3) Cyclin+Cdc2~{p1}=>Cdc2~{p1}-Cyclin~{p1}
(4) Cyclin+Cdc2~{p1}=[Cdc2~{p1}]=>Cdc2~{p1}-Cyclin~{p1}
```

It is worth noting that in these algorithms, the use of types [25] specifying
the protein functions for instance, has the effect of reducing the number of possibilities and improving the performances in terms of both adequacy of results
and computation time.

3   Parameter Search from Quantitative Temporal Properties

For relatively small networks of less than a hundred of proteins, kinetic models
have been proved successful to perform quantitative analyses and predictions.
Since the models of most datasets are in SBML, it is quite natural to handle
the kinetic expressions provided in those models, especially for relating them to
quantitative biological properties. In this section, we recal the two most usual
semantics for those expressions, the differential semantics and the stochastic
semantics, and relate them to PILP representations. We then show that the
Linear Time Logic LTL with numerical constraints provides the expressive power
necessary to represent both qualitative and quantitative properties of biological
systems. Similarly to what is done in the boolean case, a model-checker is then
used as basis for a learning process allowing here to find parameter values fitting
a given LTL specification of the biological properties that the model is supposed
to reproduce. This is shown on example 1 and is developed in an application in
the next section.

3.1   Continuous Semantics with ODE's

The concentration semantics of BIOCHAM associates to each molecule a real number representing its concentration. Reaction rules are in fact interpreted with their kinetic expressions by a set of nonlinear ordinary differential equations (ODE)[3]. Formally, to a set of BIOCHAM reaction rules $E = \{e_i$ for S_i => $S_i'\}_{i=1,\ldots,n}$ with variables $\{x_1, \ldots, x_m\}$, one associates the system of ODEs:

$$dx_k/dt = \sum_{i=1}^{n} r_i(x_k) * e_i - \sum_{j=1}^{n} l_j(x_k) * e_j$$

where $r_i(x_k)$ (resp. l_i) is the stoichiometric coefficient of x_k in the right (resp. left) member of rule i.

Given an initial state, i.e. initial concentrations for each of the objects, the evolution of the system is deterministic, and numerical integration algorithms compute a time series describing the temporal evolution of the system variables. The integration methods actually implemented in BIOCHAM are the adaptive step-size Runge-Kutta method and the Rosenbrock implicit method for stiff systems, which both produce simulation traces with variable time steps and are implemented in Prolog.

3.2   Stochastic Semantics with SLPs

The stochastic semantics is the most realistic semantics but also the most difficult to compute. This semantics associates to each BIOCHAM object an integer representing the number of molecules in the system. Rules are interpreted as a continuous time Markov chain where transition probabilities are defined by the kinetic expressions of reaction rules.

Stochastic simulation techniques [26] compute realizations of the process. The results are generally noisy versions of those obtained with the concentration semantics. However, in models with, for instance, very few molecules of some kind, qualitatively different behaviors may appear in the stochastic simulation, and thus justify the recourse to that semantics in such cases. A classical example is the model of the lambda phage virus [27] in which a small number of molecules, promotion factors of two genes, can generate an explosive multiplication (lysis) after a more or less long period of passive wait (lysogeny).

In the stochastic semantics, for a given volume V of the location where a compound is situated, its concentration C is translated into a number of molecules $N = C \times V \times K$, where K is Avogadro's number. The kinetic expression e_i for the reaction i is converted into a transition rate τ_i by replacing all concentrations by the corresponding number of molecules multiplied by volume. After normalization on all possible transitions, this gives the transition probability $p_i = \frac{\tau_i}{\sum_{j=1}^{n} \tau_j}$.

---

[3] The kinetic expressions in BIOCHAM can actually contain conditional expressions, in which case the reaction rules are interpreted by a deterministic hybrid automaton.

This semantics is close to SLPs. Two points however render unusable the classical learning techniques, and suggest an extension of the SLP framework:

- Kinetic expressions, and thus the corresponding transition probabilities τ_i, can contain variables representing the molecular concentrations (resp. number) of the reactants in each rule. A faithful translation of those models into SLP would thus involve dynamic probabilities according to variables values, like in the stochastic semantics of BIOCHAM by continuous time Markov chains [12]. On the other hand, SLPs as defined in Section 3 of Chapter 2, are restricted to constant probabilities on each rule.
- In stochastic simulation and Gillespie algorithms [26], the time is a random variable over reals, which cannot be mixed with SLPs in the current version of the formalism.

3.3 Biological Properties in LTL with Numerical Constraints

The *Linear Time Logic*, LTL is the fragment of CTL* that uses only temporal operators. A first-order version of LTL is used to express temporal properties about the molecular concentrations in the simulation trace. A similar approach is used in the DARPA BioSpice project [11]. The choice of LTL is motivated by the fact that the concentration semantics given by ODEs is deterministic, and there is thus no point in considering path quantifiers. The version of LTL with arithmetic constraints we use, considers first-order atomic formulae with equality, inequality and arithmetic operators ranging over real values of concentrations and of their derivatives.

For instance F([A]>10) expresses that the concentration of A eventually gets above the threshold value 10. G([A]+[B]<[C]) expresses that the concentration of C is always greater than the sum of the concentrations of A and B. Oscillation properties, abbreviated as oscil(M,K), are defined as a change of sign of the derivative of M at least K times:

F((d[M]/dt>0) & F((d[M]/dt<0) & F((d[M]/dt>0)...))). The abbreviated formula oscil(M,K,V) adds the constraint that the maximum concentration of M must be above the threshold V in at least K oscillations.

For practical purposes, some limited forms of quantified first-order LTL formulae are also allowed. As an example of this, constraints on the periods of oscillations can be expressed with a formula such as period(A,75), defined as $\exists t \, \exists v \, F(Time = t \,\&\, [A] = v \,\&\, d([A])/dt > 0 \,\&\, X(d([A])/dt < 0) \,\&\, F(Time = t + 75 \,\&\, [A] = v \,\&\, d([A])/dt > 0 \,\&\, X(d([A])/dt < 0)))$ where $Time$ is the time variable. This very formula is used extensively in the example of the next section.

Note that the notion of *next state* (operator X) refers to the state of the following time point computed by the (variable step-size) simulation, and thus does not necessarily imply real-time neighborhood. Nevertheless, for computing local maxima as in the formula above for instance, the numerical integration methods do compute the relevant time points with a very good accuracy.

3.4   Parameter Search from Temporal Properties

We implemented a dedicated LTL model checker for biochemical properties over simulation traces and proceeded, as in the boolean case, to use it for a learning method. Actually, it is mostly a search method automatically evaluating the fitness of a given parameter set w.r.t. an LTL specification.

The same method could theoretically be used to sample for a probability of satisfaction of an LTL specification for the stochastic semantics, however experimental trials proved to be too computationally expensive.

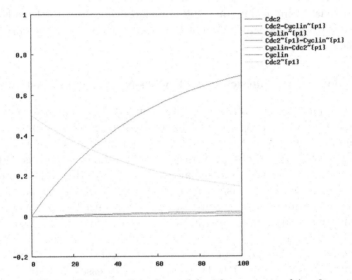

Fig. 1. Broken cell cycle model with parameter $k4 = 0$

The parameter learning method can be illustrated by changing the value of some parameter like $k4$ for instance. Figure 1 shows that the model is not oscillating as it should when $k4$ is set to zero.

The property of oscillation can be added as a temporal logic constraint to the parameter value search system as follows:

```
learn_parameters([k4],[(0,200)],20,oscil(Cdc2-Cyclin~{p1},3),100).
First values found that make oscil(Cdc2-Cyclin~{p1},3) true:
parameter(k4,200).
```

The value 200 found for $k4$ is close to the original value (180) and satisfies the experimental results formalized in LTL, as depicted in Figure 2.

Note that beacuase of the highly non-linear nature of the kinetics used in most biological models of the literature it is not possible to rely on usual tools of control theory for this kind of parameter estimation. The other available techniques are mostly local optimization based (simulated annealing and derivatives) but require to optimize with respect to a precise quantitative objective function, whereas the presented technique allows to mix qualitative and quantitative data.

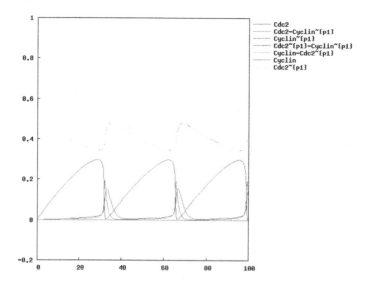

Fig. 2. Curated cell cycle model showing oscillations with the inferred parameter value $k4 = 200$

4 Application to Modelling the Synchronization in Period of the Cell Cycle by the Circadian Cycle

Cancer treatments based on the administration of medicines at different times of the day have been shown to be more efficient against malign cells and less damaging towards healthy ones. These results might be related to the recent discovery of links between the circadian clock (controlled by the light/dark cycle of a day) and the cell cycle. However, if many models have been developed to describe both of these cycles, to our knowledge none has described a real interaction between them.

In the perspective of the European Union project TEMPO[4] on temporal genomics for patient tailored chronotherapies, we developed a coupled model at the molecular level and studied the conditions of synchronization in period of these cycles, by using the parameter learning features of the modeling environment BIOCHAM. More specifically, the learning of parameter values from temporal properties with numerical constraints has been used to search how and where in the parameter space of our model the two cycles get synchronized. The technical report [28] describes the conditions of synchronization (i.e. synchronization by forcing the period of the target oscillator to be the same as that of the forcing oscillator) of the cell cycle by the circadian cycle via a common protein kinase WEE1 (see Figure 3).

---

[4] http://www.chrono-tempo.org

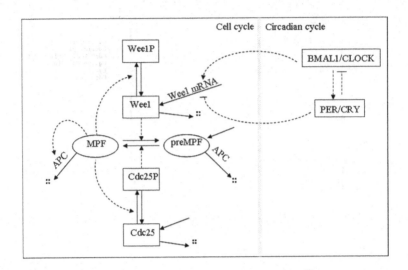

Fig. 3. Linking the circadian and the cell cycles via WEE1

The proteins chosen to illustrate the cell cycle are MPF, preMPF, the degradation factor of the cyclins, APC, the WEE1 kinase and the CDC25 phosphatase (Figure 3). Early in the cycle, MPF is kept inactive because the cyclin is not synthesized and WEE1 is present. As the cyclin is slowly synthesized, MPF activates and reaches a threshold that both inactivates WEE1 and activates CDC25 which maintains MPF in its active state. The cell enters mitosis. With a short delay, APC is activated and degrades the cyclin component of MPF. The cell exits mitosis and repeats its cycle. The model is composed of two positive feedback loops (CDC25 activates MPF which in turn activates CDC25, and WEE1 inactivates MPF which in turn inactivates WEE1) and a negative feedback loop (MPF activates APC through an intermediary enzyme X and APC degrades the cyclin component of the complex MPF). See Figure 4.

The two models describing the cell and circadian cycles are linked through the transcription of WEE1. In the model of the cell cycle alone, *wee1* mRNA was a parameter equal to 1. In the coupled model, the production of Wee1m is a function of the nuclear form of the complex BMAL1/CLOCK (BN) and the unphosphorylated nuclear form of the complex PER/CRY (PCN).

To find values for which synchronization occurs, the parameter space for each parameter has been explored using the BIOCHAM learning features from temporal properties. The values of three parameters appeared to be more significant than others: *ksweem*, *kswee* and *kimpf*. The two parameters *ksweem* and *kswee* both control the level of the WEE1 protein and show such similarities that in the following discussion, we will only report on *kswee*. The parameter values are varied in a given interval and reveal domains of synchronization reported in Figure 5. The parameters are plotted as a function of the period of three proteins that account for the behavior of the two cycles, BN (BMAL1/CLOCK nuclear) for the circadian cycle, MPF for the cell cycle, and their link, WEE1. For low

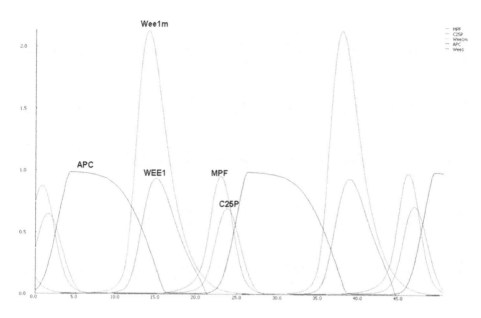

Fig. 4. Temporal simulation of a generic cell cycle

values of the parameters (region 1), MPF and BN have independent periods of oscillations of 22.4h and 23.85h respectively and no sustained oscillations are observed for WEE1. The reason for the perturbation in WEE1 oscillations in this region is that WEE1 receives simultaneously two influences: from the circadian cycle that controls the transcription of the protein mediated by the circadian transcription factors BMAL1/CLOCK and PER/CRY; and from the cell cycle that controls the activity of the protein via phosphorylation by MPF. WEE1 is produced but as soon as MPF activates, it is inactivated because WEE1 has no or little effect on MPF activation and MPF inhibits WEE1 protein. The two influences operate on WEE1 at different times as they both have different periods, perturbing WEE1 period.

For intermediate values of the parameters (region 2), WEE1 starts to play a more significant role in the cell cycle by inhibiting MPF activity, and as a result, disturbing MPF oscillations. It is only when the parameters reach a high value (either $kimpf$=1.2 or $kswee$=0.4) that the oscillations of MPF become stable again but with a period similar to that of the circadian cycle (region 3) revealing the synchronization of the cell cycle through WEE1 activity (through $kimpf$) or protein level (through $kswee$).

However, the study of $kimpf$, the parameter controlling the activity of WEE1 on MPF inactivation, shows that the synchronization does not solely depend on the value of the parameter but more particularly on the ratio $kimpf/kampf$ since both CDC25 and WEE1 are involved in the positive feedback loops that activate MPF and therefore responsible for the G2-M transition. To investigate this dual effect, the limit of synchronization is measured as the two parameters $kimpf$ and

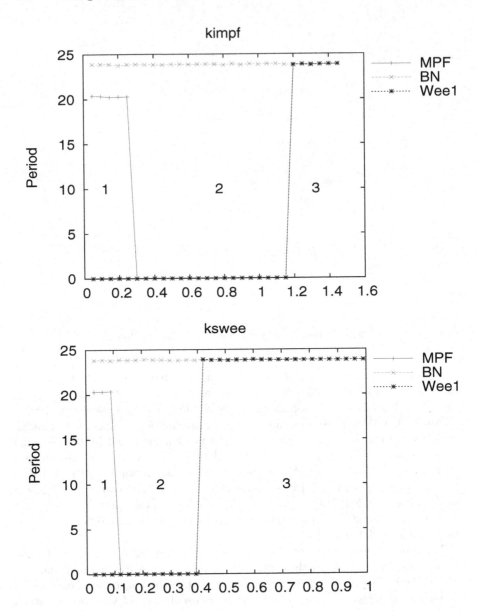

Fig. 5. Plot of the period as a function of the parameter *kimpf* from 0.01 to 1.6 and *kswee* from 0.01 to 1. The system shows synchronization for values superior to 1.2 for *kimpf* and 0.4 for *kswee*. For our purposes, constant periods are defined as follows: the last 11 peaks of the simulation over 500 time units show no more than 4% difference in their maxima and the length of the periods. MPF starts with an autonomous period of 22.4h and BN a period of 23.85h. As *kimpf* and *kswee* increase, MPF oscillations (accounting for cell cycle) lose stability and are entrained, along with WEE1, for higher values of the parameter with a period of 23.85h.

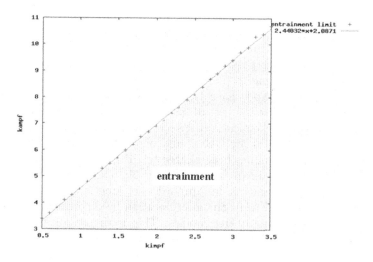

Fig. 6. BIOCHAM-generated plot of the synchronization in period of the cell cycle by the circadian cycle for different values of *kimpf* (action of WEE1 on MPF) and *kampf* (action of CDC25 on MPF). The limit of synchronization computed by BIOCHAM (red crosses) is interpolated by the linear function $kampf = 2.44832 \cdot kimpf + 2.0071$ (solid line).

kampf are varied simultaneously. A linear function of the form: $kampf = 2.44832 \cdot kimpf + 2.0071$ is obtained, the region below the line being the synchronization region (Figure 6).

These studies have been carried out thanks to the machine learning features of BIOCHAM to express the condition of synchronization in period by a first-order LTL formula with time constraints, and to explore the parameter values satisfying that formula. These methods are complementary to and have no counterpart in traditional tools for the study of dynamical systems such as the computation of bifurcation diagrams.

5   Discussion and Conclusions

Temporal logic is a powerful formalism for expressing the biological properties of a living system, such as state reachability, checkpoints, stability, oscillations, etc. This can be done both qualitatively and quantitatively, by considering respectively propositional and first-order temporal logic formulae with numerical constraints.

In the propositional case, we have given a Datalog representation of general biochemical reaction models allowing the use of ILP techniques for discovering reaction rules from given CTL properties. Because of the relative complexity of the CTL model checker however, this approach was limited to reachability properties. For general CTL properties, our approach was to use the symbolic OBDD model checker NuSMV as a black box within a model revision algorithm that searches for rule additions and deletions in order to satisfy a CTL specification. The first results are encouraging but also show some limitations concerning the boolean abstraction that simply forgets the kinetic expressions. Less crude abstractions are however possible and are currently under investigation. Furthermore, when restricting to pure reachability properties between completely defined states, ILP methods have been shown more efficient. There is thus a perspective for combining the best of both methods in this general setting.

Kinetic models of biochemical systems have been considered too with their two most usual interpretations, by ODEs, and by continuous time Markov chains. The second interpretation has been related to PILP representations with a generalized notion of SLPs involving dynamic probabilities according to variable values. However the stochastic interpretation of kinetic expressions is computationally too expensive, while the interpretation by differential equations does scale up to real-size quantitative models. This is the reason why the continuous semantics of BIOCHAM rules based on non-linear ordinary differential equations, instead of the stochastic semantics based on continuous-time Markov chains is used in these applications. We have shown that the inference of parameter values from a temporal logic specification is flexible enough to be easy to use, and provides accurate results with reasonable execution times. These functionalities are completely new and complement the other tools the modeler can use to estimate the range of parameter values. This has been illustrated by an original application to the modelling of the synchronization in period of the cell cycle by the circadian cycle.

Acknowledgements. The authors would like to thank Stephen Muggleton and Luc de Raedt for extensive discussions on the topics of this paper, and Nathalie Chabrier-Rivier and Laurence Calzone for their contributions. This work has been partly supported by the EC Sixth Framework Project Application of Probabilistic Inductive Logic Programming II (APrIL II) (Grant Ref: FP-508861).

References

1. Regev, A., Silverman, W., Shapiro, E.Y.: Representation and simulation of biochemical processes using the pi-calculus process algebra. In: Proceedings of the sixth Pacific Symposium of Biocomputing, pp. 459–470 (2001)
2. Cardelli, L.: Brane calculi - interactions of biological membranes. In: Danos, V., Schachter, V. (eds.) CMSB 2004. LNCS (LNBI), vol. 3082, pp. 257–280. Springer, Heidelberg (2005)
3. Regev, A., Panina, E.M., Silverman, W., Cardelli, L., Shapiro, E.: Bioambients: An abstraction for biological compartments. Theoretical Computer Science 325, 141–167 (2004)
4. Danos, V., Laneve, C.: Formal molecular biology. Theoretical Computer Science 325, 69–110 (2004)
5. Phillips, A., Cardelli, L.: A correct abstract machine for the stochastic pi-calculus. Transactions on Computational Systems Biology Special issue of BioConcur (to appear, 2004)

6. Eker, S., Knapp, M., Laderoute, K., Lincoln, P., Meseguer, J., Sönmez, M.K.: Pathway logic: Symbolic analysis of biological signaling. In: Proceedings of the seventh Pacific Symposium on Biocomputing, pp. 400–412 (2002)
7. Chabrier, N., Fages, F.: Symbolic model cheking of biochemical networks. In: Priami, C. (ed.) CMSB 2003. LNCS, vol. 2602, pp. 149–162. Springer, Heidelberg (2003)
8. Bernot, G., Comet, J.P., Richard, A., Guespin, J.: A fruitful application of formal methods to biological regulatory networks: Extending thomas' asynchronous logical approach with temporal logic. Journal of Theoretical Biology 229, 339–347 (2004)
9. Batt, G., Bergamini, D., de Jong, H., Garavel, H., Mateescu, R.: Model checking genetic regulatory networks using gna and cadp. In: Graf, S., Mounier, L. (eds.) SPIN 2004. LNCS, vol. 2989, Springer, Heidelberg (2004)
10. Calder, M., Vyshemirsky, V., Gilbert, D., Orton, R.: Analysis of signalling pathways using the prism model checker. In: Plotkin, G. (ed.) CMSB 2005: Proceedings of the third international conference on Computational Methods in Systems Biology (2005)
11. Antoniotti, M., Policriti, A., Ugel, N., Mishra, B.: Model building and model checking for biochemical processes. Cell Biochemistry and Biophysics 38, 271–286 (2003)
12. Calzone, L., Chabrier-Rivier, N., Fages, F., Soliman, S.: Machine learning biochemical networks from temporal logic properties. In: Priami, C., Plotkin, G. (eds.) Transactions on Computational Systems Biology VI. LNCS (LNBI), vol. 4220, pp. 68–94. Springer, Heidelberg (2006) (CMSB 2005 Special Issue)
13. Fages, F., Soliman, S., Chabrier-Rivier, N.: Modelling and querying interaction networks in the biochemical abstract machine BIOCHAM. Journal of Biological Physics and Chemistry 4, 64–73 (2004)
14. Calzone, L., Fages, F., Soliman, S.: BIOCHAM: An environment for modeling biological systems and formalizing experimental knowledge. BioInformatics 22, 1805–1807 (2006)
15. Hucka, M., et al.: The systems biology markup language (SBML): A medium for representation and exchange of biochemical network models. Bioinformatics 19, 524–531 (2003)
16. Fages, F.: From syntax to semantics in systems biology - towards automated reasoning tools. Transactions on Computational Systems Biology IV 3939, 68–70 (2006)
17. Muggleton, S.H.: Inverse entailment and progol. New Generation Computing 13, 245–286 (1995)
18. Bryant, C.H., Muggleton, S.H., Oliver, S.G., Kell, D.B., Reiser, P.G.K., King, R.D.: Combining inductive logic programming, active learning and robotics to discover the function of genes. Electronic Transactions in Artificial Intelligence, 6 (2001)
19. Angelopoulos, N., Muggleton, S.H.: Machine learning metabolic pathway descriptions using a probabilistic relational representation. Electronic Transactions in Artificial Intelligence 7 (2002) (also in Proceedings of Machine Intelligence 19)
20. Angelopoulos, N., Muggleton, S.H.: Slps for probabilistic pathways: Modeling and parameter estimation. Technical Report TR 2002/12, Department of Computing, Imperial College, London, UK (2002)
21. Bratko, I., Mozetic, I., Lavrac, N.: KARDIO: A study in Deep and Qualitative Knowledge for Expert Systems. MIT Press, Cambridge (1989)
22. Clarke, E.M., Grumberg, O., Peled, D.A.: Model Checking. MIT Press, Cambridge (1999)

23. Cimatti, A., Clarke, E., Enrico Giunchiglia, F.G., Pistore, M., Roveri, M., Sebastiani, R., Tacchella, A.: Nusmv 2: An opensource tool for symbolic model checking. In: Brinksma, E., Larsen, K.G. (eds.) CAV 2002. LNCS, vol. 2404, Springer, Heidelberg (2002)

24. Chabrier-Rivier, N., Chiaverini, M., Danos, V., Fages, F., Schächter, V.: Modeling and querying biochemical interaction networks. Theoretical Computer Science 325, 25–44 (2004)

25. Fages, F., Soliman, S.: Type inference in systems biology. In: Priami, C. (ed.) CMSB 2006. LNCS (LNBI), vol. 4210, Springer, Heidelberg (2006)

26. Gillespie, D.T.: General method for numerically simulating stochastic time evolution of coupled chemical-reactions. Journal of Computational Physics 22, 403–434 (1976)

27. Gibson, M.A., Bruck, J.: A probabilistic model of a prokaryotic gene and its regulation. In: Bolouri, H., Bower, J. (eds.) Computational Methods in Molecular Biology: From Genotype to Phenotype, MIT Press, Cambridge (2000)

28. Calzone, L., Soliman, S.: Coupling the cell cycle and the circadian cycle. Research Report 5835, INRIA (2006)

A Behavioral Comparison of Some Probabilistic Logic Models

Stephen Muggleton and Jianzhong Chen

Department of Computing, Imperial College London, London SW7 2AZ, UK
{shm,cjz}@doc.ic.ac.uk

Abstract. *Probabilistic Logic Models* (PLMs) are efficient frameworks that combine the expressive power of first-order logic as knowledge representation and the capability to model uncertainty with probabilities. *Stochastic Logic Programs* (SLPs) and *Statistical Relational Models* (SRMs), which are considered as *domain frequency* approaches, and on the other hand *Bayesian Logic Programs* (BLPs) and *Probabilistic Relational Models* (PRMs) (*possible worlds* approaches), are promising PLMs in the categories. This paper is aimed at comparing the relative expressive power of these frameworks and developing translations between them based on a behavioral comparison of their semantics and probability computation. We identify that SLPs augmented with combining functions (namely *extended SLPs*) and BLPs can encode equivalent probability distributions, and we show how BLPs can define the same semantics as complete, range-restricted SLPs. We further demonstrate that BLPs (resp. SLPs) can encode the relational semantics of PRMs (resp. SRMs). Whenever applicable, we provide inter-translation algorithms, present their soundness and give worked examples.

1 Introduction

Probabilistic Logic Models (PLMs) combine expressive knowledge representation formalisms such as relational and first-order logic with principled probabilistic and statistical approaches to inference and learning. This combination is needed in order to face the challenge of real-world learning and data mining problems in which data are complex and heterogeneous and we are interested in finding useful predictive and/or descriptive patterns.

Probabilistic logic representations and PLMs have varying levels of expressivity. As yet more effort has been put into defining new variants of PLMs than into characterising their relationships. Studying the relative expressive power of various probabilistic logic representations is a challenging and interesting task. On the one hand, there exist some theoretical study of the relations of two probabilistic approaches, i.e., possible-worlds and domain-frequency. On the other hand, it is interesting to see how both logic programming-based approaches and relational model/database-based methods have converged and to analyze relations between them. Essentially the following questions at different levels can be asked:

L. De Raedt et al. (Eds.): Probabilistic ILP 2007, LNAI 4911, pp. 305–324, 2008.

- From a semantical perspective: can we compare the semantics between PLMs in terms of the definition of probability distributions?
- From a theoretical perspective: given an expert domain, can we encode the same knowledge in different PLMs?
- More practically: can we find inter-translations between these PLMs?

This study is based on a *behavioural* approach: we analyze what the respective interests of these formulations are. In particular, how can we semantically compare PLMs and if there exist inter-translations, do the translations have the same features (computation of the probabilities, inference with or without evidence)?

The chapter is organised as follows. In section 2, we shortly introduce logical/relational models and probabilistic models as well as their extension to first-order probabilistic models. In section 3, we present a introduction to the four PLMs we choose for the study. In section 4, we explain the behavioural comparison approach and compare the semanics of the four PLMs based on the catigories of first-order probabilistic models. In the next three sections, we detail probabilistic knowledge encoding and inter-translations between the PLMs of interest. We deal with the concluding remarks and direction of future work in the last section.

2 Preliminaries

2.1 Logical/Relational Models

Knowledge encoded in PLMs are mainly based on two representation languages and their associated system of inference, namely the *Definite Clause Logic* used in logic programming model and the *Relational Algebra* used in the relational model.

Logic Programming Model. *Definite Clause Logic* (DCL), also known as *Horn clause logic*, is the subset of first-order logic whose well formed formulas are universally quantified conjunctions of disjunctions of literals. Each disjunct is called a clause. DCL further requires that clauses are definite, which means that they have exactly one positive literal each (the *head* of the clause). The list of negative literals (if any) is called the body of the clause. A list of definite clauses is called a logic program (LP). We assume our terminology discussed in the paper are within DCL.

The semantics, ie. the knowledge that can be encoded by a logic program L, can be defined with the least Herbrand model of L (noted $LH(L)$). There exists several inference algorithms to compute $LH(L)$ such as the SLD-resolution proof procedure. The success set of L, ie. the set of statements that can be refuted using SLD-resolution is exactly $LH(L)$. It is interesting to note that the use of functors and recursive clauses allows DCL programs to encode potentially infinite semantics. However, the functor-free subset of DCL (called Datalog) has a finite semantics since Datalog programs have finite least Herbrand models.

The use of DCL or Horn clause logic is supported by the wide availability and applicability of the programming language Prolog [1] and most Inductive Logic Programming systems, such as Progol [17].

Relational Model. Knowledge is often stored in relational databases which are formally defined by the so called *relational model* for data. We use the notations of [4] and [6] to introduce the relational model setting. This setting uses the abstract notion of *relational schema* to structure the data encoded in a database. Formally, a *relational schema* is a set of *classes* (also called *relations* or *tables*) $\mathcal{R} = \{R_i\}$. Each class R is associated to a *primary key* $R.K$, a set of *foreign keys* $\mathcal{F}(R) = \{F_k\}$ and a set of *descriptive attributes* $\mathcal{A}(R) = \{A_j\}$. Primary keys are unique identifiers for class instances. Descriptive attributes A_j take their respective values in the finite domains $\mathcal{V}(A_j)$. Each foreign key $F_k \in \mathcal{F}(R)$ establishes a directed relationship from its source class $Dom[F_k] = R$ onto a target class $Range[F_k]$. An instance of such a relational schema is a set of *objects* (also called *entities*, *records* or *tuples*). Instances of relational schemas are also called *databases*. Each object x is an instance of a class R: for each descriptive attribute $A_j \in \mathcal{A}(R)$, x is associated to an attribute value $x.A_j = a_j \in \mathcal{V}(A_j)$. Furthermore, class level foreign keys define binary relationships between objects: for each foreign key $F_k \in \mathcal{F}(R)$, x is associated an object $y = x.F_k$ such that y is an instance of the class $Y = Range[\rho_k]$. Databases that respect the constraints implied by foreign keys are said to respect the *referential integrity*.

Relational Algebra (RA) is a procedural language to perform queries on relational databases. RA is a procedural equivalent to declarative calculi such as the *tuple calculus* and the *domain calculus* that provides mathematical foundation for relational databases. RA is built upon six fundamental operations on sets of tuples: the selection, the projection, the Cartesian product, the set union, the set difference and the renaming of attribute names. These operations can be considered building blocks to define higher level operations such as joins, set intersections and divisions. From a more practical perspective, RA corresponds to SQL (*Structured Query Language*) without the aggregate and group operations. RA's expressive power is equivalent to non-recursive Datalog's (which is not Turing complete).

2.2   Probabilistic Models

Uncertainty is commonly modelled by defining a probability distribution over the domain $Dom(V) = \{v_i\}$ of a *random variable* V. We assume all random variables are discrete in the paper. We note $P(V = v_i) = p_i$. A set $\{v_i, p_i\}$ defines a probability distribution if and only if the $\{p_i\}$ are *normalized* ($\sum_i p_i = 1$).

Bayesian networks (BNs) [19,10] are introduced to make additional independency assumptions between random variables and to calculate the probabilities of the conjunctions. A BN is a Directed Acyclic Graph (DAG) where each vertex/node represents a random variable. Independency assumptions are encoded in the edges between chance nodes. Each variable is independent of its non-descendants given its parents. The parameters of the Bayesian networks are embedded in Conditional Probability Tables (CPTs) which specify, for each variable V_i, the probability that $V_i = v$ ($v \in Dom(V_i)$) given the values of the parents $Pa(V_i)$. Given these and based on the Bayes theorem, a joint probability distribution of a set of n random variables can be defined using the following chain rule formula:

$$P(V_1, V_2, \ldots, V_n) = \prod_{i=1}^{n} P(V_i | Pa(V_i))$$

There exists a variety of algorithms to efficiently compute this probability distribution, such as Pearl's message passing, Variable Elimination, graph based Junction Tree, etc.

2.3   First-Order Probabilistic Models

First-Order Probabilistic Reasoning. *First-order logic* alone is not suitable to handle uncertainty, while this is often required to model non-deterministic domains with noisy or incomplete data. On the other hand, *propositional* probabilistic models (such as BNs) can't express relations between probabilistic variables; such frameworks suffer from a lack of structure over the domain, and the knowledge they can encode is fairly limited.

First-order probabilistic models are models which integrate logics and probabilities; they overcome the limits of traditional models by taking the advantages of both logics and probabilities.

Categories of First-Order Probabilistic Models. A well-known categorization of first-order probabilistic models was introduced by Halpern [7], in which two types of first-order probabilistic logic are categorized, ie. probabilities on the domain (or *type 1* probability structure) and probabilities on possible worlds (or *type 2* probability structure).

Type 1 probability structure can represent statements like "The probability that a randomly chosen bird will fly is greater than .9". It provides a type of *domain-frequency* approaches, which semantically illustrates objective and 'sampling' probabilities of domains. Precisely, a type 1 probability structure is a tuple (D, π, μ), where D is a domain, π maps predicate and function symbols in alphabet to predicates and functions of the right arity over D, and μ is a discrete probability function on D. The probability here is taken over the domain D. In the logic programming setting, it is reasonable to consider the Herbrand base over a given signature to have the same function as the domain.

On the other hand, type 2 probability structure may represent statements like "The probability that Tweety (a particular bird) flies is greater than .9". It is a kind of *possible-world* approaches and illustrates the subjective and 'degree-of-belief' semantics of the probabilities of domains. Formally, a type 2 probability structure is a tuple (D, W, π, μ), where D is a domain, W is a set of states or possible worlds, for each state $w \in W$, $\pi(w)$ maps predicate and function symbols in alphabet to predicates and functions of the right arity over D, and μ

is a discrete probability function on W. The probability here is taken over W, the set of states (or possible worlds or logic models). BNs and related models are type 2 approaches.

One of the differences between the two models is that there seems to assume only one possible world in type 1 case, saying that "any bird has a probability of flying greater than 0.9" is like giving the result of some statistical analysis (by counting domain frequency) in the real world.

3 Presentation of PLMs

We choose four promising PLMs to do the comparison.

3.1 Stochastic Logic Programs

Stochastic Logic Programs (SLPs) were first introduced in [14] as a generalization of stochastic grammars.

Syntax. An SLP consists of a set of labelled clauses $p : C$, where p is from the interval $[0, 1]$, and C is a range-restricted[1] definite clause. Later in this report, the labelled clauses $p : C$ will be named *parameterized* clauses or *stochastic* clauses. This original SLP definition requires that for each predicate symbol q, the probability labels for all clauses with q in the head sum to 1. However, this can be a restrictive definition of SLPs. In other articles ([2] for instance), SLPs having this property are called *complete* SLPs, while in *uncomplete* SLPs, the probability labels for all clauses with a same predicate symbol in the head sum to less than 1. *Pure* SLPs are introduced in [2], whose clauses are all parameterized (whereas *impure* SLPs can have non-parameterized clauses, that is, definite logical clauses). Furthermore, *normalized* SLPs are like complete SLPs, but in *unnormalised* SLPs, the probability labels for all clauses with a same predicate symbol in the head can sum to any positive value other than 1.

Semantics. An SLP S has a *distributional semantics*, that is one which assigns a probability distribution to the atoms of each predicate in the Herbrand base of the clauses in S. The probabilities are assigned to atoms according to an SLD-resolution strategy that employs a stochastic selection rule[2].

Three different related distributions are defined in [2], over derivations, refutations and atoms. Given an SLP S with n parameterized clauses and a goal G, it is easy to define a *log-linear probability distribution over the set of derivations*

$$\psi_\lambda(x) = e^{\lambda.\nu(x)} = \prod_{i=1}^{n} l_i^{\nu_i(x)}$$

---

[1] C is said to be range-restricted iff every variable in the head of C is found in the body of C.

[2] The selection rule is not deterministic but stochastic; the probability that a clause is selected depends on the values of the labels (details can be found in [15]).

where x is a derivation of goal G; $\lambda = (\lambda_1, \lambda_2, ..., \lambda_n) \in \Re^n$ is a vector of log-parameters where $\lambda_i = log(l_i)$, l_i being the label of the clause C_i; $\nu = (\nu_1, \nu_2, ..., \nu_n) \in N^n$ is a vector of clause counts s.t. $\nu_i(x)$ is the number of times C_i is used in the derivation x. If we assign the probability 0 to all derivations that are not refutations of the goal G, and normalize the remaining probabilities with a normalization factor Z, we obtain the *probability distribution $f_\lambda(r)$ over the set R of the refutations of G*

$$f_\lambda(r) = Z_{\lambda,G}^{-1} e^{\lambda.\nu(r)}$$

The computed answer in the SLD-tree is the most general instance of the goal G that is refuted by r, which is also named the *yield atom*. Let $X(y)$ be the set of refutations which lead to the yield atom y, we can finally define a *distribution of probabilities over the set of yield atoms*

$$p_{\lambda,G}(y) = \sum_{r \in X(y)} f_\lambda(r) = Z_{\lambda,G}^{-1} \sum_{r \in X(y)} \left(\prod_{i=1}^{n} l_i^{\nu_i(r)} \right)$$

Given an SLP S, a query G and a (possibly partial) instantiation of G noted G_a, if λ is the vector of log-parameters associated to S and Y the set of yield atoms appearing in the refutations of G_a, we define $\mathbf{P_S^{SLP}}(G_a) = \sum_{y \in Y} p_{\lambda,G}(y)$.

3.2   Bayesian Logic Programs

Bayesian Logic Programs (BLPs) were first introduced in [13], as a generalization of Bayesian networks (BNs) and Logic Programs.

Syntax. A Bayesian logic program has two components – a *logical* one, which is a set of *Bayesian clauses*), and a *quantitative* one, which is a set of conditional probability distributions and *combining rules* corresponding to that logical structure. A *Bayesian clause* is an expression of the form: $A \mid A_1, ..., A_n$ where $n \geq 0$ and the A_i are *Bayesian atoms* which are (implicitly) universally quantified. The difference between a logical definite clause and a Bayesian clause is that: the sign $\mid$ is employed instead of $:-$; Bayesian atoms are assigned a (finite) *domain*, whereas first order logic atoms have binary values. Following the definitions in [13], we assume that atom domains in BLPs are discrete.

In order to represent a probabilistic model, each Bayesian clause c is associated with a conditional probability distribution $cpd(c)$ which encodes the probability that $head(c)$ takes some value, given the values of the Bayesian atoms in $body(c)$, ie. $P(head(c)|body(c))$. This conditional probability distribution is represented with a conditional probability table. As there can be many clauses with the same head (or non-ground heads that can be unified), *combining rules* are introduced to obtain the distribution required, i.e. functions which map finite sets of conditional probability distributions onto one *combined* conditional probability distribution. Common combining rules include the *noisy-or* rule, when domains are boolean, and the *max* rule, which is defined on finite domains.

Semantics. The link of BLPs to BNs is straightforward: each ground Bayesian atom can be associated to a chance node (a standard random variable), whose set of states is the domain of the Bayesian atom. The links (influence relations) between chance nodes are given by the Bayesian clauses, and the link matrices by the conditional probability distributions associated to these Bayesian clauses. The set of ground Bayesian atoms in the least Herbrand model together with the structure defined by the set of ground instances of the Bayesian clauses define a global (possibly infinite) dependency graph.

The semantics of BLPs can be discussed in a *well-defined* BLP. A range restricted BLP B is *well-defined* if:

1. Its least Herbrand model not empty: $LH(B) \neq \emptyset$. There must be at least one ground fact in B.
2. The induced dependency graph is acyclic;
3. Each random variable is only influenced by finite set of random variables.

Any such *well-defined* BLP B defines a unique probability distribution over the possible valuations of a ground query $G_a \in LH(B)$ [13]. The query-answering procedure actually consists of two parts: first, given a ground query and some evidence, the Bayesian network (namely the support network) containing all *relevant* atoms is computed, using *Knowledge Based Model Construction* (KBMC). Then the resulting Bayesian network can be queried using any available inference algorithm, the results we were looking for being the probability of the initial ground query over its domain.

Let B be a *well-defined* BLP and G_a a ground query. The Bayesian network constructed with KBMC is denoted by BN_{B,G_a}. The probability of a chance node Q taking the value v in BN_{B,G_a} (i.e. the probability of the set of possible worlds of BN_{B,G_a} in which Q has the value v) is denoted $\mathbf{P}^{\mathbf{BLP}}_{\mathbf{B},\mathbf{G_a}}(Q = v)$.

3.3 Statistical Relational Models

Statistical Relational Models (SRMs) were introduced in [6] in order to provide ways to infer statements over the success of some relational databases queries.

Syntax. SRMs are defined with respect to a given relational schema $\mathcal{R}$. Furthermore, SRMs require $\mathcal{R}$ to be *table stratified*, that is there must exist a partial ordering $\prec$ over classes in $\mathcal{R}$ such that for any $R.F \in \mathcal{F}(\mathcal{R})$ and $S = Dom[R.F]$, $S \prec R$ holds. Given such a *table stratified* relational schema $\mathcal{R}$, an SRM ψ is a pair $(\mathcal{S}, \theta)$ that defines a local probability model over a set of variables $\{R.A\}$ (for each class R and each descriptive attribute $A \in \mathcal{A}(R)$) and a set of boolean join indicator $\{R.J_F\}$ (for each foreign key $F \in \mathcal{F}(R)$ with $S = Dom[R.F]$). For each random variable of the form $R.V$, $\mathcal{S}$ specifies a set of parents $Pa(R.V)$ where each parents has the form $R.B$ or $R.F.B$, and θ specifies a CPT $\theta_{R.V} = P(R.V|Pa(R.V))$. $\mathcal{S}$ is further required to be a directed acyclic graph.

Semantics. Any SRM ψ defines a unique probability distribution $\mathbf{P}^{\mathbf{SRM}}_{\psi}$ over the class of so called *inverted-tree-foreign-key-join* queries (or *legal* queries) of a table stratified relational schema $\mathcal{R}$.

A *legal* query Q has form: $\bowtie_Q (\sigma_Q(R_1 \times R_2 \times \ldots \times R_n))$. The set $T = \{t_1, \ldots, t_n\}$ of tuple variables occurring in Q must be *closed* with respect to the *universal foreign key closure* of $\mathcal{R}$ as defined in [6] so that:

$$\bowtie_Q = \{t.F \bowtie s.K \mid t \in T, \; s \in T \text{ is associated to } t.F\}$$

The select part of Q occurs on some subset of $\mathcal{A}(T)$, $\sigma_Q = \{A_i = a_i \mid A_i \in \mathcal{A}(T)\}$

Given an SRM $\psi = (\mathcal{S}, \theta)$, $\mathcal{S}$ induces a Bayesian network B over the attributes of tuples variables in T (joint indicators included). The parameters of B are set according to θ. $\mathbf{P}_\psi^{\mathbf{SRM}}$ is then defined as the probability distribution induced by B over possible instantiations of attributes of T that correspond to σ_Q of any *legal* query Q over T:

$$\mathbf{P}_\psi^{\mathbf{SRM}}(Q) = \prod_{t.V_i \in Q} \theta(t.V_i | Pa_B(t.V_i))$$

SRMs can thus be used to estimate the probability P_D of success of legal queries against a database D that implements the relational schema $\mathcal{R}$. For any select-join query Q over D, P_D is defined as follows:

$$P_D(Q) = \frac{|\bowtie_Q (\sigma_Q(R_1 \times R_2 \times \ldots \times R_n))|}{|R_1| \times |R_2| \times \ldots \times |R_n|}$$

A *table stratified* database D is said to be a model of an SRM ψ if ψ's estimations are correct, ie. for any legal query Q, $\mathbf{P}_\psi^{\mathbf{SRM}}(Q) = P_D(Q)$. In this case, we note $D \models \psi$.

3.4   Probabilistic Relational Models

Probabilistic Relational Models (PRMs) were introduced in [4], which extends the relational model presented in section 2.1 by introducing *reference slots* and *relational skeletons*. For a given class R, the set of reference slots $\varrho(R) = \{\rho_k\}$ is the union of R's foreign keys $\mathcal{F}(R)$ with the set of foreign keys $R'.F$ that point to $R.K$ (ie. reverse foreign keys). Such a reference slot ρ may thus establish a one-to-many relationship between R and $R' = R.\rho$. A relational skeleton σ is a set of objects respecting the constraints of a given relational schema. The difference between a relational skeleton σ and a complete instance is that the values of some descriptive attributes of objects in σ are unknown. However σ specifies the values for the foreign keys.

Syntax. PRMs with attribute uncertainty consider each class-level descriptive attribute as a random variable. PRMs make some independency assumptions in order to shrink the model size. As with BNs, these independency assumptions are encoded in an dependency structure $\mathcal{S}$ where the vertices represent the descriptive attributes. $\mathcal{S}$ is defined with respect to a given relational structure $\mathcal{R}$. For each descriptive attribute $R.A$ in $\mathcal{R}$, $\mathcal{S}$ specifies a set of parents $Pa(R.A)$. A parent takes either the form $R.A'$ (another descriptive attribute of the same class) or $\gamma(R.\tau.A')$ where $\tau = \rho_{k_1}.\rho_{k_2}. \ldots .\rho_{k_n}$ is a chain of n reference slots and

γ is an aggregate function. Indeed, for a given object $r \in R$, $r.\tau$ is potentially a multi-set of objects of class $R.\tau$. In such a case S uses an aggregate function γ to map the different values in $r.\tau.A'$ to a single value in the domain $\mathcal{V}(\gamma)$. Aggregate functions can be any of those traditionally used in SQL: *min, max, average*, etc. A dependency structure S is said to be *legal* with respect to a given relational skeleton if it is *guaranteed-acyclic* at the object-level: an object's descriptive attribute cannot be its own ancestor.

A PRM quantifies the probabilistic dependencies encoded in S through a set of parameters θ_S. For each attribute $r.A$, $\theta_S(r.A) = P(r.A|Pa(r.A))$. The CPTs are identical for every objects of the same class. However, as the aggregate functions might compute different values for two different objects, the resulting probability can change from an object to another. A PRM Π is fully defined by a dependency structure S and its associated CPTs θ_S (parameters), $\Pi = (S, \theta_S)$.

Semantics. Given a relational skeleton σ, every PRM $\Pi = (S, \theta_S)$, with S legal w.r.t. σ, defines a coherent probability distribution over $\mathcal{I}^\sigma$, the set of possible instances of σ, by the following chain-rule formula

$$\mathbf{P}^{PRM}_{\Pi,\sigma}(i) = \prod_{x \in \sigma} \prod_{A \in \mathcal{A}(x)} P(i(x.A)|i(Pa(x.A)))$$

where $i(x.A)$ and $i(Pa(x.A))$ are the respective representations of the random variables $x.A$ and $Pa(x.A)$ in the instance i.

4 Behavioural Comparison of Expressive Knowledge Representations

Suppose that A, B represent two Herbrand bases[3] over given signatures Σ, Ω and that p, q represent probability functions over sets. Halpern's two types of probabilistic logic can be characterised as classes of probability functions with the following forms: $p : A \rightarrow [0, 1]$ (type 1) and $q : 2^B \rightarrow [0, 1]$ (type 2). Here 2^B represents the set of all possible worlds over the Herbrand base B. In this paper the approach taken to establishing relationships between type 1 and type 2 probabilistic logics involves demonstrating the existence of mappings between the logics. Suppose R_1, R_2 denote particular type 1, 2 logics respectively.

We say that R_1 is behaviourally weaker than R_2, or simply $R_1 \preceq_b R_2$ in the case that for every probability function p in R_1 with Herbrand base A there exists a probability function q in R_2 with Herbrand base B and a function f such that $f : A \rightarrow 2^B$ where $\forall a \in A \cdot q(f(a)) = p(a)$. Similarly, R_2 is behaviourally weaker than R_1, or simply $R_2 \preceq_b R_1$ when for every q in R_2 with Herbrand base B there exists a probability function p in R_1 with Herbrand base A and a function g such that $g : 2^B \rightarrow A$ where $\forall b \in 2^B \cdot p(g(b)) = q(b)$. As usual we say that R_1 is behaviourally equivalent to R_2, or simply $R_1 \equiv_b R_2$, in the case that $R_1 \preceq_b R_2$ and $R_2 \preceq_b R_1$.

---

[3] As stated before, we treat the Herbrand base of a logic model as its domain.

Halpern's work [7] provides good clarifications about what respective kinds of knowledge can be captured with probabilities on the domain (such as those defined by SLPs and SRMs) and probabilities on possible worlds (BLPs and PRMs). Links between these probabilities are also provided. However, the conclusions that can be drawn from a *behavioral* approach differ from the results obtained in previous *model-theoretical* studies (such as that of [7]): our aim is to provide ways in which knowledge encoded in one framework can be transferred into another framework. We focus on inter-translations, their features and limits.

We have adopted the following methodology:

- We first demonstrate relations between semantics: for a pair of frameworks (say, SLPs and BLPs), we define *equivalent programs* and *equivalent set of queries*. For instance, the fact that a k-ary Bayesian atom G_a takes the value v in a BLP can be represented in an *equivalent* SLP with a $(k+1)$-ary logical atom G having the same predicate and k first arguments as G_a, and the value v as last argument. We then say that a k-ary atomic BLP query is equivalent to the associated $(k+1)$-ary atomic SLP query.
- We say that the semantics are *equivalent* when equivalent (set of) queries on equivalent programs infer the same (set of) probability distributions.
- Hence our goal is eventually to provide algorithms that transform a program into an *equivalent* program in another framework, (such algorithms are referred to as *inter-translations*) and to analyze their features and their limits.

From the semantics perspective, we compare the four PLMs in terms of the following categories (as presented in section 2)

- Logic programming (LP) vs. relational models (RM): SLPs and BLPs are LP-based, while SRMs and PRMs are RM-based.
- Possible-world vs. domain-frequency: SLPs and SRMs are type 1 / domain-frequency approaches, in contrast type 2 / possible-world perspective is dominant in BLPs and PRMs.

In addition, SLPs are considered to be grammar-based models, while BLPs, PRMs and SRMs are classified to be graph-based models. In the rest sections, we detail the inter-translations between the PLMs of interests: SLPs-BLPs, SLPs-SRMs and BLPs-PRMs respectively.

5 A Behavioral Comparison of SLPs and BLPs

We first claim that a BLP B and an SLP S define *equivalent semantics* if the probability that any ground Bayesian atom G_a in the Herbrand model of the BLP takes some value v is identical to the probability of the associated logical atom G in S, ie. $\mathbf{P}_{\mathbf{S}}^{\mathbf{SLP}}(G_a) \equiv_b \mathbf{P}_{\mathbf{B},\mathbf{G_a}}^{\mathbf{BLP}}(G_a = v)$. There is an intuitive and global approach to find an inter-translation: any BLP B can be represented by a (possibly infinite) Bayesian network BN_B, and the KBMC stage consists in finding the Bayesian variables relevant to the query (hence leading to a finite BN -a subpart of BN_B- that can be queried). Provided that the least Herbrand model of the BLP is finite, BN_B will be finite, and it is possible to use the method in [2] to translate BN_B into SLPs. But this approach cannot be extended to general BLPs.

To solve the problem, we need either restrict BLPs or extend SLPs. Therefore we developed a *standard translation* [20], which exists in two versions: one translates restricted BLPs (which do not make use of combining rules) into SLPs; and the other one translates general BLPs into extended SLPs (which are augmented with combining functions). One remaining drawback is that the standard translations do not handle evidence, that is, some prior knowledge about the domain in BNs. The reason is that SLPs and e-SLPs define semantics on tree structure, whereas KBMC in BLPs permits the union of several trees and the computation of probabilities in singly connected networks.

We summarize the translation approaches and theorems presented in [20] without examples and proofs, and provide some revisions with examples.

5.1 Restricted BLPs and Extended SLPs

If S is an SLP, the subset S_h of clauses in S with predicate symbol h in the head is called the definition of h. A *restricted BLP* is a BLP whose predicate definitions contain one single stochastic clause each. A ground query G_a is said to be safe with regards to a BLP B if the and-or tree rooted at G_a does not contain 2 identical nodes (no merging of nodes takes place during KBMC). $\mathcal{N}_n$ is the set of natural numbers from 1 to n.

An *extended SLP* (e-SLP) is an SLP S augmented with a set of *combining functions* $\{CR_h\}$, for all predicates h appearing in the head of some stochastic clause in S. A combining function is a function that maps a set of possible resolvents of h (obtained using one clause in S_h) and associated real numbers in $[0,1]$ to a real number in $[0,1]$, $CR_h : ((r_1, p_1), ..., (r_n, p_n)) \mapsto r \in [0,1]$.

Given an e-SLP S_e consisting of the SLP S and the combining functions $\{CR_h\}$, and a query Q (consisting of a predicate h), the probability $\mathbf{P}_{S_e}^{eSLP}(Q)$ is the probability of the *pruned* and-or tree T rooted at the or-node Q. The probability of a pruned and-or tree is defined by structural induction:

- Base case: if T is a single or-node, $\mathbf{P}_{S_e}^{eSLP}(Q)$ is $\mathbf{P}_S^{SLP}(Q)$, the probability of S at query Q.
- If the root of T is an or-node with n branches leading to the resolvents (and-nodes) $(r_i)_{i \in \mathcal{N}_n}$, then $\mathbf{P}_{S_e}^{eSLP}(Q) = CR_h((r_i, p_i)_{i \in \mathcal{N}_n})$, where p_i is the probability of the pruned and-or subtree rooted at the and-node r_i.
- If the root of T is an and-node leading to the resolvents (or-nodes) $(r_i)_{i \in \mathcal{N}_n}$, then $\mathbf{P}_{S_e}^{eSLP}(Q) = \prod_{i=1}^{n} p_i$, where p_i is the probability of the pruned and-or subtree rooted at the or-node r_i.

5.2   Standard Translation from Restricted BLPs to SLPs

Let B denote a restricted BLP.

- Identify each k-ary Bayesian atom b, which appears in B and has the value domain V, to the $(k+1)$-ary (logical) atom $b(v_b)$ having the same k first arguments and a value v_b of V as last argument.
- For each Bayesian clause $head|b_1, ..., b_n$ in B, for each value in the associated CPT, which indicates the probability $p_{v_h, v_{b1}, ..., v_{bn}}$ that the Bayesian atom $head$ takes the value v_h given that the $\{b_i : i \in \mathcal{N}_n\}$ take the values $(v_{b1}, ..., v_{bn})$, construct the stochastic clause consisting of the parameter $p_{v_h, v_{b1}, ..., v_{bn}}$, and the definite clause $head(v_h) \leftarrow b_1(v_{b1}), ..., b_n(v_{bn})$.
- The standard translation of B consists of the n stochastic clauses constructible in that way, n being the sum of the numbers of coefficients in the CPTs. This SLP is pure and unnormalised (the parameters of the clauses in $S_h \subseteq S$ sum to the product of the domain sizes of the Bayesian atoms in the body of the Bayesian clause with head h).

Theorem. *Given a restricted BLP B, its standard translation S obtained as defined above, and a ground Bayesian query G_a which is safe with regards to B. Let us associate to G_a the logical query $G(v)$, $v \in dom(G_a)$. Then* $\mathbf{P}_S^{\mathrm{SLP}}(G(v)) \equiv_b \mathbf{P}_{B,G_a}^{\mathrm{BLP}}(G_a = v)$.

5.3   Standard Translation from BLPs to e-SLPs

Let B denote a BLP. The standard translation of B is the extended SLP S_e defined by the following stochastic clauses and combining functions:

- The stochastic clauses (which form the set S) are obtained in the same way as the stochastic clauses obtained from a restricted BLP.
- Let us take a ground predicate h in the head of some clause in S and assume that it can be unified with the heads of some clauses in S_h, leading to the resolvents $\{r_{i,j}\}$ with probabilities in S equal to $\{p_{i,j}\}$. A resolvent can contain several atoms. The clauses in S_h come from z different Bayesian clauses with the same predicate in the head. These original clauses can be indexed with a number that corresponds to the first index $i \in \mathcal{N}_z$ in the name of the resolvents. The second index $j \in \mathcal{N}_{n_i}$ refers to one of the n_i different distributions of values over the Bayesian atoms in the body of the Bayesian clause i. We define CR_h by:

$$CR_h = \sum_{j_1 \in \mathcal{N}_{n_1}, ..., j_z \in \mathcal{N}_{n_z}} CR(h, r_{1,j_1}, ..., r_{z,j_z}) \times \prod_{t=1}^{z} p_{t,j_t}$$

where CR is the combining rule defined in B.

Theorem. *Given any BLP B, its standard translation S_e obtained as defined above, and a ground Bayesian query G_a which is safe with regards to B. Let us associate to G_a the logical query $G(v)$, $v \in dom(G_a)$. Then* $\mathbf{P}_{S_e}^{\mathrm{eSLP}}(G(v)) \equiv_b \mathbf{P}_{B,G_a}^{\mathrm{BLP}}(G_a = v)$.

5.4   Translation from SLPs to BLPs

Let S denote a complete, range-restricted and non-recursive SLP[4].

- For each stochastic clause $p : head \leftarrow b_1, ..., b_n$ in S, identify each atom to a Bayesian atom whose domain is $\{true, false\}$.
- Construct the Bayesian clause having the same head, the same body, and the following conditional probability table:

			head	
b_1	...	b_n	true	false
true	true	true	p	$1 - p$
true	true	false	0	1
?	?	?	0	1
false	false	false	0	1

- To complete the definition of the BLP, we need to define a *combining rule* CR. Suppose that we have to combine n conditional probability tables CPT_i $(1 \leq i \leq n)$. Each CPT_i defines the probabilities $P(head \mid \mathcal{B}_i)$, where $\mathcal{B}_i$ is the set of ground Bayesian atoms in the body of the associated clause. Thus to define $CR((CPT_i)_{1 \leq i \leq n})$, and by using normalization, we only have to set the values of $P(head = true \mid \cup_{i=1}^{n} \mathcal{B}_i)$ for all possible instantiations of the ground Bayesian atoms in $(\cup_{i=1}^{n} \mathcal{B}_i)$. The value of $P(head = false \mid \cup_{i=1}^{n} \mathcal{B}_i) = 1 - P(head = true \mid \cup_{i=1}^{n} \mathcal{B}_i)$ can then be deduced.
- For each possible instantiation $(\cup_{i=1}^{n} Inst_i)$ of $(\cup_{i=1}^{n} \mathcal{B}_i)$, we take the sum $\sum_{i=1}^{n} P(head = true \mid \mathcal{B}_i = Inst_i)$ and assign it to $P(head = true \mid \cup_{i=1}^{n} \mathcal{B}_i)$. Since the SLP is complete, this sum will never be greater than 1, and the CR is well defined.

Theorem. *Given a complete, range-restricted and non-recursive SLP S, its translation into a BLP B obtained as defined above , and a ground query G. Let us associate to G the Bayesian atom G_a, whose domain is $\{true, false\}$, and which is itself associated to a chance node in the Bayesian net BN_{B,G_a}. If G_a is safe with regards to B then $\mathbf{P}_S^{SLP}(G) \equiv_b \mathbf{P}_{B,G_a}^{BLP}(G_a = true)$.*

5.5   A Revised Translation from BLPs to SLPs

There exists a potential 'contradictory refutation' problem in BLPs-SLPs translation, which is illustrated in Figures 1, 2 and 3 for an example. The error lies in the potential inconsistent value settings (or substitutions) between atoms in a clause, eg. in clause $\mathbf{d(tom,y)} \leftarrow \mathbf{b(tom,y)}, \mathbf{c(tom,y)}$, $\mathbf{b(tom,y)}$ may be set to $\mathbf{a(tom,y)}$ while $\mathbf{c(tom,y)}$ might be set to a contradictory value $\mathbf{a(tom,n)}$ simultaneously. To solve the problem, we introduce an extra data structure of list to 'set and remember' values instead of just setting values. Translations from the

---

[4] A clause C is said to be *non-recursive* iff the head of C is not found in the body of C.

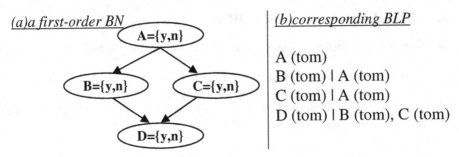

Fig. 1. An example of a first-order BN and corresponding BLP representation

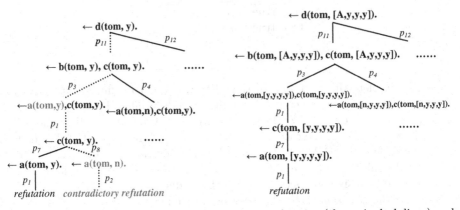

Fig. 2. (a) Stochastic SLD-tree with contradictory refutation (shown in dash lines) and (b) Resolved SSLD-tree without contradictory refutation

p_1: a(tom, y) ← .

p_2: a(tom, n) ← .

p_3: b(T, y) ← a(T, y).

p_4: b(T, y) ← a(T, n).

p_5: b(T, n) ← a(T, y).

p_6: b(T, n) ← a(T, n).

p_7: c(T, y) ← a(T, y).

p_8: c(T, y) ← a(T, n).

p_9: c(T, n) ← a(T, y).

p_{10}: c(T, n) ← a(T, n).

p_{11}: d(T, y) ← b(T, y), c(T, y).

p_{12}: d(T, y) ← b(T, y), c(T, n).

p_{13}: d(T, y) ← b(T, n), c(T, y).

p_{14}: d(T, y) ← b(T, n), c(T, n).

p_{15}: d(T, n) ← b(T, y), c(T, y).

p_{16}: d(T, n) ← b(T, y), c(T, n).

p_{17}: d(T, n) ← b(T, n), c(T, y).

p_{18}: d(T, n) ← b(T, n), c(T, n).

p_1: a(tom, [y,B,C,D]) ← .

p_2: a(tom, [n,B,C,D]) ← .

p_3: b(T, [y,y,C,D]) ← a(T, [y,y,C,D]).

p_4: b(T, [n,y,C,D]) ← a(T, [n,y,C,D]).

p_5: b(T, [y,n,C,D]) ← a(T, [y,n,C,D]).

p_6: b(T, [n,n,C,D]) ← a(T, [n,n,C,D]).

p_7: c(T, [y,B,y,D]) ← a(T, [y,B,y,D]).

p_8: c(T, [n,B,y,D]) ← a(T, [n,B,y,D]).

p_9: c(T, [y,B,n,D]) ← a(T, [y,B,n,D]).

p_{10}: c(T, [n,B,n,D]) ← a(T, [n,B,n,D]).

p_{11}: d(T, [A,y,y,y]) ← b(T, [A,y,y,y]), c(T, [A,y,y,y]).

p_{12}: d(T, [A,y,n,y]) ← b(T, [A,y,n,y]), c(T, [A,y,n,y]).

p_{13}: d(T, [A,n,y,y]) ← b(T, [A,n,y,y]), c(T, [A,n,y,y]).

p_{14}: d(T, [A,n,n,y]) ← b(T, [A,n,n,y]), c(T, [A,n,n,y]).

p_{15}: d(T, [A,y,y,n]) ← b(T, [A,y,y,n]), c(T, [A,y,y,n]).

p_{16}: d(T, [A,y,n,n]) ← b(T, [A,y,n,n]), c(T, [A,y,n,n]).

p_{17}: d(T, [A,n,y,n]) ← b(T, [A,n,y,n]), c(T, [A,n,y,n]).

p_{18}: d(T, [A,n,n,n]) ← b(T, [A,n,n,n]), c(T, [A,n,n,n]).

Fig. 3. Previous and revised translations from the above BLP to an SLP

BLP to an SLP by applying previous method (in [20]) and a revised method (this paper) are shown in Fig. 3(a) and (b) respectively, and the resolved stochastic SLD-tree can be seen in Fig.2(b). More precisely, a revised BLP-SLP translation algorithm is shown in the following steps. Let B denote a restricted BLP and S denote its translation SLP.

- Identify each k-ary Bayesian atom b, which appears in B and has the value domain V_b, to the $(k+1)$-ary (logical) atom $b(v_b)$ having the same k first arguments and a value $v_b \in V_b$ as last argument.
- Construct a list lv_b to replace v_b. The length of lv_b is the number of all Bayesian atoms. Each element of lv_b corresponds to an arbitrary Bayesian atom b' and is set to a fresh variable if $b' \neq b$ or a value $v_{b'} \in V_{b'}$ if $b' = b$.
- For each Bayesian clause $head \mid b_1, ..., b_n$ in B, for each value in the associated CPD, which indicates the probability $p_{v_h, v_{b1}, ..., v_{bn}}$ that the Bayesian atom $head$ takes the value v_h given that the $\{b_i : i \in \mathcal{N}_n\}$ take the values $(v_{b1}, ..., v_{bn})$, construct a list lv_h for $head$ as done in step 2, then construct the stochastic clause consisting of the parameter $p_{v_h, v_{b1}, ..., v_{bn}}$, and the definite clause: $head(lv_h) \leftarrow b_1(lv_{b1}), ..., b_n(lv_{bn})$.
- For $lv_h, lv_{b1}, ..., lv_{bn}$, update the value for each element in lists with respect to $v_h, v_{b1}, ..., v_{bn}$ respectively.
- The standard translation of B consists of the n stochastic clauses constructible in that way, n being the sum of the numbers of coefficients in the CPD tables. This SLP is pure and unnormalised (the parameters of the clauses in $S_h \subseteq S$ sum to the product of the domain sizes of the Bayesian atoms in the body of the Bayesian clause with head h).

Note that, in a definite clause (eg. **b(tom,[n,y,C,D])** $\leftarrow$ **a(tom,[n,y,C,D])**), all atoms have the same list values (eg. **[n,y,C,D]**), in which the elements corresponding to the atoms occurred in the clause are value set (eg. **[n,y, ,]** corresponds to atoms **a,b**) and other elements are assigned to be variables (eg. **[, ,C,D]** correspond to atoms **c,d**). The introduction of lists with variables will guarantee atoms to propagate consistent values to their predecessors in stochastic SLD-trees.

6 A Behavioral Comparison of SRMs and SLPs

SRMs naturally have a type 1 semantics with domain frequency over the rows of a database. This section will show how to build SLPs that encode the same class of semantics. Our approach is to find for any SRM ψ, a SLP S whose least Herbrand model is a contraction of a minimal *table stratified* database D such that $\psi \models D$ and such that the probabilities of success of legal queries against D or ψ match the probabilities induced by S on a set of corresponding SLP queries.

Let $\mathcal{R}$ be a *table stratified* relational schema and ψ an associated SRM. In order to translate ψ into an equivalent SLP S, we need to translate the model itself on one hand and the associated set of *legal* queries on the other hand.

For every descriptive attribute $R_i.A_j$ and for each ground instantiation with $\vec{k}$ as tuple of indices, we assert the parameterised ground fact:

$$\theta_{R_i.A_j}^{\vec{k}} \ : \ r_{i,j}(a_{i,j}^{k_1}, \overrightarrow{pa}(R_i, A_j)^{\vec{k}})$$

where $\theta_{R_i.A_j}^{\vec{k}}$ is the corresponding ψ parameter in which all joint indicator variables in the parents are set to $true$.

For each class R_i, we recursively define the key-validator predicate g_i as follows:

$$1 \ : \ g_i(k_i(\overrightarrow{A_{i_1}}, \overrightarrow{F_i}, \overrightarrow{J_i})) \ \leftarrow$$
$$g_{i_1}(F_{i_1}), \ldots, g_{i_1}(F_{i_n}),$$
$$ga_{i,1}(A_{i,1}), \ldots, ga_{i,n_i}(A_{i,n_i}),$$
$$gj_{i,1}(J_{i,1}), \ldots, gj_{i,m_i}(J_{i,m_i}).$$

For each class R_i in ψ we can build a clause that defines the predicate r_i/n_i by using the above predicates and by introducing additional helper predicates[5].

$$1 \ : \ r_i(k_i(\overrightarrow{A_i}, k_j(\overrightarrow{A_{j_1}}, \overrightarrow{F_j}, true(\overrightarrow{A_{i|J_{j,i}}}), \overrightarrow{J_j}), \overrightarrow{A_i}, \overrightarrow{F_i}, \overrightarrow{J_i}), \overrightarrow{A_i}, \overrightarrow{F_i}) \ \leftarrow$$
$$g_j(k_j(\overrightarrow{A_{j_1}}, \overrightarrow{F_j}, true(\overrightarrow{A_{i|J_{j,i}}}), \overrightarrow{J_j})),$$
$$r_{i,1}(A_{i,1}, \overrightarrow{Pa}(A_{i,1})), \ldots, r_{i,n}(A_{i,n}, \overrightarrow{Pa}(A_{i,n})),$$
$$j_{i,1}(J_{i,1}, \overrightarrow{A_{i|J_{i,1}}}), \ldots, j_{i,m}(J_{i,m}, \overrightarrow{A_{i|J_{i,m}}}).$$

Let Q be a *legal* query with respect to ψ. The following shows how to build a corresponding SLP query G_Q. Initialise G_Q to an empty formula. For each tuple variable t_j in of class R_i occurring in Q, add a literal to G_Q with predicate symbol r_i and the free variable K_j as primary key. Descriptive attributes arguments are set to the constants $a_{i,k}^l$ corresponding to the values $t_j.a_k^l$ specified in σ_Q or to some new free variable if no value is specified. Foreign key arguments are set bound variables K_j' according to the join $\bowtie_Q$.

Theorem. *The previous procedure translates any SRM ψ into a SLP S that computes the same probability distribution over legal queries; that is, for any legal query Q, the corresponding SLP query G_Q, such that $\mathbf{P}_\psi^{SRM}(Q) \equiv_b \mathbf{P}_S^{SLP}(G_Q)$.*

7   A Behavioral Comparison of PRMs and BLPs

PRMs naturally have a type 2 semantics with possible worlds which correspond to possible instances of a given relational skeleton. This section will show how to build BLPs that can capture the same class of semantics where unary Bayesian predicates represent descriptive attributes and aggregate functions, binary predicates represent reference slots and constants represent objects of the relational skeleton.

Given a relational skeleton σ and a PRM $\Pi = (\mathcal{S}, \theta_S)$, Table 1 defines a translation procedure prm2blp that builds a BLP B inferring a probabilistic distribution on $\mathcal{I}^\sigma$, the set of complete instances of the relational skeleton σ.

---

[5] Definition and translation of helper predicates are omitted.

Table 1. The `prm2blp` translation procedure from PRMs to BLPs

proc prm2blp(σ, $\mathcal{S}$, $\theta_{\mathcal{S}}$):

1. for each class R_i:
   (a) for each descriptive attribute $A_j \in A(R_i)$:
       define the unary Bayesian predicate $\mathtt{p}_{\mathtt{i,j}}/1$ with
       $dom(\mathtt{p}_{\mathtt{i,j}}/1) = \mathcal{V}(A_j)$
   (b) for each reference slot $\rho_k \in R(R_i)$:
       define the binary Bayesian predicate $\mathtt{r}_{\mathtt{i,k}}/2$ with
       $dom(\mathtt{r}_{\mathtt{i,k}}/2) = \{true, false\}$
   (c) for each aggregate function γ_i in $\theta_{\mathcal{S}}$:
       define the unary Bayesian predicate $\mathtt{g}_{\mathtt{i}}/1$ with
       $dom(\mathtt{g}_{\mathtt{i}}/1) = \mathcal{V}(\gamma_i)$
2. let B be an empty BLP
3. for each class R_i:
   (a) for each object $o \in \mathcal{O}^\sigma(R_i)$:
       i. for each reference slot $\rho_k \in \varrho(R_i)$ and each
          $o' \in \rho_k(o)$:
          assert in B the ground Bayesian fact $\mathtt{r}_{\mathtt{i,k}}(o,o')$. with associated
          (instantiated) CPT: $[1,0]$
   (b) for each descriptive attribute $A_j \in A(R_i)$:
       – if $(Pa(R_i.A_j) = \emptyset)$ according to $\mathcal{S}$ then:
         for each object $o \in \mathcal{O}^\sigma(R_i)$:
         i. assert in B the ground Bayesian fact $\mathtt{p}_{\mathtt{i,j}}(o)$. with
            associated CPT: $\theta_{\mathcal{S}}(R_i.A_j)$
       – else:
         i. let C be a Bayesian clause with head $\mathtt{p}_{\mathtt{i,j}}(V)$
         ii. for each $U_l \in \mathbf{U} = Pa(R_i.A_j)$
             • if $U_l = R_i.A_m$ then:
               add the literal $\mathtt{p}_{\mathtt{i,m}}(V)$ to the body of C
             • else $U_l = \gamma_k(R_i.\tau.A_m)$ where τ a chain of
               reference slots of the form $\tau = \rho_{k_1}, \rho_{k_2}, \ldots, \rho_{k_n}$:
               * add the literal $\mathtt{g}_{\mathtt{k}}(V)$ to the body of C
               * let $R_{i'} = R_i.\tau$
               * assert in B the following helper Bayesian clause :
                 $\mathtt{g}_{\mathtt{k}}(V) \quad | \quad \mathtt{r}_{\mathtt{i,k_1}}(V, V_1), \ldots, \mathtt{r}_{\mathtt{k_{n-1},k_n}}(V_{n-1}, V_n), \mathtt{p}_{\mathtt{i',m}}(V_n).$
         iii. let CPT_C be $\theta_{\mathcal{S}}(R_i.A_j | Pa(R_i.A_j))$
         iv. assert C in B
4. build the Combining Rules for each predicate in B by applying the
   build_cr procedure
5. return B

Theorem. *For any PRM Π that is* guaranteed-acyclic *w.r.t. some non-empty relational skeleton σ:*

$$\forall i \in \mathcal{I}^\sigma : \mathbf{P}^{PRM}_{(\sigma,\Pi)}(i) \equiv_b \mathbf{P}^{BLP}_{prm2blp(\sigma,\Pi)}(i'),$$

where i' is the ground Bayesian query corresponding to the database instance i.

8   Discussion and Conclusions

The first result we achieved in the study is BLPs $\equiv_b$ e-SLPs. We argue that SLPs augmented with combining functions (namely extended SLPs) and BLPs can encode the same knowledge, in that they encode equivalent probability distributions for equivalent set of queries. Since SLPs need to be augmented with combining rules in order to be as expressive as BLPs, and BLPs are able to encode complete, range-restricted and non-recursive SLPs, we are tempted to conclude that BLPs are more expressive than strict SLPs. However, SLPs' and BLPs' formalisms are more or less intuitive, depending on the kind of knowledge we want to model. It should be noted that BLP's query-answering procedure benefits from different frameworks, say logic programming and Bayesian networks, while inference mechanisms in SLPs are straightforward using only logic programming.

Another finding is also shown in the study, denoting as PRMs $\preceq_b$ BLPs and SRMs $\preceq_b$ SLPs. When considering models within the same probabilistic model category (type 1 or type 2), BLPs (resp. SLPs) can naturally express PRMs (resp. SRMs), i.e., translated models and queries can be forged, which compute the same probability distributions.

We believe this study to be a formal basis for further research. Several learning algorithms have been devised for SLPs [2,15,16], BLPs [11,12], PRMs [4,5] and SRMs [5,6]. Further work thus includes the study of how inter-translations between those frameworks can help devising better learning algorithms for PLMs depending on the kind of knowledge we want to model. For instance, inter-translations of e-SLPs and BLPs can be used to extend learning techniques designed for BLPs to the learning of e-SLPs (and vice-versa). Investigating such extensions could be interesting. We also hope this study provide a bridge to developing an integrated theory of probabilistic logic learning.

As the related studies, one may find an intuitive approach of translating SLPs into BNs, Markov networks and stochastic context free grammars in [2]; a simpler scheme for mapping PRMs to BLPs is contained in [3]; a theoretical comparison of BLPs and Relational Markov Models (RMMs) is presented in [18]; and another approach of analysing the expressive power of different probabilistic logic languages could be found in [9,8] as well as in this volume.

Acknowledgement

We are very grateful to Aymeric Puech and Olivier Grisel for their initial contributions on the topic when they were studying in Imperial College London. This work was supported by the Royal Academy of Engineering/Microsoft Research Chair on 'Automated Microfluidic Experimentation using Probabilistic Inductive Logic Programming'; the BBSRC grant supporting the Centre for Integrative Systems Biology at Imperial College, Grant Reference BB/C519670/1; the BBSRC grant on 'Protein Function Prediction using Machine Learning by Enhanced Novel Support Vector Logic-based Approach', Grant Reference BB/E000940/1;

ESPRIT IST project 'Application of Probabilistic Inductive Logic Programming II (APRIL II)', Grant reference FP-508861.

References

1. Bratko, I.: Prolog for artificial intelligence. Addison-Wesley, London (1986)
2. Cussens, J.: Parameter estimation in stochastic logic programs. Machine Learning 44(3), 245–271 (2001)
3. De Raedt, L., Kersting, K.: Probabilistic Logic Learning. ACM-SIGKDD Explorations: Special issue on Multi-Relational Data Mining 5(1), 31–48 (2003)
4. Friedman, N., Getoor, L., Koller, D., Pfeffer, A.: Learning probabilistic relational models. In: Dean, T. (ed.) Proceedings of the Sixteenth International Joint Conferences on Artificial Intelligence (IJCAI 1999), Stockholm, Sweden, pp. 1300–1309. Morgan Kaufmann, San Francisco (1999)
5. Getoor, L.: Learning Statistical Models from Relational Data. PhD thesis, Stanford University (2001)
6. Getoor, L., Koller, D., Taskar, B.: Statistical models for relational data. In: Wrobel, S. (ed.) MRDM 2002, University of Alberta, Edmonton, Canada, July 2002, pp. 36–55 (2002)
7. Halpern, J.Y.: An analysis of first-order logics of probability. Artificial Intelligence 46, 311–350 (1989)
8. Jaeger, M.: Type extension trees: A unified framework for relational feature construction. In: Gärtner, T., Garriga, G.C., Meinl, T. (eds.) Working Notes of the ECML 2006 Workshop on Mining and Learning with Graphs (MLG 2006), Berlin, Germany (September 2006)
9. Jaeger, M., Kersting, K., De Raedt, L.: Expressivity analysis for pl-languages. In: Fern, A., Getoor, L., Milch, B. (eds.) Working Notes of the ICML 2006 Workshop Open Problems in Statistial Relational Learning (SRL 2006), Pittsburgh, USA, June 29 (2006)
10. Jensen, F.V.: Introduction to Bayesian Networks. Springer, New York (1996)
11. Kersting, K., De Raedt, L.: Adaptive Bayesian Logic Programs. In: Rouveirol, C., Sebag, M. (eds.) ILP 2001. LNCS (LNAI), vol. 2157, Springer, Heidelberg (2001)
12. Kersting, K., De Raedt, L.: Towards Combining Inductive Logic Programming and Bayesian Networks. In: Rouveirol, C., Sebag, M. (eds.) ILP 2001. LNCS (LNAI), vol. 2157, Springer, Heidelberg (2001)
13. Kersting, K., De Raedt, L.: Bayesian logic programs. In: Cussens, J., Frisch, A. (eds.) Proceedings of the Work-in-Progress Track at the 10th International Conference on Inductive Logic Programming, pp. 138–155 (2000)
14. Muggleton, S.H.: Stochastic logic programs. In: de Raedt, L. (ed.) Advances in Inductive Logic Programming, pp. 254–264. IOS Press, Amsterdam (1996)
15. Muggleton, S.H.: Learning stochastic logic programs. Electronic Transactions in Artificial Intelligence 4(041) (2000)
16. Muggleton, S.H.: Learning structure and parameters of stochastic logic programs. In: Matwin, S., Sammut, C. (eds.) ILP 2002. LNCS (LNAI), vol. 2583, Springer, Heidelberg (2003)
17. Muggleton, S.H., Firth, J.: CProgol4.4: a tutorial introduction. In: Dzeroski, S., Lavrac, N. (eds.) Relational Data Mining, pp. 160–188. Springer, Heidelberg (2001)

18. Muggleton, S.H., Pahlavi, N.: The complexity of translating blps to rmms. In: Muggleton, S., Otero, R., Tamaddoni-Nezhad, A. (eds.) ILP 2006. LNCS (LNAI), vol. 4455, pp. 351–365. Springer, Heidelberg (2007)
19. Pearl, J.: Probabilistic Reasoning in Intelligent Systems: Networks of Plausible Inference. Morgan Kaufmann, Los Altos (1988)
20. Puech, A., Muggleton, S.H.: A comparison of stochastic logic programs and Bayesian logic programs. In: IJCAI 2003 Workshop on Learning Statistical Models from Relational Data, IJCAI (2003)

Model-Theoretic Expressivity Analysis

Manfred Jaeger

Institut for Datalogi, Aalborg University
Selma-Lagerlöfs Vej 300, 9220 Aalborg Ø, Denmark
jaeger@cs.aau.dk

1 Introduction

In the preceding chapter the problem of comparing languages was considered from a behavioral perspective. In this chapter we develop an alternative, model-theoretic approach.

In this approach we compare the expressiveness of probabilistic-logic (pl-) languages by considering the models that can be characterized in a language. Roughly speaking, one language L' is at least as expressive as another language L, if every model definable in L also is definable in L'. Results obtained in the model-theoretic approach can be somewhat stronger than results obtained in the behavioral approach in that equivalence of models entails equivalent behavior with respect to any possible type of inference tasks. On the other hand, the model-theoretic approach is somewhat less flexible than the behavioral approach, because only languages can be compared that define comparable types of models. A comparison between Bayesian Logic Programs (defining probability distributions on possible worlds) and Stochastic Logic Programs (defining probability distributions over derivations), therefore, is already quite challenging in a model-theoretic approach, as it requires first to define a unifying semantic framework. In this chapter, therefore, we focus on pl-languages that exhibit stronger semantic similarities (Bayesian Logic Programs (BLPs) [6], Probabilistic Relational Models (PRMs) [1], Multi-Entity Bayesian Networks [7], Markov Logic Networks (MLNs) [12], Relational Bayesian Networks (RBNs) [4]), and first establish a unifying semantics for these languages. However, the framework we propose is flexible to enough (with a slightly bigger effort) to also accommodate languages like Stochastic Logic Programs [9] or Prism [13].

The focus of this chapter is expressivity analysis. Clearly, expressivity is only one relevant aspect in the comparison of languages. Further highly important issues are compactness of representation, efficiency of inference, and learnability in different languages. A meaningful comparison of these issues, however, requires concepts of equivalence of models and inferences, which is just what our expressivity analysis provides. Thus, this analysis is to be understood as a first step towards more comprehensive comparisons.

L. De Raedt et al. (Eds.): Probabilistic ILP 2007, LNAI 4911, pp. 325–339, 2008.
© Springer-Verlag Berlin Heidelberg 2008

2   PL-Models

In this chapter the word *model* is used to refer to unique distributions over some state space. This is consistent with the usage in logic, where "model" refers to a unique structure. It is different from the usage in statistics, where "model" refers to a parametric class of distributions. Specifically, when we talk about the model represented by some BLP, RBN or MLN, for example, we are referring to a fully quantified BLP, etc., i.e. all numeric parameters set to specific values.

As a first step towards a unifying semantics for different pl-languages, we have to find a common structure of the state spaces on which distributions are defined. A sufficiently general class of state spaces consists of the spaces that are generated by a set of random variables that can be written in the syntactic form of ground atoms, e.g. *blood_pressure(tom)*, *sister(susan,tom)*, *genotype(mother(tom))*,... These random variables take values in finite sets of states that are associated with the relation symbol, e.g. *states(genotype)*={*AA, Aa, aa*}. At this point we do not consider continuous variables. We call any assignment of states to the set of all ground atoms constructible over a given vocabulary S of relation, function and constant symbols (the *Herbrand base of S, $HB(S)$*) a *Multi-valued Herbrand interpretation*.

To reason about identity we allow that $=\in S$. The symbol $=$ is seen as a binary Boolean relation symbol. Interpretations of $=$ are constrained to be consistent with identity relation on domain elements (i.e. they must satisfy the axioms of equality). Some languages (including RBNs and MLNs) use the $=$ relation to define models, but do not provide probabilistic models of $=$ itself. Some approaches have been proposed to model "identity uncertainty", i.e. to build probabilistic models for $=$ [10,8,11].

The set of all multi-valued Herbrand interpretations for S is denoted $MVHI(S)$. We use $\omega, \omega', \dots$ to denote individual multi-valued Herbrand interpretations. In the case where all relations in S are Boolean, then these ω are also referred to as *possible worlds* (in agreement with standard logic terminology). When $\omega \in MVHI(S)$, and $S' \subseteq S$ then $\omega[S']$ denotes restriction of ω to the symbols in S'. Similarly, when $\boldsymbol{\alpha}$ is an arbitrary vector of ground S-atoms, then $\omega[\boldsymbol{\alpha}]$ denotes the state assignment in ω to the ground atoms in $\boldsymbol{\alpha}$. Another notational convention we will use is to refer by r/k to a relation symbol of arity k. Specifically, $r/k \in S$ is to be read as "r is a k-ary relation symbol in S". Similarly for function symbols.

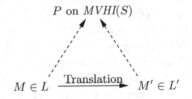

Fig. 1. Preliminary translation schema

We always assume that probability distributions P on $MVHI(S)$ are defined on the σ-algebra $\mathcal{A}(S)$ generated by all elementary events of the form $\alpha = s$, where $\alpha \in HB(S)$, and s is a value in the state space of α.

Figure 1 gives a preliminary view of the model-theoretic language comparison: a language L' is at least as expressive as a language L, if for every model $M \in L$ defining a distribution P on $MVHI(S)$ there exists a model $M' \in L'$ defining the same distribution. The schema in Figure 1 is not yet fully adequate, however. The first problem with this schema is that we cannot expect the model M' to define a distribution on exactly the same state space $MVHI(S)$ as M. For example, language L' might only permit Boolean relations, whereas L operates with multi-valued relations. The translation from M to M' then will involve a "binarization" of the vocabulary S, leading to a new vocabulary S', and hence a different probability space $MVHI(S')$. We must therefore allow that M' does not represent exactly the distribution P defined by M, but only that M' defines some P' that encodes all the information contained in P. In the following definition we formalize this scenario. The definition also provides for the case where the model M' does not encode all of M, but only as much as is needed to answer a restricted class of queries.

Definition 1. *Let* P, P' *be probability distributions over* $\mathrm{MVHI}(S)$, *respectively* $\mathrm{MVHI}(S')$. *Let* $\mathcal{Q} \subseteq \mathcal{A}(S)$. *A* $\mathcal{Q}$-embedding *of* P *in* P' *is a mapping*

$$h : \mathcal{Q} \to \mathcal{A}(S') \qquad (1)$$

such that for all $Q \in \mathcal{Q}$:

$$P(Q) = P'(h(Q)).$$

We write $P \preceq_{\mathcal{Q}} P'$ *if there exists a* $\mathcal{Q}$-embedding *of* P *in* P'.

A conditional $\mathcal{Q}$-embedding *of* P *in* P' *is a mapping (1) together with a subset* $C \in \mathcal{A}(S)$, *such that for all* $Q \in \mathcal{Q}$:

$$P(Q) = P'(h(Q) \mid C).$$

We write $P \preceq_{\mathcal{Q},c} P'$ *if there exists a conditional* $\mathcal{Q}$-embedding *of* P *in* P'.
If $\mathcal{Q} = \mathcal{A}(S)$ *we just write* $\preceq, \preceq_c$ *instead of* $\preceq_{\mathcal{Q}}, \preceq_{\mathcal{Q},c}$.

An important example for $\mathcal{Q}$ is the set of all events of the form $\alpha = s$. If then $P \preceq_{\mathcal{Q}} P'$, we can retrieve from P' all single variable marginals of P, but not necessarily joint or conditional distributions of P.

We now turn to a second, more subtle and fundamental deficiency of the schema in Figure 1. Consider the situation where $MVHI(S)$ is finite (which happens when S contains only finitely many constant and no function symbols). In this case basically every pl-language will be able to represent any distribution P on $MVHI(S)$ (P could be expressed by a Bayesian network with one node for each $\alpha \in HB(S)$; for essentially all pl-languages it is known that they can encode any standard Bayesian network). Thus, it would immediately follow that for purely relational vocabularies all pl-languages are equally expressive, and that they have the same expressive power as standard Bayesian networks.

To see why this argument misses the point, consider a typical pl-model for genotypes in a pedigree. Such a model would be given by two distinguishable elements: on the one hand, there are general probabilistic rules that specify, for example, that each of the two alleles of one gene is passed from parent to child with equal probability, or that specify the probability of a random mutation. On the other hand, there are basic facts that describe the structure of the pedigree, e.g. that John and Mary are the parents of Paul. The power and usefulness of pl-languages derives from this modularity that separates generic underlying probabilistic rules from domain-specific information.

The modularity in the model specification is most clearly expressed in PRMs, where the specification of the *skeleton structure* is distinguished from the actual probabilistic model, and in RBNs, where the specification of an *input structure* is distinguished from the specification of the actual RBN model. BLPs make a distinction between the *intensional* and the *extensional* model part, where the extensional part mostly is expressed in terms of special *logical relations*, roughly corresponding to the *predefined relations* of RBNs.

In the following we adopt the extensional/intensional terminology (originating in database theory), and by the following definition demand that a pl-model has a modular structure that separates the generic, high-level (intensional) part of the model from a specific, non-probabilistic (extensional) domain specification.

Definition 2 *(PL-model). A PL-model M for a vocabulary S is a specification in a formal language L of a probability distribution $P[M]$ on $\mathrm{MVHI}(S)$. The model M can be decomposed as $M = (M_{\mathrm{int}}, M_{\mathrm{ext}})$, such that*

(i) *For a given M_{int} there exist infinitely many different $M_{\mathrm{ext}}^{(1)}, M_{\mathrm{ext}}^{(2)}, \ldots$, such that $(M_{\mathrm{int}}, M_{\mathrm{ext}}^{(i)})$ defines a distribution on some $\mathrm{MVHI}(S_i)$, where for $i \neq j$ the vocabularies S_i, S_j contain different constant symbols.*

(ii) *If $\alpha \in \mathrm{HB}(S)$ with $0 < P[M](\alpha) < 1$, then there exists $M' = (M'_{\mathrm{int}}, M_{\mathrm{ext}})$ with $P[M'](\alpha) \neq P[M](\alpha)$.*

Definition 2 requires that a model M has a modular structure (M_{int}, M_{ext}). Moreover, conditions (i) and (ii) make certain minimal requirements for the components: condition (i) ensures that the generic, intensional part of the model gives rise to infinitely many concrete model instances obtained by exchanging the extensional part, and that these changes permit a change of the underlying domain as represented by the constants in S. Condition (i) alone would permit the trivial decomposition $M = (\emptyset, M_{ext})$. Condition (ii), therefore, requires that M_{int} actually contains the essential probabilistic information, and that by changes to M_{int} (typically just by change of numerical parameter values) one can change the quantitative aspects of the model.

It must be emphasized that the partitioning of a model into intensional and extensional part may not be unique. For some languages there exists a canonical decomposition that is also reflected in a syntactic distinction between the two parts. For other languages, several meaningful partitions satisfying Definition 2 can exist.

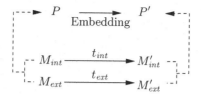

Fig. 2. Translations and Embeddings

We now arrive at the refined schema in Figure 2: to show that L' is at least as expressive as L, we have to find translations between L and L' models that respect the modular structure of the models, i.e. we need separate translations for the intensional and extensional parts. For the following precise definition we take it for granted that for L and L' decompositions into intensional and extensional parts have been defined, and emphasize that modifications to how intensional and extensional parts are identified can lead to changes in the partial expressivity order $\preceq$ here defined.

Definition 3. *Language L' is at least as expressive as L with respect to queries $\mathcal{Q}$, $L \preceq_{\mathcal{Q}} L'$, if $\exists t_{\text{int}} \forall M_{\text{int}} \exists t_{\text{ext}} \forall M_{\text{ext}}$*

$$P[M_{\text{int}}, M_{\text{ext}}] \preceq_{\mathcal{Q}} P[t_{\text{int}}(M_{\text{int}}), t_{\text{ext}}(M_{\text{ext}})] \tag{2}$$

If (2) is only satisfied by a conditional embedding $\preceq_{\mathcal{Q},c}$, we write $L \preceq_{\mathcal{Q},c} L'$.

The quantifier string $\exists t_{int} \forall M_{int} \exists t_{ext} \forall M_{ext}$ in Definition 3 requires some explanation. According to the definition, the exact translation used for the extensional part may depend on the concrete intensional part. This reflects to some extent the "primacy" of the intensional model part, which is supposed to contain the essential probabilistic specifications, whereas the extensional part contains ancillary domain information. The following example illustrates how the possible dependence of t_{ext} on M_{int} can become relevant in practice.

Example 1. One special application of comparisons of the form $L \preceq L'$ is the case where L' is a fragment of L. In such a case, a relation $L \preceq L'$ is basically a normal form theorem: every model M is equivalent to a model M' in a normal form characterized by the syntactic restrictions of L'. As an example, let L be the language of BLPs, and L' the BLP fragment in which Bayesian clauses are not allowed to contain constant symbols.

Consider the following BLP (here not showing the probability annotation of the intensional clauses):

M_{ext} :

```
father(peter,paul)
mother(mary,paul)
```

M_{int} :

```
bloodtype(X)|father(thomas,X)
bloodtype(X)|father(Y,X),mother(Z,X),bloodtype(Y),bloodtype(Z)
```

Here an intensional probabilistic rule contains the constant 'thomas'. In order to eliminate the occurrence of this constant, we can introduce a new unary relation symbol *thomas_rel*/1, and translate the original model into

M'_{ext} :
```
father(peter,paul)
mother(mary,paul)
thomas_rel(thomas)
```
M'_{int} :
```
bloodtype(X)|father(Y,X),thomas_rel(Y)
bloodtype(X)|father(Y,X),mother(Z,X),bloodtype(Y),bloodtype(Z)
```

In general, t_{int} replaces constants cons in M_{int} with new variables, and adds cons_rel() atoms to the clauses. This translation is independent of M_{ext}. The translation t_{ext} adds clauses cons_rel(cons) to M_{ext}. This depends on M_{int}, because we first have to inspect M_{int} in order to find the constant symbols in need of elimination.

3 Case Study: MLNs and RBNs

In this section we apply the general framework established in the previous section to compare the expressiveness of Markov Logic Networks [12] with Relational Bayesian Networks [4]. We begin by briefly reviewing the essential concepts and definitions for both languages.

3.1 Markov Logic Networks

In the following we give a definition of MLNs following [12]. Notation and presentation are somewhat adapted to our general conceptual setting. In particular, we make explicit an intensional/extensional division in MLN models, which is only implicit in the original definitions. Our definitions are based on the *known functions assumption* stated in [12], which basically stipulates that all function symbols in the language have a fixed and known interpretation, and are not modeled probabilistically. The general MLN paradigm allows to relax or eliminate this assumption. The translation we present in this chapter can be generalized also to MLN versions without the known functions assumption.

Under the known function assumption, MLNs contain a domain specification given by a set of constant symbols S_C^{mln}, and interpretations over this domain of a set of function symbols S_F^{mln}. This domain specification is the extensional part of the model, i.e.

$$M_{ext}^{mln} \in MVHI(S_{ext}^{mln}),$$

where $S_{ext}^{mln} := S_C^{mln} \cup S_F^{mln} \cup \{=\}$.
    The intensional part of an MLN is given by a set of pairs

$$M_{int}^{mln} = \{(\phi_i(x_1, \ldots, x_{k_i}), w_i) \mid i = 1, \ldots, n\}, \tag{3}$$

Here the ϕ_i are first-order logic formulas in $S^{mln} := S_{ext}^{mln} \cup S_R^{mln}$, where S_R^{mln} is a set of Boolean relation symbols. The w_i are numbers in $\mathbb{R} \cup \{\infty\}$. $M^{mln} = (M_{ext}^{mln}, M_{int}^{mln})$ defines a distribution on $MVHI(S^{mln})$ as follows:

$$P[M^{mln}](\omega) = \begin{cases} 0 & \text{if } \omega[S_{ext}^{mln}] \neq M_{ext}^{mln}, \\ & \quad \text{or } \omega \not\models \forall \boldsymbol{x}\phi_i(\boldsymbol{x}) \text{ for some } i \text{ with } w_i = \infty \\ \frac{1}{Z} exp(\displaystyle\sum_{\substack{i=1 \\ w_i \neq \infty}}^{n} n_i(\omega)w_i) & \text{otherwise} \end{cases}$$

(4)

where $n_i(\omega)$ is the number of true instances of ϕ_i in ω obtained by grounding ϕ_i with constants from S_C^{mln}. Z is a normalizing constant.

Table 1. MLN: friends and smokers example

ϕ_i	w_i
$Fr(x,y) \wedge Fr(y,z) \Rightarrow Fr(x,z)$	0.7
$\neg\exists y\, Fr(x,y) \Rightarrow Sm(x)$	2.3
$Sm(x) \Rightarrow Ca(x)$	1.5
$Fr(x,y) \Rightarrow (Sm(x) \Leftrightarrow Sm(y))$	1.1
$Fr(Anna,Bob)$	∞

Example 2. (adapted from [12]) Table 1 shows a small intensional model using relation symbols $S_R^{mln} = \{Fr(iend), Sm(okes), Ca(ncer)\}$. The model consists of four weighted formulas expressing, respectively, that the friends relation is transitive, friendless people smoke, smoking causes cancer, and friends will either both smoke or both not smoke. Furthermore, there is a hard constraint saying that *Anna* is friends with *Bob* (not necessarily implying the converse). This intensional model is to be combined with domain specifications given by a set of constants, including the constants *Anna,Bob*, e.g $S_C^{mln} = \{Anna, Bob, Paul\}$. There are no function symbols, so this set of constants (together with the unique names assumption) defines $M_{ext}^{mln} \in MVHI(S_C^{mln} \cup \{=\})$.

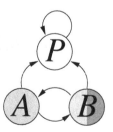

Fig. 3. A small $\omega \in MVHI(S_C \cup S_R \cup \{=\})$

Let $\omega \in MVHI(S_{ext}^{mln})$ as shown in Figure 3, where arrows indicate the interpretation of the Fr relation, light shading indicates the objects for which Sm is true, and dark shading indicates the objects for which Ca is true. In ω there are 26 groundings that satisfy $\phi_1(x, y, z)$ (there are 3^3 possible groundings, and only the grounding $x = A, y = B, z = A$ does not satisfy ϕ_1), i.e. $n_1(\omega) = 26$. Similarly, $n_2(\omega) = 3$ (the condition $\neg\exists y Fr(x, y)$ is not satisfied for any x), $n_3(\omega) = 2$, and $n_4(\omega) = 7$ (this example highlights some potential difficulties with calibrating the weights for material implications, which can have a large number of true groundings simply because most groundings do not satisfy the antecedent).

3.2   Relational Bayesian Networks

We here give a condensed summary of all relevant technical definitions for syntax and semantics of RBNs. For more detailed explanations and motivating examples the reader is referred to [5].

In RBN models the vocabulary is partitioned into predefined (extensional) and probabilistic (intensional) symbols: $S^{rbn} = S_{ext}^{rbn} \cup S_{int}^{rbn}$, where S_{ext}^{rbn} consists of relation and constant symbols (including the $=$ relation), and S_{int}^{rbn} of relation symbols only. The extensional part of a RBN model consists of $M_{ext}^{rbn} \in MVHI(S_{ext}^{rbn})$, where the interpretation of $=$ follows the unique names assumption (in the original RBN terminology, M_{ext}^{rbn} is called an *input structure*).

The intensional part (i.e. the RBN proper) consists of a collection of *probability formulas* F_r for the intensional relations:

$$M_{int}^{rbn} = \{F_r(x_1, \ldots, x_k) \mid r/k \in S_{int}^{rbn}\}. \tag{5}$$

Probability formulas are formal expressions generated by a syntax which can be seen as a probabilistic counterpart of the syntax of predicate logic: the probability formula constructs of *atoms*, *convex combinations* and *combination functions* (cf. Table 2) correspond to predicate logic constructs of *atomic formulas*, *Boolean connectives*, and *quantification*, respectively. A first-order formula $\phi(x_1, \ldots, x_k)$ evaluates for particular domain elements $c_1, \ldots, c_k$ from some possible world ω to a truth value $\phi(c_1, \ldots, c_k)[\omega] \in \{true, false\}$ (note that $\phi(c_1, \ldots, c_k)[\omega] = true$ is synonymous with $\omega \models \phi(c_1, \ldots, c_k)$).

A probability formula $F(x_1, \ldots, x_k)$ evaluates to a probability value

$$F(c_1, \ldots, c_k)[\omega] \in [0, 1].$$

Both the first-order and the probability formula depend for their evaluation usually not on the whole possible world ω, but only on the truth values of a set of ground atoms $\alpha(\phi, c)$, respectively $\alpha(F, c)$. For example, the evaluation of the first-order formula $\phi(x) = \exists y(r(x, y) \wedge t(y))$ depends for $x = c$ on the atoms $r(c, c'), t(c')$ for all c' in the domain of ω.

In the case of probability formulas we will be mostly interested in the dependence on S_{int}^{rbn}-atoms. The set of S_{int}^{rbn}-atoms that the evaluation of $F(c)[\omega]$ depends on is determined by c and the extensional part of ω, i.e. $\omega[S_{ext}^{rbn}]$. We write $\alpha(F, c, \omega[S_{ext}^{rbn}])$ for this set of ground S_{int}^{rbn}-atoms.

Table 2 now summarizes syntax and semantics of probability formulas. Shown are the syntactic form of the formulas $F(\boldsymbol{x})$ constructed via the four different construction rules, a specification of the sets $\boldsymbol{\alpha}(F, \boldsymbol{c}, \omega[S_{ext}^{rbn}])$, and the computation rule for the probability value $F(\boldsymbol{c})[\omega]$. The combination function construct here is only shown for the *noisy-or* combination function, which is the only one we will need for MLN encodings.

Table 2. RBN syntax and semantics – F_1, F_2, F_3, F' are any probability formulas; $\psi(\boldsymbol{x}, \boldsymbol{y})$ is any Boolean combination of S_{ext}^{rbn}-atoms

	$F(\boldsymbol{x})$	$\boldsymbol{\alpha}(F, \boldsymbol{c}, \omega[S_{ext}^{rbn}])$
Constant	$p \; (p \in [0, 1])$	$\emptyset$
Atom	$r(\boldsymbol{x}) \; (r \in S_{int}^{rbn})$	$r(\boldsymbol{c})$
Convex Combination	$(F_1 : F_2, F_3)$	$\cup_{i=1}^{3} \boldsymbol{\alpha}(F_i, \boldsymbol{c}, \omega[S_{ext}^{rbn}])$
Combination Function	$noisy\text{-}or\{F'(\boldsymbol{x}, \boldsymbol{y}) \mid \boldsymbol{y} : \psi(\boldsymbol{x}, \boldsymbol{y})\}$	$\displaystyle\bigcup_{\substack{\boldsymbol{c}' \\ \omega[S_{ext}^{rbn}] \models \psi(\boldsymbol{c}, \boldsymbol{c}')}} \boldsymbol{\alpha}(F, (\boldsymbol{c}, \boldsymbol{c}'), \omega[S_{ext}^{rbn}])$

	$F(\boldsymbol{c})[\omega]$
Constant	p
Atom	$\begin{cases} 1 \text{ if } r(\boldsymbol{c})[\omega] = true \\ 0 \text{ otherwise} \end{cases}$
Convex Combination	$F_1(\boldsymbol{c})[\omega]F_2(\boldsymbol{c})[\omega] + (1 - F_1(\boldsymbol{c})[\omega])F_3(\boldsymbol{c})[\omega]$
Combination Function	$\displaystyle 1 - \prod_{\substack{\boldsymbol{c}' \\ \omega \models \psi(\boldsymbol{c}, \boldsymbol{c}')}} (1 - F(\boldsymbol{c}, \boldsymbol{c}')[\omega])$

A pair $M^{rbn} = (M_{ext}^{rbn}, M_{int}^{rbn})$ induces a dependency relation between S_{int}^{rbn}-atoms:

$$r(\boldsymbol{c}) \preceq r'(\boldsymbol{c}') \; :\Leftrightarrow \; r'(\boldsymbol{c}') \in \boldsymbol{\alpha}(F_r, \boldsymbol{c}, M_{ext}^{rbn}).$$

If this dependency relation is acyclic, then we obtain a well-defined probability distribution on $MVHI(S)$ via

$$P[M^{rbn}](\omega) = \begin{cases} 0 & \text{if } \omega[S_{ext}^{rbn}] \neq M_{ext}^{rbn} \\ \displaystyle\prod_{r \in S_{int}^{rbn}} \prod_{\substack{\boldsymbol{c} \\ \omega \models r(\boldsymbol{c})}} F_r(\boldsymbol{c})[\omega] \prod_{\substack{\boldsymbol{c} \\ \omega \not\models r(\boldsymbol{c})}} (1 - F_r(\boldsymbol{c})[\omega]) & \text{otherwise} \end{cases}$$

Probability formulas can encode standard first-order formulas in the following sense: for all first-order $\phi(\boldsymbol{x})$ there exists a probability formula $F_\phi(\boldsymbol{x})$, such that for all ω, c: $\phi(c)[\omega] = \textit{true}$ iff $F_\phi(c)[\omega] = 1$, and $\phi(c)[\omega] = \textit{false}$ iff $F_\phi(c)[\omega] = 0$ [4]. This encoding will be the cornerstone for our MLN to RBN translation in the next section.

3.3 MLN to RBN Translation

Let $M^{mln} = (M_{ext}^{mln}, M_{int}^{mln})$ be a MLN model with M_{int}^{mln} as in (3). We begin by defining the vocabulary for the target RBN model:

$$S_{ext}^{rbn} = S_C^{mln} \cup \{r_f/(k+1) \mid f/k \in S_F^{mln}\}$$
$$S_{int}^{rbn} = S_R^{mln} \cup \{r_{\phi_i}/k_i \mid i = 1, \ldots, n\}.$$

Thus, S_{ext}^{rbn} is $S_C^{mln} \cup S_F^{mln}$ with relations instead of functions, and S_{int}^{rbn} adds to S_R^{mln} new relation symbols r_{ϕ_i} corresponding to the formulas in M_{int}^{mln}.

The translation t_{ext} is independent of the intensional model part M_{int}^{mln}, and simply consists of a transformation of $M_{ext}^{mln} \in MVHI(S_{ext}^{mln})$ into $M_{ext}^{rbn} \in MVHI(S_{int}^{rbn})$ by replacing interpretations of $f/k \in S_F^{mln}$ with corresponding interpretations of $r_f/(k+1) \in S_{ext}^{rbn}$.

To define $M_{int}^{rbn} := t_{int}(M_{int}^{mln})$, we have to define for each relation $r \in S_{int}^{rbn}$ a probability formula. For $r/k \in S_R^{mln}$ we simply define:

$$F_r(x_1, \ldots, x_k) = 0.5. \tag{6}$$

The formulas F_r ($r \in S_R^{mln}$) together with the input structure M_{ext}^{rbn} define a uniform distribution over $\{\omega \in MVHI(S_{ext}^{rbn}, S_R^{mln}) \mid \omega[S_{ext}^{rbn}] = M_{ext}^{rbn}\}$.

The core of the translation lies in the definition of probability formulas F_{ϕ_i} for the new relation symbols r_{ϕ_i}. The main component in the construction of the F_{ϕ_i} are sub-formulas H_{ϕ_i} that are essentially encodings of the formulas ϕ_i, as mentioned at the end of the preceding section. More precisely, we construct probability formulas H_{ϕ_i} with the following properties:

(i) $\alpha(H_{\phi_i}, c, \omega[S_{ext}^{rbn}])$ only contains atoms in relations from S_R^{mln}.
(ii) $H_{\phi_i}(c)[\omega] \in \{0, 1\}$ for all ω, c.
(iii) For all $\omega \in MVHI(S^{mln}), c$, and $\omega' \in MVHI(S_{ext}^{rbn}, S_R^{mln})$ with $\omega'[S_{ext}^{rbn}] = t_{ext}(\omega[S_{ext}^{mln}])$ and $\omega'[S_R^{mln}] = \omega[S_R^{mln}]$:

$$\phi_i(c)[\omega] = \textit{true} \Leftrightarrow H_{\phi_i}(c)[\omega'] = 1.$$

The formulas H_ϕ are defined inductively in the manner described in [4]. Some additional provisions are necessary for dealing with the transformation of function symbols into a relational representation.

Case 1a: ϕ is a relational atom. This is the most difficult case, as it involves the elimination of function. We demonstrate the construction of H_ϕ by a generic example: let $\phi = r(f(d), x)$, with $r \in S_R^{mln}$, $f \in S_F^{mln}$, $d \in S_C^{mln}$. Define

$$H_\phi(x) := \textit{noisy-or}\{r(y, x) \mid y : r_f(d, y)\}.$$

According to the semantics of probability formulas, the evaluation $H_\phi(c)[\omega']$ performs all substitutions $r(y, c)[y/c']$ for $c' \in S_C^{mln}$ that satisfy $r_f(d, c')$, by evaluating the resulting ground probability formulas, and by combining all values so obtained by noisy-or. We first observe that this evaluation does not require truth values of any atoms in the relations r_{ϕ_i}, so that (i) is satisfied. By the definition of M_{ext}^{rbn}, the condition $r_f(d, y)$ is satisfied exactly for $y = c'$ with $c' = f(d)$ in M_{ext}^{mln}. Thus, only this substitution is performed, and the evaluation of $H_\phi(c)[\omega']$ reduces to the evaluation of $noisy\text{-}or\{r(c', c)\}$. The evaluation of $r(c', c)$ returns 1, respectively 0, according to whether $r(c', c)$ is true, respectively false, in ω', or, equivalently $r(f(d), c)$ is true, respectively false, in ω. Since, finally $noisy\text{-}or\{0\} = 0$ and $noisy\text{-}or\{1\} = 1$, we obtain (ii) and (iii).

Case 1a: ϕ is an equational atom. This case is similar. The formula for the equational atom $f(c) = x$ is given by $noisy\text{-}or\{1 \mid y : r_f(c, y) \wedge y = x\}$. This construction utilizes the convention that $noisy\text{-}or\emptyset := 0$.

Case 2a (Negation): $\phi(x) = \neg\psi(x)$. Define $H_\phi(x) := (H_\psi(x) : 0, 1)$, using the convex combination construct for probability formulas.

Case 2a (Conjunction): $\phi(x) = \psi(x) \wedge \chi(x)$: Define $H_\phi(x) := (H_\psi(x) : H_\chi(x), 0)$, again using convex combinations.

Case 3 (Existential quantifiers): $\phi(x) = \exists y \psi(x, y)$. Define $H_\phi(x) := noisy\text{-}or$ $\{H_\psi(x, y) \mid y : \tau\}$, where τ stands for a tautological constraint (e.g. $y = y$). Thus, the sub-formula $H_\psi(x, c)$ will be evaluated for all $c \in C$, and $H_\phi(x)$ returns 1 iff $H_\psi(x, c)$ evaluates to 1 for at least one $c \in C$.

In all cases condition (i) is immediate from the syntactic form of the constructed probability formulas (they do not contain any ϕ_i-atoms), and (ii),(iii) follow from the evaluation rules for probability formulas.

Given the formulas H_{ϕ_i}, we define the final probability formulas F_{ϕ_i} as follows:

$$F_{\phi_i}(x) := \begin{cases} (H_{\phi_i} : 1, 0) & \text{if } w_i = \infty \\ (H_{\phi_i} : 1, 1/e^{w_i}) & \text{if } \infty > w_i \geq 0 \\ (H_{\phi_i} : e^{w_i}, 1) & \text{if } w_i < 0 \end{cases}$$

Example 3. Table 3 shows the formulas H_{ϕ_i} and F_{ϕ_i} for $\phi_1, \ldots, \phi_5$ from Table 1. Here we have translated implications $\phi \Rightarrow \psi$ directly into probability formulas $(H_\phi : H_\psi, 1)$, rather than applying the translation given above for $\neg$ and $\wedge$ to $\neg(\phi \wedge \neg\psi)$. Note, too, that we need to encode $Fr(Anna, Bob)$ in the roundabout way shown in the table, because the RBN syntax does not allow constants from S_{ext}^{rbn} as arguments in atomic relation formulas (cf. Table 2).

Table 3. Translation of M_{int}^{mln} of Table 1

H_{ϕ_i}	F_{ϕ_i}
$((Fr(x, y) : Fr(y, z), 0) : Fr(x, z), 0)$	$(H_{\phi_1} : 1, 0.496)$
$(noisy\text{-}or\{(Fr(x, y) : Sm(x), 1) \mid y : y = y\} : 0, 1)$	$(H_{\phi_2} : 1, 0.1)$
$(Sm(x) : Ca(x), 1)$	$(H_{\phi_3} : 1, 0.223)$
$(Fr(x, y) : (Sm(x) : Sm(y), (Sm(y) : 0, 1)))$	$(H_{\phi_4} : 1, 0.332)$
$noisy\text{-}or\{Fr(x, y) \mid x, y : x = Anna \wedge y = Bob\}$	$(H_{\phi_5} : 1, 0)$

Having defined the translations t_{ext}, t_{int}, we have to show that

$$P[M^{mln}] \preceq_c P[M^{rbn}]. \tag{7}$$

where $M^{mln} = (M^{mln}_{int}, M^{mln}_{ext})$ and $M^{rbn} = (t_{int}(M^{mln}_{int}), t_{ext}(M^{mln}_{ext}))$. For this we have to find a suitable embedding, and a conditioning set C.

Since both $MVHI(S^{mln})$ and $MVHI(S^{rbn})$ are finite, we need to define the embedding $h(Q)$ only for singleton $Q = \{\omega\}$ ($\omega \in MVHI(S^{mln})$). First define $\tilde{h}(\omega) \in MVHI(S^{rbn}_{ext}, S^{mln}_R)$ as the unique $\tilde{\omega}$ with $\tilde{\omega}[S^{rbn}_{ext}] = t_{ext}(\omega[S^{rbn}_{ext}])$, and $\tilde{\omega}[S^{mln}_R] = \omega[S^{mln}_R]$. Now let

$$h(\omega) := \{\omega' \in MVHI(S^{rbn}) \mid \omega'[S^{rbn}_{ext}, S^{mln}_R] = \tilde{h}(\omega)\}. \tag{8}$$

Thus, $h(\omega)$ contains all possible extensions of $\tilde{h}(\omega)$ with interpretations of the relations $r_{\phi_i}/k_i \in S^{rbn}_{int} \setminus S^{mln}_R$.

Now let

$$C := \{\omega' \in MVHI(S^{rbn}) \mid \forall i = 1, \ldots, n \forall c : r_{\phi_i}(c)[\omega'] = true\}. \tag{9}$$

To show (7) it is sufficient to show that for all $\omega \in MVHI(S^{mln})$:

$$P[M^{mln}](\omega) = 0 \iff P[M^{rbn}](h(\omega) \mid C) = 0, \tag{10}$$

and for all $\omega_1, \omega_2 \in MVHI(S^{mln})$ with $P[M^{mln}](\omega_i) > 0$:

$$log\frac{P[M^{mln}](\omega_1)}{P[M^{mln}](\omega_2)} = log\frac{P[M^{rbn}](h(\omega_1) \mid C)}{P[M^{rbn}](h(\omega_2) \mid C)}. \tag{11}$$

It is quite straightforward to verify (10) from the definitions of $P[M^{mln}]$ and $P[M^{rbn}]$. We therefore only show (11), for which we then can make the simplifying assumption that $w_i < \infty$ for all i. By the semantics of MLNs, we obtain for the left-hand side of (11):

$$log\frac{P[M^{mln}](\omega_1)}{P[M^{mln}](\omega_2)} = \sum_{i=1}^{n} w_i(n_i(\omega_1) - n_i(\omega_2)), \tag{12}$$

For the right-hand side, we first obtain

$$log\frac{P[M^{rbn}](h(\omega_1) \mid C)}{P[M^{rbn}](h(\omega_2) \mid C)} = logP[M^{rbn}](h(\omega_1) \cap C) - logP[M^{rbn}](h(\omega_2) \cap C)$$

$$= logP[M^{rbn}](h(\omega_1))P[M^{rbn}](C \mid h(\omega_1)) - logP[M^{rbn}](h(\omega_2))P[M^{rbn}](C \mid h(\omega_2))$$

$$= logP[M^{rbn}](C \mid h(\omega_1)) - logP[M^{rbn}](C \mid h(\omega_2)). \tag{13}$$

The last equality follows from $P[M^{rbn}](h(\omega_1)) = P[M^{rbn}](h(\omega_2))$, which holds because $P[M^{rbn}](h(\omega_i))$ is equal to the marginal probability $P[M^{rbn}](\tilde{h}(\omega_i))$ defined by M^{rbn}_{ext} and the probability formulas for $r \in S^{mln}_R$ alone. According to (6), these probabilities are uniform over the ω that have nonzero probability.

We now determine

$$P[M^{rbn}](C \mid h(\omega_i)) = \bigcap_{i=1}^{n} \bigcap_{c} P[M^{rbn}](r_{\phi_i}(c) = true \mid h(\omega_i)).$$

Since $F_\phi(c)$ only depends on relations in S_R^{mln}, we have that the random variables $r_{\phi_i}(c)$ are conditionally independent given an interpretation of all relations in S_R^{mln}. Furthermore, since all $\omega' \in h(\omega)$ have the same interpretation of S_R^{mln}, we obtain

$$P[M^{rbn}](r_{\phi_i}(c) = true \mid h(\omega_i)) = F_{\phi_i}(c)[\tilde{h}(\omega)].$$

This gives us

$$logP[M^{rbn}](C \mid h(\omega)) = \sum_{i=1}^{n} \sum_{c} logF_{\phi_i}(c)[\tilde{h}(\omega)]$$

$$= \sum_{i=1}^{n} \left(\sum_{\substack{c \\ w_i \geq 0 \ \ \tilde{h}(\omega) \models \phi_i(c)}} log(1) + \sum_{\substack{c \\ \tilde{h}(\omega) \not\models \phi_i(c)}} log(1/e^{w_i}) \right)$$

$$+ \sum_{i=1}^{n} \left(\sum_{\substack{c \\ w_i < 0 \ \ \tilde{h}(\omega) \models \phi_i(c)}} log(e^{w_i}) + \sum_{\substack{c \\ \tilde{h}(\omega) \not\models \phi_i(c)}} log(1) \right)$$

$$= \sum_{i:w_i \geq 0} -w_i(N_i - n_i(\omega)) + \sum_{i:w_i < 0} w_i n_i(\omega), \quad (14)$$

where N_i is the total number of possible groundings c of $\phi_i(x)$, and, thus, $N_i - n_i(\omega)$ is the number of groundings with $\tilde{h}(\omega) \not\models \phi_i(c)$. The terms $-w_i N_i$ cancel when taking the difference in (13), so that we finally obtain for the right-hand side of (11) the same expression as in (12) for the left-hand side.

We have now shown that RBNs are at least as expressive as MLNs. It is an open question whether the converse also holds.

Beyond the pure expressivity result, our MLN to RBN translation provides some additional insights: first, it is clear that the size of the RBN encoding of a MLN model is linear in the size of the MLN, so that compactness of representation is preserved. Second, one can see that MLN models and their RBN encodings will exhibit very similar behavior in terms of inference complexity: inference for MLNs is conducted on a *ground Markov network* [12] whose nodes are ground atoms in the relations from S_R^{mln} with constants from S_C^{mln}. Inference for RBNs (usually) is conducted on a *ground Bayesian network*, whose nodes are ground atoms in the relations from S_{int}^{rbn} with constants from S_C^{mln}. For inference, this Bayesian network will first be transformed into its *moral graph*. This moral graph turns out to have essentially the same structure as the ground Markov network from the MLN, only that to the cliques of S_R^{mln}-nodes are attached nodes with ground r_ϕ-atoms. Since for inference these nodes are all instantiated to true, they can easily be eliminated, and one ends up with a graphical support structure for inference in the RBN model that is identical to the ground Markov network. Thus, the commonly used inference techniques (exact or approximate) that operate on ground graphical models will show very similar behavior for MLN and RBN encodings. This does not preclude the possibility, however, that for one language one might find a more sophisticated inference technique, which does not readily translate into a corresponding inference technique for the other language.

4   Conclusion

In this chapter we have developed a model-theoretic framework for comparisons of probabilistic logic languages. The framework is based on the key hypothesis that the essential feature of pl-languages is their modularity: they allow to represent general, high-level probabilistic specifications (the intensional model part), that is combined with the specification of concrete domains (the extensional model part).

Within this framework we have shown that the RBN language can encode MLN models. This result is based on basic versions of RBNs and MLNs. Both languages can be extended in various ways, e.g. to provide probabilistic models of functions in addition to probabilistic relations, or to provide probabilistic models for infinite domains [2,14]. For some of the simpler extensions the basic translation method described in this chapter will also be applicable. For more complex extensions (notably infinite domains), however, some substantial additional effort may be required to determine whether MLN models can be translated into RBN models, or vice-versa.

Turning our attention to other languages, we conjecture that BLPs and RBNs are equally expressive when both languages are restricted to the noisy-or combination function. Since only noisy-or is required for the RBN encodings of MLNs, this would also mean that $MLN \preceq_c BLP$.

Acknowledgments

The author wants to thank Kristian Kersting and Luc De Raedt for many fruitful discussions on the topic of this paper. A preliminary account of some of the material in this chapter was given in [3]. This work was supported in part by the EU IST program: FP6-508861, Application of Probabilistic ILP II (April-II).

References

1. Friedman, N., Getoor, L., Koller, D., Pfeffer, A.: Learning probabilistic relational models. In: Proceedings of the 16th International Joint Conference on Artificial Intelligence (IJCAI 1999) (1999)

2. Jaeger, M.: Reasoning about infinite random structures with relational bayesian networks. In: Cohn, A.G., Schubert, L., Shapiro, S.C. (eds.) Proceedings of the 6th International Conference on Principles of Knowledge Representation and Reasoning (KR 1998), Trento, Italy, pp. 570–581. Morgan Kaufmann, San Francisco (1998)

3. Jaeger, M., Kersting, K., De Raedt, L.: Expressivity analysis for pl-languages (position paper). In: Online Proceedings of the Workshop on Statistical Relational Learning (SRL 2006) (2006)

4. Jaeger, M.: Relational bayesian networks. In: Geiger, D., Shenoy, P.P. (eds.) Proceedings of the 13th Conference of Uncertainty in Artificial Intelligence (UAI-13), Providence, USA, pp. 266–273. Morgan Kaufmann, San Francisco (1997)

5. Jaeger, M.: Complex probabilistic modeling with recursive relational Bayesian networks. Annals of Mathematics and Artificial Intelligence 32, 179–220 (2001)

6. Kersting, K., De Raedt, L.: Towards combining inductive logic programming with bayesian networks. In: Rouveirol, C., Sebag, M. (eds.) ILP 2001. LNCS (LNAI), vol. 2157, pp. 118–131. Springer, Heidelberg (2001)

7. Laskey, K.B., da Costa, P.C.G.: Of starships and klingons: Bayesian logic for the 23rd century. In: Proceedings of UAI 2005 (2005)

8. Milch, B., Marthi, B., Russell, S., Sontag, D., Ong, D.L., Kolobov, A.: Blog: Probabilistic logic with unknown objects. In: Proc. 19th International Joint Conference on Artificial Intelligence (IJCAI), pp. 1352–1359 (2005)

9. Muggleton, S.: Stochastic logic programs. In: de Raedt, L. (ed.) Advances in Inductive Logic Programming, pp. 254–264. IOS Press, Amsterdam (1996)

10. Pasula, H., Marthi, B., Milch, B., Russell, S., Shpitser, I.: Identity uncertainty and citation matching. In: Proceedings of NIPS 2003 (2003)

11. Poole, D.: Logical generative models for probabilistic reasoning about existence, roles and identity. In: Proceedings of AAAI 2007 (2007)

12. Richardson, M., Domingos, P.: Markov logic networks. Machine Learning 62(1-2), 107–136 (2006)

13. Sato, T.: A statistical learning method for logic programs with distribution semantics. In: Proceedings of the 12th International Conference on Logic Programming (ICLP 1995), pp. 715–729 (1995)

14. Singla, P., Domingos, P.: Markov logic in infinite domains. In: Proceedings of UAI 2007 (2007)

Author Index

Lecture Notes in Artificial Intelligence (LNAI)